START
PROGRAMMING
USING HTML, CSS,
AND JAVASCRIPT

CHAPMAN & HALL/CRC
TEXTBOOKS IN COMPUTING

Series Editors

John Impagliazzo
Professor Emeritus, Hofstra University

Andrew McGettrick
Department of Computer
and Information Sciences
University of Strathclyde

Aims and Scope

This series covers traditional areas of computing, as well as related technical areas, such as software engineering, artificial intelligence, computer engineering, information systems, and information technology. The series will accommodate textbooks for undergraduate and graduate students, generally adhering to worldwide curriculum standards from professional societies. The editors wish to encourage new and imaginative ideas and proposals, and are keen to help and encourage new authors. The editors welcome proposals that: provide groundbreaking and imaginative perspectives on aspects of computing; present topics in a new and exciting context; open up opportunities for emerging areas, such as multi-media, security, and mobile systems; capture new developments and applications in emerging fields of computing; and address topics that provide support for computing, such as mathematics, statistics, life and physical sciences, and business.

Published Titles

Paul Anderson, Web 2.0 and Beyond: Principles and Technologies

Henrik Bærbak Christensen, Flexible, Reliable Software: Using Patterns and Agile Development

John S. Conery, Explorations in Computing: An Introduction to Computer Science

John S. Conery, Explorations in Computing: An Introduction to Computer Science and Python Programming

Iztok Fajfar, Start Programming Using HTML, CSS, and JavaScript

Jessen Havill, Discovering Computer Science: Interdisciplinary Problems, Principles, and Python Programming

Ted Herman, A Functional Start to Computing with Python

Pascal Hitzler, Markus Krötzsch, and Sebastian Rudolph, Foundations of Semantic Web Technologies

Mark J. Johnson, A Concise Introduction to Data Structures using Java

Mark J. Johnson, A Concise Introduction to Programming in Python

Lisa C. Kaczmarczyk, Computers and Society: Computing for Good

Mark C. Lewis, Introduction to the Art of Programming Using Scala

Efrem G. Mallach, Information Systems: What Every Business Student Needs to Know

Bill Manaris and Andrew R. Brown, Making Music with Computers: Creative Programming in Python

Uvais Qidwai and C.H. Chen, Digital Image Processing: An Algorithmic Approach with MATLAB®

David D. Riley and Kenny A. Hunt, Computational Thinking for the Modern Problem Solver

Henry M. Walker, The Tao of Computing, Second Edition

CHAPMAN & HALL/CRC
TEXTBOOKS IN COMPUTING

START PROGRAMMING USING HTML, CSS, AND JAVASCRIPT

Iztok Fajfar

University of Ljubljana
Slovenia

CRC Press
Taylor & Francis Group
Boca Raton London New York

CRC Press is an imprint of the
Taylor & Francis Group, an **informa** business

A CHAPMAN & HALL BOOK

CRC Press
Taylor & Francis Group
6000 Broken Sound Parkway NW, Suite 300
Boca Raton, FL 33487-2742

© 2016 by Taylor & Francis Group, LLC
CRC Press is an imprint of Taylor & Francis Group, an Informa business

Printed on acid-free paper
Version Date: 20150904

International Standard Book Number-13: 978-1-4987-3144-7 (Paperback)

Library of Congress Cataloging-in-Publication Data

Fajfar, Iztok, author.
   Start programming using HTML, CSS, and JavaScript / author, Iztok Fajfar.
     pages cm -- (Chapman & Hall/CRC textbooks in computing)
   Includes bibliographical references and index.
   ISBN 978-1-4987-3144-7 (alk. paper)
   1. Internet programming--Textbooks. 2. Web site development--Textbooks. 3. HTML (Document markup language)--Textbooks. 4. Cascading style sheets--Textbooks. 5. JavaScript (Computer program language)--Textbooks. I. Title.

   QA76.625.F355 2015
   006.7'6--dc23
                                                       2015020329

Visit the Taylor & Francis Web site at
http://www.taylorandfrancis.com

and the CRC Press Web site at
http://www.crcpress.com

To my family

Contents

Acknowledgments

A huge thank you goes to the guys at Taylor and Francis, especially to my editor Randi Cohen for her enthusiasm for the whole project, my project coordinator Ashley Weinstein, who oversaw production attentively, and technical reviewers for their detailed comments making the whole book more enjoyable. Many thanks also to the proofreader for correcting typos and grammar. Indeed, it was a great pleasure to work with such a professional team.

Honestly, all this wouldn't have happened were it not for Igor and the other guys from the morning-coffee crew, who suggested that I should really write a book. Thanks, chaps, it cost me a year of my life. Thank you to all my amazing students for sitting through my programming lectures and asking nasty questions. Man, how should I know all that? I shall not forget to also thank the other teaching staff from the team. The joy of working together is immeasurable. I'm deeply indebted to Žiga, who had painstakingly read the whole manuscript before releasing it to the wild. (I sincerely hope you spotted all the silly mistakes so I don't make a fool of myself.) Thank you, Andrej, for technical advice on preparing the camera-ready PDF. Those are really details that make a difference. A thousand thanks go to Tanja and Tadej for that little push that did the trick. You are terrific!

I also wish to extend my considerable gratitude to everyone that gave away their precious time, energy, and invaluable expertise answering questions on forums, posting on blogs, and writing all those wonderful LATEXpackages. It's impossible to list you all by name because I'm contracted for only 400 or so pages.

A colossal thank you goes out to my mom and dad for instantiating and personalizing me. It wasn't the easiest assignment in the world but you did a marvelous job! Many thanks to my second parents, Dana and Ivo, for telling me that I should also eat if I am ever to finish the book. A zillion thanks go out to my close family. Thank you, Erik, for patiently checking which page I am on with an I-want-my-daddy-back determination; and thank you, Monika, for tons of understanding and supportive coffee mugs. I love you!

I am also thankful for the support of the Ministry of Education, Science, and Sport of the Republic of Slovenia within the research program P2-0246—Algorithms and Optimization Methods in Telecommunications, which made possible some of the research for this book.

And, of course, thank *you*, the reader. Without you, this book wouldn't make much sense, would it?

—Iztok Fajfar (iztok.fajfar.eu), March 2015

Introduction

Easy to Use

Normally, putting honey in my tea is not a particularly demanding task, but that morning my hand was paralyzed in astonishment, trying to do its routine job of pouring some honey in the steaming cup. Honey labels usually say things like "All Natural," "Contains Antioxidants," or "With Grandma's Recipe Book." Over time, I've got used to more absurd labels like "Improved New Flavor" or "Gathered by Real Bees." The label that knocked me out was surprisingly plain, with an award-winning message printed on it: "Easy to Use." I don't recall honey ever being hard to use, except maybe when it crystallizes, or when I was six months old, but that's probably not exactly what the author of the message had in mind.

You can also buy programming books that promise easy and quick learning, even as fast as in 24 hours. An average adult can read a novel in 24 hours. But let's face it, no one can read—let alone understand and learn—a 500-page technical book in 24 hours. While using honey is not difficult even when it doesn't explicitly say so, learning to program is not easy. It can be fun if you're motivated and have decent material to study from, but it's also an effort. If you're not ready to accept that, then this book is not for you. Otherwise, I invite you to join Maria, Mike, and me at exploring the exciting world of computer programming. It's going to be fun but it's also going to be some work.

About the Book

This handbook is a manual for undergraduate students of engineering and natural science fields written in the form of a dialog between two students and a professor discovering how computer programming works. It is organized in 13 thematic meetings with explanations and discussions, supported by gradual evolution of engaging working examples of live web documents and applications using HTML, CSS, and JavaScript. You will see how the three mainstream languages interact, and learn some of the essential practices of using them to your advantage. At the end of each meeting there is a practical homework, which is always discussed at the beginning of the next meeting. There is also a list of related keywords to help you review important topics of each meeting.

The general structure of the book is multilayered: the basic language syntax and rules are fleshed out with contents and structure while still keeping things simple and manageable, something that many introductory textbooks lack.

The main body of the text is accompanied by five appendices. The first of them contains a solution of the last homework, the second summarizes (also with examples) some major directions in which you can continue your study, including hints on some of the relevant sources. The last three appendices are abbreviated references of the three languages used in the book.

There will be situations when you need to use yet more languages and technologies in order to get the job done. Some such situations are gently dealt with in this book. For example, you will learn just enough about a Server Side Includes language to be able to include external HTML code, which will save you a tremendous amount of time and energy.

Is This Book for Me?

If you know absolutely nothing about computer programming and want to learn, this is the book for you. It has been written with a complete beginner in mind in the first place.

If you have been exposed to programming before, you might find the book useful as well. Today, many people learn from examples and forums, and thus acquired knowledge is mostly skills and not much theory. If you ever want to build more serious software, you need a firm and systematic understanding of what is going on. You need a framework to which you can systematically attach your partial skills to form a sound structure of connected knowledge. Hopefully, this book can give you this as well.

Last but not least, if you're a teacher of an introductory programming course, you might find a handful of useful examples and approaches for your classes on the few hundred pages that follow.

But most likely, as there are as many learning styles as there are learners, you will have to find out for yourself whether or not this book is for you.

How to Avoid Reading the Whole Book

Don't panic! If you are only up to JavaScript programming, you can just read Meeting 1 to get a basic idea of what HTML is (you need this in order to be able to run the JavaScript examples in this book), and then you can immediately skip to Meeting 6—more specifically, Section 6.3. There are some examples involving CSS in the JavaScript part but they won't stand in the way of your learning JavaScript. Later, if you feel like it, you can just as well skim over Meeting 3, where you can get the basic idea of what CSS is all about.

For Your Safety

This book is not about cutting-edge web technologies, so you don't need any protective equipment. It is more about general computer programming and some web-related principles using the mainstream web languages HTML, CSS, and JavaScript as examples. Some of the principles are over 40 years old, but are extremely important because they allow you to write cleaner and more easily maintainable code, and they will not go away just like that.

It's a busy world, and the sixth edition of ECMAScript standard (the standardized version of JavaScript) has just entered the official publication process. The good news is that it only introduces additions to its predecessor, so the essential concepts stay. Also, while CSS3 isn't completely finished yet, there already exist some so-called "level 4" CSS modules. Fortunately, they are also just additions to the CSS standard and there are no serious plans for a single CSS4 specification on the horizon. This book pays attention to the basic concepts that have matured with the latest HTML5, CSS3, and ECMAScript 5 standards to the point where it seems these concepts are going to persist for some time.

The Software Used

In researching this book, I used Google Chrome and Notepad++ v6.5.3 (*notepad-plus-plus.org*) on a Windows 7 Professional SP1 64bit operating system. I also used the EasyPHP DevServer 13.1 VC11 web development server (*www.easyphp.org*). However, you will be able to follow most of the examples and experiments in this book using any modern browser and plain text editor. They are already installed on your computer, so you can start experimenting right away.

Conventions Used in This Book

The following typographical conventions are used in this book:

A `monospaced font` is used for all code listings and everything that you normally type on a keyboard, including keys and key combinations.

A `monospaced italic font` is used as a general placeholder to mark items that you should replace with an actual value or expression in your code.

An italic font is used to indicate the first appearance of a term, or as an emphasis.

A sans serif font is used to indicate a menu item.

A sans serif italic font is used to indicate URLs and file names and extensions.

Feedback and Supporting Online Material

I deeply appreciate having any comments, suggestions, or errors found brought to my attention at the email address *start-programming@fajfar.eu*. You will find source code of the examples in this book and some additional materials and problems for each chapter at *fajfar.eu/start-programming*.

About the Author

Iztok Fajfar got his first computer in the early 1980s, a ZX Spectrum with an amazing 48 KB of RAM. Computers soon turned into a lifelong fascination and an indispensable companion, assisting him in his professional work and hobbies alike. Iztok has a PhD degree in electrical engineering from the University of Ljubljana, Slovenia, where he is currently Associate Professor at the Faculty of Electrical Engineering. His research topics include evolutionary algorithms, in particular, genetic programming. He teaches computer programming at all levels, from assembly to object-oriented, and to all kinds of audience. Now and then he even ventures to explain to his mother-in-law how to forward an email, and he hasn't given up yet. He is also a programmer and writer. Iztok lives with his family in Ljubljana, and when he is not programming, or teaching, or researching weird stuff, he makes the most yummy pancakes, not to mention the pizza.

Content and Structure

1.1 Opening

Professor: I'm thrilled that you accepted my invitation to help me with a new book I am researching. There are three languages awaiting us in this course: HTML, CSS, and JavaScript.

Mike: Why three? You'll just confuse us, won't you?

Professor: The languages have been designed for quite specific purposes and work very differently, so there is little danger in confusing them. At the same time, the three languages nicely complement each other: HTML holds the structure and content of a web page, CSS takes care of presentation, and JavaScript is responsible for action. I like to say that HTML is bones, CSS is flesh, and JavaScript is the brain and muscles of web programming.

Maria: How much of a chance is there of us learning three languages to the level that we can use any of them to our advantage?

Professor: You don't have to be a guru in any of them to start using them effectively. It's only important that you know the basic principles. The good news is you don't have to install or learn to use any new software. All you need to start off is already installed on your computer.

Do you have any programming experience?

Maria: Actually, I use a computer a lot but not for programming. I have never written a computer program before.

Mike: Neither have I.

Professor: In a way, programming is like speaking. You speak English, right?

Mike: Yes...?

Professor: I even know people who have learned Finnish. Quite well, to be honest.

English and Finnish are examples of *natural languages*, which people learn to communicate with other people. However, if you want to talk to computers, you have to learn *artificial languages* so that computers understand and *obey* you. It's very similar. The only difference is that people won't obey you if you lack charm, while computers won't obey you if you're not accurate. Accuracy is crucial. Similar to both is that it takes a certain amount of practice before your interlocutor understands you. I won't lie to you on this one.

Maria: I'm just starting to learn Spanish and I must use a sign language a lot. I suppose you cannot use a sign language with a computer.

Professor: That's true. In natural languages, people use context and even a sign language to guess what others have to say even though what they say may not be grammatically correct. Computers don't do that, though, and that's the difficult part of programming. You have to be *exact*.

All right. Let's start programming, shall we?

1.2 Introducing HTML

Professor: To be precise, HTML is not a programming language but it is a so-called *markup language*. That's what the acronym HTML stands for: Hypertext Markup Language. Markup is a modern approach for adding different annotations to a document in such a way that these annotations are distinguishable from plain text. Markup instructions tell the program that displays your text what actions to perform while the instructions themselves are hidden from the person that views your text. For example, if you want a certain part of your text to appear as a paragraph, you simply mark up this part of the text using appropriate *tags*:

```
<p>But it's my only line!</p>
```

Maria: It looks quite straightforward. Are those p's in the angle brackets like commands?

Professor: You could say that. They are called *tags* and they *instruct* or *command* a browser to make a paragraph out of the text between them.

Mike: That's like formatting, isn't it?

Professor: In a way, yes. Tags are like commands in a word processor that allow you to format paragraphs, headings, and so forth. However, they only specify *what* to format, not *how* to do it.

The above code fragment is an example of an HTML *element*—the basic building block of an HTML document. An HTML document is composed exclusively of elements. Each element is further composed of a *start tag* and *end tag*, and everything in between is the *content*:

Start Tag	Content	End Tag
`<p>`	`But it's my only line!`	`</p>`

The start tag is also called the *opening tag* while the end tag is also called the *closing tag*. By the way, the name, or the abbreviation of the name of the element is written inside the tags. In particular, p stands for a paragraph. The closing tag should have an additional slash (/) before the element's name.

In order for a paragraph to show in the browser, we need to add two more things to get what is generally considered the *minimum HTML document*. The first line should be a special declaration called *DOCTYPE*, which makes a clear announcement that HTML5 content follows. The DOCTYPE declaration is written within angle brackets with a preceding exclamation mark and the `html` keyword after it: `<!DOCTYPE html>`. Although it looks like a tag, this is actually the only part of an HTML document that isn't a tag or an element. As a matter of fact, this code is here for historical reasons. I don't want to kill you with details, but you have to include it if you want your document to be interpreted by the browser correctly.

One more thing that the minimum document should contain is a `<title>` element. This element is necessary as it identifies the document even when it appears out of context, say as a user's bookmark or in search results. The document should contain no more than one `<title>` element.

Putting it all together, we get the following code:

```
<!DOCTYPE html>
<title>The Smallest HTML Document</title>
<p>But it's my only line!</p>
```

Maria: You just showed us what the document code should look like. But I still don't know where to type the code and how to view the resulting page.

1.3   The Tools

Professor: You can use any plain text editor you like. For example, you can use the Windows Notepad, which is already on your computer if you use Windows.

Mike: What if I don't use Windows?

Professor: It doesn't matter. Just about any operating systems contains a plain text editor. Personally, I use Notepad++, a programmer-friendly free text editor (*notepad-plus-plus.org*).

After you type the code, it is important that you save the file with a *.htm* or *.html* extension. While it doesn't really matter which one you use, it is quite important that you choose one and stick to it consistently. Otherwise, you could throw yourself into a real mess. For example, you could easily end up editing two different files (same names, different extensions) thinking they're one and the same file.

Now we open the file in a browser and voilà!

Notice how the content of the `<title>` element appears at the top of the browser tag.

Mike: How did you open the file in the browser?

Professor: Oh yes, sorry about that. Inside Notepad++, I chose Run→Launch in Chrome. If you use another browser, it will automatically appear under the Run menu item in your Notepad++. You can of course also simply double-click the file or drag and drop it into the browser. Once the file is open in the browser, you don't have to repeat this operation. If you modify the source code—the original HTML code, that is—you simply refresh the browser window. If you use Chrome like I do, you can do that by pressing F5. Later, you will use more than a single file to build a page. In that case, you will sometimes have to *force reload* all files of a page, which you can do by pressing Ctrl+F5 on Chrome. On Windows, to switch between the text editor and browser quickly, you press Alt+Tab, a standard key combination for switching between running tasks.

Maria: What would happen if we forgot to include the `<title>` element?

Professor: Nothing fatal, to be honest. One of the basic rules of rendering web pages is that the browser always tries its best to show the content. Of course, if the document isn't fully formatted according to the recommendations, the results are sometimes not in our favor. If you forget the title, then the name of the file containing the document usually takes over its role. If nothing else, that looks ugly and unprofessional.

1.4   Minimal HTML Document

Professor: One of the general prerequisites to good technical design is simplicity, which should not be confused with minimalism. In our last example, we saw a truly minimal HTML document, which you will rarely see in practice. Even with no extra content it is normally a good idea to flesh out this skeleton HTML document. For instance, most web developers share the belief that the traditional `<head>` and `<body>` elements can contribute to clarity, by cleanly separating your document into two sections. You pack all the content into the `<body>` section, while the other information about your page goes to the `<head>` section. Sometimes it is also a good idea to wrap both these sections in the traditional `<html>` element:

```
<!DOCTYPE html>
<html>
  <head>
    <title>The Smallest HTML Document</title>
  </head>
```

```
  <body>
    <p>But it's my only line!</p>
  </body>
</html>
```

Mike: I noticed that an element can contain not only text but another element as well. For example, you placed the `<title>` element within the `<head>` element.

Professor: Good observation! The content of an element can in fact be any valid HTML conforming to the rules of that specific element. We call putting one element into another *nesting*. When an element is nested (contains other HTML elements), it is important that it contains whole elements, including start and end tags. So if, for example, an `<elementA>` starts *before* an `<elementB>`, then it must by all means end *after* the `<elementB>`:

```
<elementA> ... <elementB> ... </elementB> ... </elementA>
```

The element that is contained inside another element *inherits* some of its behavior, and we often say that the contained element is a *descendant* of its owner, which is in turn its *parent*. The direct descendant is also called a *child*. This concept will become especially important when we come to styling elements with CSS. Now I only mention it so that later the terms will already sound familiar to you.

Maria: What are those periods inside?

Professor: Oh, yes. A set of three periods is an *ellipsis*. An ellipsis indicates the omission of content that is not important for understanding the explanation.

We will soon come back to our last example and furnish it with a little more. For that purpose we need another element called `<meta>`. This element is used to provide additional page description (so-called metadata), which is not displayed on the page, but can be read by a machine. The information stored in the `<meta>` element includes keywords, author of the document, character encoding, and other metadata. The `<meta>` element has neither content nor the closing tag:

```
<meta>
```

An element that is composed only of the opening tag is called an *empty* or *void* element.

Mike: I don't understand that. Where do you put all the information you talked about if there is no content?

Professor: That's the job for *attributes*. An attribute is the means of providing additional information about an HTML element. For example, by using the `src` attribute on the `<img>` element, one can tell the browser where to find the image to display. There are two things you should know about attributes: they are always specified after the element name in the start tag, and they come in name/value pairs like this one:

```
name="value"
```

The quotes (double style are the most common, but single style quotes are also allowed) around the value are not necessary under HTML5, as long as the value doesn't contain some restricted character (mostly =, >, or a space). That said, it is still a good practice that you always use quotes. Your code will look cleaner and more readable, which in turn lessens the possibility of errors. Likewise, it is not necessary that you use lower-case names and values, although it is recommended that you do so.

Some attributes do not have values. All that counts is their presence: they are either present or not, similar to an electrical switch, which can be put in on or off position. That type of attribute is called a *Boolean attribute* after an English mathematician, George Boole, the inventor of the so-called Boolean logic based on only two values. Because Boolean attributes have no value (the value is implied by their presence or absence), we omit the equals sign as well:

```
name
```

There are often cases when an element has more than one attribute. In that case the attributes are separated by spaces.

If I confused you with all this theoretical talking, don't worry. I will now show you how this works in practice.

Maybe you've already heard about the *character encoding*. Basically, that's a system that tells you how each character of a given repertoire is represented physically. In a computer, this physical representation consists of a series of ones and zeros, called bits. Relax, I'm not going into more detail with this explanation. The important thing is that the browser must know what encoding has been used to store the document text so it can read it back and show it properly. If you don't provide the encoding information to your markup, a browser will of course try to guess it, which may drive it into an obscure security issue. You should provide this information through the *charset* (that's short for character set) attribute of a `<meta>` element, assigning it the value `utf-8`. Today, more than half of all web pages are encoded in UTF-8 and honestly, I don't think you will ever need to use a character that is not included in UTF-8.

Let's check if you followed me. Can you add a `<meta>` element containing a character encoding attribute to our previous example without my help?

Maria: You didn't tell us in which section to put it. Let me think.... You said that the content went into the `<body>` and all the other information in the `<head>`. Something like the following code, perhaps?

```
<head>
  <meta charset="utf-8">
  <title>The Smallest HTML Document</title>
</head>
```

Professor: Exactly! One more thing here: when saving your work, don't forget to save it in UTF-8. For example, if you edit your document with Notepad (on Windows) the Save As dialog box lets you choose UTF-8 from the Encoding list at the bottom. In Notepad++, there is a top-level menu Encoding, under which you select Encode in UTF-8.

Maria: Is the order of the above elements within `<head>` important?

Professor: Not really. HTML5 only specifies that the character encoding declaration must be within the first 1024 bytes of the document. Translated into English, always include it as early as possible.

Mike: What if I wanted to write a page in some other language than English?

Professor: The encoding has nothing to do with the natural language of your web page. It only defines the set of characters you will use and, as I said before, you will hardly ever need a character (English or non-English) that is not a part of UTF-8. I'm glad you asked that, though. Your question takes us to the one last thing we should add to our basic document. Specifying a web page's natural language is considered a good style by many as it can be a useful piece of information for many users—for example, for filtering web search results by language. Interestingly, as far I as I know, Google Translate ignores this tag, relying on its own language-recognition algorithms. Anyway, you specify the language using the `lang` attribute on any element. For English, the attribute value will be `en`, and you can find codes for many other languages at *www.w3schools.com/tags/ref_language_codes.asp*. In a most likely scenario that your whole web page uses a single language (in our case English) you simply add the `lang` attribute to the `<html>` element:

```
<html lang="en">
```

By putting it all together we arrive at a decent starting point for any modern web page you want to build:

```
<!DOCTYPE html>
<html lang="en">
  <head>
    <meta charset="utf-8">
    <title>The Smallest HTML Document</title>
  </head>
  <body>
    <p>But it's my only line!</p>
  </body>
</html>
```

Mike: I noticed that you indent the code, and even some lines more than others. How do you know how much to indent which line?

Professor: You don't have to know that. The indentation you see is in fact completely optional and without any effect on how the browser interprets the code. At the same time it is of the utmost importance for the person writing the code. Notice how the start and end tags of nested elements "see" each other. This "visibility" is made possible by indenting the content of an element. That way, the structure of the document stands out more clearly before the writer. In such a small document the advantage may not be evident at first glance, but believe me, when a document's size reaches several hundred lines, you want to have some order in your code.

Another friendly feature that helps programmers find their way through the chaos of computer code are *comments*. In HTML, comments are enclosed in a so-called *comment tag*, which starts with a left angle bracket, an exclamation mark, and two hyphens (<!--), and ends with two hyphens and a right angle bracket (-->). For example:

```
<!-- This is a comment and will not be visible in the browser. -->
```

Comments are completely ignored by the browser but quite useful for writing a short human-readable summary of code, for example. With proper comments, you don't need to decipher the code every time you need to upgrade it. Even if it is your own code, human readable remarks will help you tremendously to understand it later. Another practical use of comments is to temporarily switch off parts of code during testing and experimenting.

If you work in a team, or plan to make your code public, it is a good idea to include your name, contact information, and a licensing notice in comments at the top of your code. That way, people will have a chance to contact you in case they have questions about the code.

While comments are not visible in the browser window, be aware that they are accessible to the web page viewer through the View page source menu command. So, don't use comments to write filthy remarks about your boss or mother-in-law.

1.5 Formatting a Page

Professor: When I was a little boy, my mother used to read me fairy tales from the book by Joseph Jacobs, an Australian folklorist, and I still can't do without them. That's why we're going to mark up the beginning of *The Rose-Tree* as our next example. We will need three more elements for the job: a main heading (the `<h1>` element) for the title, an image (the `<img>` element) to include a fancy decorative capital letter, and quotes (the `<q>` element) for quoted speech.

In HTML, you can use six different levels of headings defined by the elements from `<h1>` to `<h6>`. `<h1>` defines the most important heading while `<h6>` defines the least important heading. Since our fairy tale only has one heading, we will of course use the `<h1>`:

```
<h1>The Rose-Tree</h1>
```

With the image element, as you might have guessed, you can include a picture into a page. You do that by specifying the file in which the picture is stored using the `src` attribute: the value of the `src` attribute is the name of the file. Note that `<img>` is an empty element so it only has the opening tag. If the image is not a key part of the content, then it is a good idea to include the `alt` attribute as well. This attribute provides a textual equivalent to show in case the image cannot be displayed or until it is downloaded.

Our fairy tale begins with a letter T. You can use some graphic software to draw a decorative T and save it to the image file named *T.jpg*. The final code for our image element looks like this:

```
<img src="T.jpg" alt="T">
```

Because the decorative capital letter is not an essential part of the content, I provided alternative text, which is obviously the letter T.

Incidentally, not all image formats are supported by browsers. The most commonly used formats that are supported are *.gif*, *.jpg* and *.png*. In the above code, I used only the file name without any path as the value of the `src` attribute. That means that the file resides in the same folder as the HTML file that contains the above `<img>` element. Should the image file be stored elsewhere, I would have to prepend the corresponding path to it.

The `<q>` element is used to represent some quoted content and is usually rendered as a pair of quotes around the marked content.

OK, let's put it all together. The following code goes into the `<body>` element:

```
<!-- From the e-book English Fairy Tales, collected and
  -- edited by Joseph Jacobs. Belongs to the public domain.
  -- Source: www.authorama.com
  -->
<h1>The
    Rose-Tree</h1><p><img src="T.jpg" alt="T">here
was once upon a time a good man who had two children:

a girl by a first wife, and a boy by the second. The girl was
as
    white as milk, and her lips were like cherries. Her hair
  was like golden silk, and it hung to the ground. Her brother
  loved her dearly, but her wicked stepmother hated her.
  <q>Child,</q> said the
  stepmother one day, <q>go to the grocer's shop and buy me
   a pound of candles.

   </q> She gave her the money; and the little girl
  went, bought the candles, and started on her return. There
  was a stile to cross. She
                  put down the candles whilst she
   got over the
  stile. Up came a dog and ran off with the candles.</p><p>She
   went back to the grocer's, and she got a second bunch.
```

```
She came to the stile, set down the candles, and proceeded to
climb
        over. Up came the dog and ran off with the candles.</p>
        <p>
She went again to the grocer's,

                and she got a third bunch; and
just the same
happened. Then she came to her stepmother crying, for she had
spent all the money and had lost three bunches of candles.</p>
```

Deliberately, I made a mess out of the text and tags so that the structure of the document is not obvious at first glance. The rendering, however, is quite appealing.

Maria: I notice that the browser does not obey your original text formatting.

Professor: Well.... Yes and no. Actually it obeys the rules perfectly, it's just that the rules are a little different from what you might have expected.

The first thing you may have noticed is that the browser ignores spaces, tabs and newlines. OK, it does not ignore them completely. For example, if there is a separation between words, it is replaced by a single space regardless of what I have actually put there: a space, tab, newline, or even more of them. A rule of thumb is that spaces, tabs, and newlines are ignored unless they represent the only separation between two entities—words, for example. Even then they are replaced by a single space. If there exists some other separator like, for example, a tag or equals sign (=), then spaces are not needed at all and it doesn't matter whether they are there or not—the result is always the same.

That said, there are a few cases where you should be careful about spaces:

- Do not put any spaces *before* the element name in the opening or closing tag. It would be wrong to write `< p>` or `< /p>` or `</ p>`.

- Do not put any unnecessary spaces between double quotes when writing attribute values. In that sense, `lang="en"` is quite different from `lang=" en "` while `lang = "en"` is still OK while it does not change the meaning.

- You should always put spaces between two attributes of the same element, even when there's a quotation mark at the end of a value. For example, it would be incorrect to write `<img src="T.jpg"alt="T">` instead of `<img src="T.jpg" alt="T">`.

It is important at this point that you start seeing HTML as a language that gives a document a *structure* and *meaning* rather than a specific look. How a page looks is taken care of by the browser. Later, we will control the document's look by means of CSS.

Mike: I have a question. If spaces and newlines do not affect the text formatting, how can you tell the browser to start, for example, some text on a new line?

Professor: This kind of formatting is implied in the element meaning or *semantics*. For example, it is usual practice that a heading or a paragraph is displayed as a block of text occupying the whole line—nothing else can be positioned on the same line in a browser window. In other words, a line break is inserted before and after a heading or paragraph. We say that such elements have a *block display*. On the other hand, an image can happily inhabit a line together with other elements, if there is enough space, of course. Such elements are said to have an *inline display*.

You'll see later that it is possible to change a type of display for an element using CSS. Doing that, however, does not change the intrinsic HTML categorization of elements. For this reason we will use the terms *inline element* and *block element* to denote elements that have an inline or block display by default.

Mike: But what if I simply want to break a line without making a paragraph or using any other block element? Is there a way to do that?

Professor: There's an element for breaking a line called `<br>`, which is short for break. Paradoxically, that element should not be used for breaking lines unless you are breaking lines because you want to convey some meaning. Typical examples are breaking lines in poems or postal addresses. If you want to break a line because you are introducing a new paragraph, then you shouldn't use the `<br>` element. A paragraph is a meaning by itself and doesn't need other elements to promote line breaks. If you don't like the amount of space between paragraphs, however, that's not a matter of semantics. You take care of stuff like that by using CSS.

Mike: I'll try to remember that, but I think I'll have to wait to get some experience to fully understand it.

Professor: That's true.

Maria: What happens with text that is written outside of any element?

Professor: Oh yes, that is extremely important. I'm glad you asked. It is strongly recommended that you do not put any plain text directly into the body. It is a good practice that you mark up every text and that you mark it up properly. I believe that's easier said than done, but as we go along you will get some experience and feeling about how to get it right. All you have to be careful about is to think of the role of the text you are writing: is it a heading, a paragraph, a caption, a sidenote? When you decide upon the role of the text, your next step is to pick up an appropriate HTML element to mark up that text. If none seems right, then generic `<span>` and `<div>` elements come in handy, but more on these later.

Mike: Is HTML case sensitive?

Professor: For the most part, no. However, just to stay on the safe side, I strongly recommend that you only use lower-case letters. If nothing else, it will save you the trouble of remembering whether case matters or not.

1.6 Homework

Professor: Before we call it a day, let me just give you lots of homework. Note that it's not important that you do everything right. The important thing is that you do it on your own and that you ask yourselves questions about why things work (or don't work) the way they do. Write down any unresolved questions, which we will use in our next meeting.

I encourage you to search and use, apart from the material we covered today, other sources as well. That's important. Because of the constant progress of technology, you'll always be searching for new answers. For your convenience, I prepared some reference material you will find at the end of this book. However, this is not a complete reference. I have only put into it what I thought would be important for our course. So if something is not there, that doesn't mean it doesn't exist. For the time being, I recommend that you use the site *www.w3schools.com*, which offers a decent learning experience for a beginner. One more good thing about this site is that it includes lots of working examples ready for you to play with. When you level up, however, I suggest that you start using other, more profound sources to learn from.

And now the homework. I have prepared for you a short document on wombats, which I want you to reproduce as faithfully as you can while using only HTML. If you don't like the picture, you can use another one. Personally, I like this old drawing of the now-extinct wombats of King Island, Tasmania, by Charles-Alexandre Lesueur from 1807.

Here's how the document should look:

Common Wombat

The **common wombat** (*Vombatus ursinus*), also called **bare-nosed wombat** or **coarse-haired wombat**, is one of three living wombat species. The common wombat reaches an average of 98 cm in length and a weight of 26 kg. It prevails in colder and wetter parts of South East Australia. The common wombat was first described by George Shaw in 1800.

There exist three subspecies of the common wombat:

- *V. ursinus hirsutus* on the Australian Mainland.
- *V. ursinus tasmaniensis* in Tasmania.
- *V. ursinus ursinus* on Flinders Island to the north of Tasmania.

For some more homework, here is a short list of keywords covering today's meeting. Draw a mind map using all the given keywords, adding some more if you feel like it.

In this meeting: element, start tag, end tag, content, indentation, comment, nested element, DOCTYPE, descendant, parent, child, markup, empty element, void element, attribute, attribute name, attribute value, Boolean attribute, <html>, <head>, keywords, author information, <meta>, <title>, <body>, <p>, <img>, <h1>, <br>, alt, src, charset, lang, block display, inline display, block element, inline element, spaces, generic elements, semantics

Building a Sound Structure

2.1 Homework Discussion

Professor: I am anxious to see what you have written since our last meeting. Did you have any trouble?

Mike: At the beginning, yes. I had to read about lists and tables to complete the homework.

Professor: No offense, but I am sort of glad you mention tables because that's a sure signal you did it wrong. Don't panic, though. That mistake will help you tremendously with mastering basic HTML principles. What about you, Maria?

Maria: I don't know.... It wasn't as easy as it looked and I have some questions.

Professor: Good. So we have material to talk about. Mike, what did you use a table for?

Mike: I haven't found any other way to wrap text around the picture. And honestly, I had that annoying feeling that this is not the way to go. Namely, I had to split the text between two cells of the table in order to wrap it around the image. The whole thing only looked good till I resized the window. So how can this be done properly?

Professor: As I already mentioned in our previous meeting, HTML is used to give a document a structure rather than a look. So there is always one very important question to bear in mind when constructing an HTML document: is what I'm trying to do in any way affecting the *structure* of the document or is it merely a matter of *presentation*? Indeed, wrapping a text around a picture has nothing to do with structure. A paragraph is a paragraph and a picture is a picture, no matter how they are positioned and formatted. You don't have to worry about whether a paragraph is wrapped around a picture when writing an HTML. That's work for CSS. I've deliberately presented you with a document that will tempt you to think about how it looks, which is completely irrelevant at this point.

Mike: So it's not possible to wrap text using pure HTML?

Professor: No.

Maria: Why not? I managed to find an attribute called `align`, which aligns a picture to one side (left or right) and wraps text around it nicely. Have I missed something?

Professor: You missed one thing. The `align` attribute is deprecated since HTML4.01 and obsolete since HTML5, and not without a reason. Nearly all elements and attributes that were historically used for presentational purposes are considered obsolete in HTML5. That doesn't mean they don't work, though. Worse still, they will work for quite some time for backward compatibility. Browsers are really forgiving when rendering a markup. You won't notice that you did something wrong because the mistake won't show. It is entirely up to you whether you conform to the rules or not. If you want to build sound web pages, the HTML is the most important thing to do right because it serves as a foundation of your whole web page. Sticking to plain, simple HTML is also important because it keeps search engines happy. The old HTML's approach of using formatting elements like `<font>`, or tables to lay out a page, stands in the way of a search engine's job.

Mike: Is there really no way of testing whether your HTML is written properly?

Professor: In fact there is. When in doubt, you can resort to one of many on-line HTML validators (for example, *validator.w3.org*). You can upload your document to a validator and it will alert you to any errors you may have in your HTML document.

Maria: I think I see things more clearly now and meanwhile you also answered some of my questions. I even discovered another mistake I made: I used the `<h2>` element for the main heading (Common Wombat) since the `<h1>` seemed too big to me. If I understand correctly, I shouldn't have done this because I was actually concerned about presentation. In fact I should have used `<h1>` because this is the main heading.

Professor: Exactly! When writing HTML, you shouldn't worry about font size.

Maria: What still puzzles me are the elements `<b>` (bold) and `<i>` (italic), which I found in the reference you gave us. Is this not presentation?

Professor: If you use them properly, then the answer is no. Those two elements survive from the past but with a different, more semantic interpretation. In modern HTML, it would be wrong to use `<b>` and `<i>` elements merely for the purpose of making text bold or italic. It is also wrong to think of `<b>` and `<i>` elements as being old elements now replaced by the `<strong>` and `<em>` elements only because they are rendered in the same way by many browsers. The `<strong>` element is used to express strong importance of its content and the `<em>` element is used to emphasize its content. The `<b>` and `<i>` elements are not used to stress importance or emphasis. Rather, they are considered as generic bold and italic elements used in cases when normally—but not necessarily—bold and italic fonts are used. Because the meaning of these two elements is not evident from their names, authors often use the `class` attribute to clearly identify their semantic meaning. For example:

```
<i class="latin-taxonomy">Vombatus ursinus ursinus</i>
<b class="english-definition">bare-nosed wombat</b>
```

The `class` attribute not only gives a clear meaning to both elements but also allows CSS to access these elements for formatting purposes—but more on that later.

Mike: There's one more question bothering me. I have found an element for writing unordered lists (`<ul>`) and I'm not quite sure whether I should put it inside a paragraph or not. A list, in my opinion, is a part of a paragraph but visually it is represented as a separate block. I know that I am mixing two concepts, which you've just clearly separated, but in this case my thinking is a bit blurred.

Professor: I agree that it is a matter of debate whether a list is a separate paragraph or not. In practice, things are sometimes not as clear as in theory. Luckily, in this special case we have recourse to the additional rules concerning *context* in which certain HTML elements can appear, and the *content* that they are allowed to include. One or both of the terms also appear with some of the element descriptions in the concise reference at the end of this book. For example, you will find that the `<p>` element should only contain text and inline elements. Because the `<ul>` element is a block element, you shouldn't put it inside a `<p>` element.

Mike: You have already told us about the block-inline categorization of elements based on their default display setting, but I'm still confused. When you are talking about display, isn't that presentation?

Professor: You couldn't be more right about that. In HTML5, a display has become purely a CSS term since it defines the visual behavior of an element. In the past, the categorization of block and inline elements helped authors in deciding which element is allowed as a content (descendant) of the other. In HTML5, this binary categorization has been replaced by a more complex one and you will hear terms like the *flow content*, *sectioning content*, *phrasing content*, and so on. If you are interested in studying these (which I don't actually recommend at this stage), I suggest that you visit the site *developer.mozilla.org*, which is written on a higher technical level than *www.w3schools.com*, but still more understandable for a beginner than the World Wide Web Consortium's (W3C) page, *www.w3.org*, which publishes original web standards and is quite a demanding read.

To keep things simple, some authors equate the flow content category roughly with the block display category and the phrasing content category roughly with the inline display category. This block-inline categorization is easier to understand. That's why it is still used by some authors.

As a rule of thumb, an inline element can only contain inline elements while a block element can contain inline elements as well as block elements. A notable exception to the rule are elements `<p>` and `<h1>` to `<h6>`, which are block elements but cannot contain block elements. Again, when in doubt, you can use an HTML validator.

OK, that's it. Here is a possible final solution of your homework:

```
<!DOCTYPE html>
<html lang="en">
  <head>
    <meta charset="utf-8">
```

```
      <title>Common Wombat</title>
  </head>
  <body>
    <h1>Common Wombat</h1>
    <p><img src="King Island_wombats.jpg">
      The <b class="english-definition">common wombat</b>
      (<i class="latin-taxonomy">Vombatus ursinus</i>), also called
      <b class="english-definition">bare-nosed wombat</b> or
      <b class="english-definition">coarse-haired wombat</b>, is
      one of three living wombat species. The common wombat reaches
      an average of 98 cm in length and a weight of 26 kg. It
      prevails in colder and wetter parts of South East Australia.
      The common wombat was first described by George Shaw in 1800.
    </p>
    <p>
      There exist three subspecies of the common wombat:
    </p>
    <ul>
      <li><i class="latin-taxonomy">V. ursinus hirsutus</i>
        on the Australian Mainland.</li>
      <li><i class="latin-taxonomy">V. ursinus tasmaniensis</i>
        in Tasmania.</li>
      <li><i class="latin-taxonomy">V. ursinus ursinus</i>
        on Flinders Island to the north of Tasmania.</li>
    </ul>
  </body>
</html>
```

Note that text will not wrap around the picture because, as I already pointed out, this cannot be done using only HTML.

2.2 Lists and Tables

Professor: I'm glad that you succeeded in finding and using unordered list on your own. Still, I would like to point out some things about elements <ul> (unordered list) and <ol> (ordered list). Basically, they are the same, only the first one is used when the order in which the items are listed is completely irrelevant, while the second one is used when the order is important. The items in an unordered list are usually preceded by bullets while in an ordered list they come with ordinal numbers or letters. The use of lists is quite straightforward. There's one thing, however, you need to be careful about.

If you look at element descriptions in the reference at the end of this book, you will notice that the elements <ul> and <ol> alike can only contain <li> (list item) elements and nothing else. Since a list is composed of items, this is hardly a surprise. Still, many people find themselves completely at a loss for where to put sublists. As we already discussed, browsers are quite tolerant of bad HTML code and you won't know whether you did it right unless you use a validator. So, let me ask you a question. Where do you think one should put a sublist? Or, more specifically, could you write HTML code for the following list?

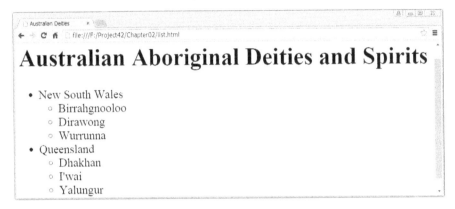

Maria: Let me think. A <ul> element can only contain <li> elements. That means that I cannot put another list inside a list. On the other hand, a sublist *should* be a part of a list. Now, the only possibility I see is to put a sublist inside an <li> element. Am I right? Like this:

```
<ul>
  <li>New South Wales
    <ul>
      <li>Birrahgnooloo</li>
      <li>Dirawong</li>
      <li>Wurrunna</li>
    </ul>
  </li>
  <li>Queensland
    <ul>
      <li>Dhakhan</li>
      <li>I'wai</li>
      <li>Yalungur</li>
    </ul>
  </li>
</ul>
```

Professor: Perfect! When you think of it, a sublist in fact always relates to a specific list item rather than a list as a whole. It is therefore the only logical solution to put a sublist inside a list item.

Lists allow authors to organize document data in a specific way. Another such element is <table>. Just like lists, the <table> element also has its content limited to a small number of allowed direct descendants. Amongst them you'll find an optional <caption> element and an obligatory <tr> element. Each <tr> (table row) element represents a row in a table and its only direct descendants can be <td> (table data) and <th> (table header) elements. The former represent table data cells and the latter table header cells. You can stretch any data or header cell over more rows or columns using their rowspan and colspan attributes, respectively. The <caption> element holds the table caption and should appear before any <tr> elements.

Here's a complicated example:

```
<table border="1">
  <tr>
    <td rowspan="3"></td>
    <th colspan="4">Imports</th><th colspan="4">Exports</th>
  </tr>
  <tr>
    <th>Value</th><th colspan="3">Annual change (%)</th>
    <th>Value</th><th colspan="3">Annual change (%)</th>
  </tr>
  <tr>
    <th>2010</th><th>2008</th><th>2009</th><th>2010</th>
    <th>2010</th><th>2008</th><th>2009</th><th>2010</th>
  </tr>
  <tr>
    <td>Mozambique</td>
    <td>1,200</td><td>10.6</td><td>-2.5</td><td>19.0</td>
    <td>2,600</td><td>5.1</td><td>1.5</td><td>25.4</td>
  </tr>
  <tr>
    <td>Kyrgyz Republic</td>
    <td>300</td><td>1.9</td><td>0.5</td><td>6.7</td>
    <td>500</td><td>2.1</td><td>-1.5</td><td>2.9</td>
  </tr>
</table>
```

And the result in the browser.

	Imports				Exports			
	Value	Annual change (%)			Value	Annual change (%)		
	2010	2008	2009	2010	2010	2008	2009	2010
Mozambique	1,200	10.6	-2.5	19.0	2,600	5.1	1.5	25.4
Kyrgyz Republic	300	1.9	0.5	6.7	500	2.1	-1.5	2.9

Mike: I think I will study this when it's quiet and do some experimenting to see how things behave.

Professor: I just wanted to suggest the same. There's really not much to explain, you'll simply have to try it out by yourselves. However, before you start pulling your hair out figuring out why I used an obsolete `border` attribute, let me tell you that I haven't had any other choice.

Maria: That's true. Border is a matter of presentation, not content.

Professor: Precisely. However, browsers by default don't draw any table borders and the above table would be completely illegible without them. The only way to draw borders properly according to HTML5 standard is to use CSS. In the above example, I resorted to the possibility of limited use of the `border` attribute in the case of absence of CSS. W3C allows that but you can only use one of the two possible values for this purpose: the empty string or the value 1.

Maria: What about `rowspan` and `colspan`? Aren't they also a matter of presentation?

Professor: Think again.

Maria: Oops! I get it. We stretched a header across many columns because they share the same *meaning*. For example, there are four columns about export.

Professor: Exactly!

2.3 Generic `<div>` and `<span>` Elements

Professor: By now, you already understand that writing HTML code is thinking primarily about meaning and structure.

Equally important is that you use the right element for assigning a part of a document the right meaning. For example, it would be wrong to use any of the six headings elements `<h1>` to `<h6>` for a table caption. For that purpose, you should use the `<caption>` element and nothing else.

The difficulty with the older HTML standards was that there really weren't many semantic elements available and authors had to resort to the generic `<div>` and `<span>` elements. In this sense, the HTML5 standard has made a significant step forward by introducing many new semantic elements.

Before I introduce you those newbies, I think it's important that you understand the role of the `<div>` and `<span>` elements. There are two good reasons for that: First, designers have used them in the past very often to organize their documents, which will continue to live on the Web for years to come. Second, they are still very useful in special cases when you don't find a more appropriate element.

`<div>` and `<span>` are called *generic* elements. That means there really isn't any inherent meaning attached to them nor is there provided any default browser rendering for them (except that `<div>` has block display while `<span>` has inline display). Rather, they can be used generally wherever needed. Authors use the `class` or `id` attributes to attach to them a specific semantic meaning, and CSS to make them look any way they want. For example, you can use the `<div>` element to divide a page into logical pieces like headers, footers, banners, sidenotes, and so forth. Later, you can design and position each piece to create page layouts to your liking using CSS.

Say you want to include in your website a copyright notice, which usually comes at the bottom of every page of your website. In typography, a piece of information that is separated from the main body of text and appears at the bottom of the page is called a *footer*. Using the `<div>` element, the code for a copyright notice could look something like this:

```
<div class="footer">
  (C) ACME 2015 All Rights Reserved
</div>
```

With the help of a `<div>` element with the `class` attribute set to `footer`, I have clearly marked the copyright text as belonging to the footer of a page. The same `class` attribute could help me later to access and visually design the element by CSS.

The generic `<div>` and `<span>` elements are useful for giving parts of a document an arbitrary, user-defined meaning. However, one of the ambitions of HTML5 is to bring into play other, more semantic elements to enable authors to easily and more accurately describe the content of their documents. For example, there actually exists the `<footer>` element and you can rewrite the above code as:

```
<footer>
  (C) ACME 2015 All Rights Reserved
</footer>
```

Maria: I see. And now the browser will already know that this is a footer and will render it smaller, won't it?

Professor: Actually, no. The `<footer>` element doesn't do anything on its own except render its content as a separate block. As a matter of fact, none of the new semantic elements will do anything special except display themselves as a block of text. Apart from that, the four sectioning elements that we are going to meet shortly, `<article>`, `<aside>`, `<nav>`, and `<section>` make their headings smaller and that's about it.

Maria: What's the use, then?

Professor: There are many reasons, actually. One of them is code readability. The limitation of the generic `<div>` and `<span>` elements is that they don't carry any information about the structure of the page. OK, authors label their generic elements using `class` and `id` attributes, but the values given to them are quite arbitrary. Sometimes it takes a lot of digging through HTML and CSS files to unscramble what the author actually had in mind. Even if a human can do that unscrambling, a search bot or a screen reader surely can't. The page structure remains a complete mystery to them.

Mike: Are you saying that search engines can profit from the sound page structure?

Professor: Right now the concept is maturing. Designers already use HTML5 semantic elements to build clearer document structures. They can be used to advantage by search engines to build a better page preview, or by screen readers to better guide visually impaired users through a deep forest of sections and subsections.

Let's take another example. Imagine you want to add to your page an enhanced heading, which is not just a title but also includes a subtitle, a byline, and a teaser. For that purpose, the `<header>` and `<hgroup>` elements come in handy:

```
<header>
  <hgroup>
    <h1>Twelfth Night</h1>
    <h2>Or what you will</h2>
  </hgroup>
  <div class="byline">by William Shakespeare</div>
  <div class="teaser">Things get complicated as Lady Olivia flips
  over Cesario (who is really Viola in disguise) and Viola secretly
  loves the Duke, who thinks she is a man.</div>
</header>
```

The `<hgroup>` element can only contain headings (`<h1>` to `<h6>`) and its purpose is to hide all the lower-rank headings and expose only the highest-rank heading to the document structure. That way you can use the `<h2>` heading to include a subtitle without a danger that the rest of the text becomes subordinate to the `<h2>` heading. If that happened, the entire play would end up as one big subsection, which wouldn't make much sense. For a byline and teaser there's no semantic elements; that's why I've used `<div>` elements with the `class` attribute. Again, this attribute could help me visually design both elements later using CSS. I wrapped my complete enhanced heading into a `<header>` element, signifying that everything is in fact the heading of the play.

2.4 Sectioning Elements

Professor: Although the `<header>` and `<footer>` elements are important for structuring a document, they don't contribute to the *outline* of the page. An outline is an important notion in HTML. You can think of it as a table of contents, where each section or subsection has its own title and hierarchical position within a document. In HTML, only the so-called *sectioning elements* contribute to the document outline. There are four of them: `<article>`, `<section>`, `<aside>`, and `<nav>`. While each of the six heading elements (`<h1>` to `<h6>`) also starts a new section, they are not considered sectioning elements per se because they do not *contain* a section. Rather, they only mark the *beginning* of a section. Therefore, their control over sections is somewhat limited.

Mike: Is an outline like a mind map?

Professor: As a matter of fact, it is. A mind map or table of contents—it's all the same. A mind map helps you organize your thoughts, and the structure (big picture) is just as important as the content itself. An outline is like a mind map not so much helping humans directly as aiding different software like screen readers or search engines.

Maria: I'm curious how our last example looks in a browser.

Professor: This is not relevant right now. Although we *will* try and shape it into a pleasing form, that comes later on our menu.

Maria: OK, I understand that the look is not important right now. You mentioned, however, that the <hgroup> element hides all the lower-rank headings. Does that mean that we won't see the <h2> heading?

Professor: The heading is not hidden from the document in the browser window but from the *document outline*. Put differently, if you wanted to create a table of contents, then the <h2> heading from our last example wouldn't be in it.

I would like to show you how the document outlining works with the next example:

```
<body>
  ...
  <article>
    <header>
      <hgroup>
        <h1>10 Things You Should Do Before The Deluge</h1>
        <h2>You Better Hurry Up</h2>
      <hgroup>
    </header>
    <p>...</p>
    <section>
      <h2>Pay Taxes</h2>
      <p>...</p>
      <aside>...</aside>
    </section>
    <section>
      <h2>Visit People</h2>
      <p>...</p>
      <section>
        <h3>Your Mom</h3>
        ...
      </section>
      <section>
        <h3>Your Best Friend From Childhood</h3>
        ...
      </section>
    </section>
    ...
    <footer>
      ...
    </footer>
  </article>
  ...
</body>
```

There are lots of things going on in the code. Yet it is not as complicated as it seems. Every sectioning element defines its own section, and whenever one sectioning element comes nested within another, the corresponding sections become hierarchically related. The first heading element inside a sectioning element takes the role of the heading for that section. If there's no heading element within a sectioning element, then the section is considered untitled. Visualization can probably help you understand the outlining concept, even more so if you can try it by yourselves. I suggest that you take one of the many free outlining tools or *outliners*, and experiment at home to see how it works. For example, *gsnedders.html5.org/outliner*. You can either upload a file or simply copy and paste HTML code into the provided text box. Unlike a validator, an outliner does not require a complete HTML document. You only need to include elements that contribute to the document outline.

After feeding the outliner with our last example, we get the following outline:

1. *Untitled Section*
    1. 10 Things You Should Do Before The Deluge
        1. Pay Taxes
            1. *Untitled Section*
        2. Visit People
            1. Your Mom
            2. Your Best Friend From Childhood

Notice the *Untitled Section* subordinate to the *Pay Taxes* section, which was generated by the `<aside>` element without a heading. By the way, the `<aside>` element is normally used for a content that is related to the main text but not essential for its understanding. So if you remove the `<aside>` element, the main text should still make sense. From the document outline point of view, however, the `<aside>` element does exactly the same as `<article>` and `<section>` elements do.

Mike: What about the main *Untitled Section*? Where does that come from? If I get it right, then the `<article>` element is the main section, which should be assigned the title of the first heading element inside the `<hgroup>` element, shouldn't it?

Professor: Why do you think so?

Mike: Simply because `<header>` and `<hgroup>` elements do not contribute to the document's outline, as you said before. You also said that the `<hgroup>` element hides all its contained headings except the first one, which is exposed to the outline. That's OK because *You Better Hurry Up* doesn't appear in the outline. Still, the main title should be the title of the top-level section, which it is not. Have I missed something?

Professor: No, you're right. I see that you understand the concept, so we are ready to move on.

There is another group of elements called *sectioning roots*. A sectioning root defines a root, or top-level section. The sections inside a sectioning root do not contribute to the outlines of other sectioning roots. Rather, they form a new outline of their own. At this moment, that's not very important, though. The only reason I'm telling you that is because `<body>` is one of the sectioning roots. Therefore, the main *Untitled Section* you see in the above outline belongs to the `<body>` element, which acts as a sectioning root for the entire page (provided there are no other sectioning roots). If you want your page outline to have a title, then you should include an appropriate heading element outside of any other sectioning element (but inside the sectioning root). At home, try to add a heading element outside of the `<article>` element in the above example (either before or after it) to see what happens. As far as the outline is concerned, it doesn't matter where inside an element a heading appears as long as it is the first one—although it is a little weird to write the page title at the end, or even in the middle of the document.

Maria: What happens if you include more headings inside a section? Are they hidden the same way as within the `<header>` element?

Professor: No. Each subsequent heading within a section implicitly starts a new section. If a heading has a lower rank than the one before it, then a new section is created, which is subordinate to the enclosing section. If, however, a heading has the same or higher rank than the one before it, then a new section is created on the same level in the hierarchy as the enclosing section of that heading. The enclosing section is closed even though its closing tag has not been reached yet because the newly created section is on the same hierarchical level. Needless to say, that can cause some confusion. I suggest that for homework you add some headings of different ranks to different places in our last example and observe what happens. That way, if you don't get lost, you will understand things a little better.

Maria: If a heading itself already starts a new section, why then do we need additional explicit section definitions?

Professor: There are numerous reasons for that, many of which may not be quite evident at first glance. Consider, for example, that a website reuses some material from other sites, say in the form of a news feed. Such a process is called *web syndication*. It can easily happen that an article from another web page is pulled under a heading with a rank that is lower than that of the included article. Although the page will still work fine, the hierarchy will become disturbed, which could make the page more difficult for search engines and other software to process. However, if you use HTML5 sectioning elements, the hierarchy of the sections is defined by the placing of the sectioning elements regardless of the ranks of headings within them.

2.5   Hyperlinks

Professor: All our discussion so far was related to the meaningful structure of a single web page. A modern website is, however, more than just a single page. It is important that you can build from different pages an ordered and logical structure where your visitor will promptly and effortlessly find the right information. That's an important part of what we call *web design*, which is unfortunately not the main focus of our course. We concentrate more on how things are done technically, although not in complete oblivion of web design considerations.

Technically, for connecting one web page with another document, we can establish a *hyperlink* by means of an `<a>` element and its `href` attribute:

```
Another species endemic to
<a href="regions/antarctica.html">Antarctica</a> is the
<a href="species/emperor_penguin.html">Emperor Penguin</a>.
```

By default, the content of the `<a>` element with the supplied `href` attribute is usually displayed blue and underlined, which is a sign for a visitor that it can be clicked. Apart from that, the mouse pointer turns into a hand when hovered over it.

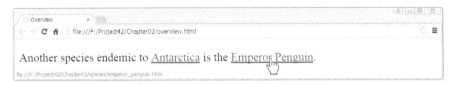

When a visitor clicks a hyperlink, the browser loads a new web resource. The destination of the hyperlink is specified as the string value of the `href` attribute in the form of a *uniform resource locator* or URL, also known as a web address. The URL can be given either in an *absolute* or *relative* form. An absolute URL requires the complete path to the destination page to be given, starting with the communications protocol (for example HTTP, HTTPS, or FTP). After the protocol name, there should be a colon and double slash (://). Next comes the domain name, which identifies a website. For example, *www.taylorandfrancis.com* or *en.wikipedia.org*. Finally, there comes another slash and the full path of the linked resource:

```
protocol://domain/path_of_the_resource
```

The URL of the web page currently displayed in the browser window is usually shown on top of the browser, inside the address bar.

This is an example of an absolute URL used as a hyperlink destination:

```
<a href="http://www.w3schools.com/tags/att_a_href.asp">Read more</a>
```

Mike: What is a communications protocol?

Professor: This is a set of message formats and rules for exchanging messages in or between computer systems. A browser must know how to communicate with its environment in order to be able to load and display documents. In the same way, people hold to certain protocol rules when communicating with other people. For example, the Japanese are very formal and have some strict rules about behavior in public:

- Bow when greeting someone.
- Remove shoes before entering homes and restaurants.
- If someone offers you sake, drink it.

Do not tell anyone, but the last one is actually my favorite.

Don't worry, though. You don't have to know much about protocols to use them in your absolute URLs. You simply copy from the browser's address bar the complete address of the page to which you want to establish a link to and the job's done.

Mike: Now when you mention it, in your penguin example I have noticed a word *file* at the beginning of the URL in the browser's address bar. Is that also a protocol name? It also bothers me that there is no protocol name in your example code. Or did I miss something?

Professor: Yes and no. Yes, FILE is a communications protocol and no, you didn't miss anything. My penguin example really doesn't include a protocol because it is not an absolute URL.

The FILE protocol is shown in the address bar because we are accessing a file directly through the computer's local file system rather than through a web server. Whenever you load a local file in the browser, you will see the FILE protocol displayed in the address bar.

The reason why you don't see any protocol name in the penguin example source code is that I used a *relative* URL in that example. A relative URL is used to locate a resource relative to the context in which it exists. To help you understand a difference between a relative and an absolute URL, I will give you an example from everyday life. If you want to write a letter to a friend in Australia, you have to write on the envelope the whole address beginning with the friend's name and ending with *Australia*. That is an absolute address and will always be the same no matter where you are sending a letter. However, if you already were in Australia, then the address probably wouldn't include *Australia*. And if you were just a few blocks away, then even the name of the street wouldn't be necessary to give someone instructions on how to reach your friend. You would simply say: "Walk down the street till the second traffic lights. Turn left and walk on to the fourth house on the right side of the street and you're there." That is a relative address because it is *relative* to your current location and would change if you moved to another place.

There are two different types of relative URLs:

- *A document-relative URL*: If a URL begins with a directory (also called a folder) or file name, then the URL is relative to the current directory; that is, the directory in which the currently displayed page resides. If, however, the document contains a `<base>` metadata element, then the URL is relative to the base URL given by this element.
- *A root-relative URL*: If a URL begins with a slash (/), then the URL is relative to the root directory of the website.

Maria: In the example with penguins on page 27, the URLs beginning with a directory name are relative URLs, aren't they?

Professor: That's right. They are document-relative URLs.

Mike: I have noticed one more thing on the screenshot on page 27. After the protocol name, there are actually three slashes while you mentioned that there are only two. Why is that?

Professor: As a matter of fact, there *are* two slashes. If a computer hosting a page is the same computer that interprets the URL—which in our case it is—then a domain name is *localhost*, which is usually omitted. A slash after a domain name, however, should not be omitted. That's why we get three slashes.

Back to relative URLs. Sometimes you will see a special notation in the form of a dot (.) or a double dot (..) at the beginning of a document-relative URL. A single dot explicitly represents the current directory. In that sense, the next code is equivalent to our previous penguin example from page 27:

```
Another species endemic to
<a href="./regions/antarctica.html">Antarctica</a> is the
<a href="./species/emperor_penguin.html">Emperor Penguin</a>.
```

With two dots, you can reach the parent directory. Consider there is another directory at the same level as our current directory, named *thingamajigs*.

Maria: I feel like I am losing the thread of your explanation. Can you please sketch the directory tree of the location of these Antarctica and penguin files?

Professor: Yes, of course. Here it is:

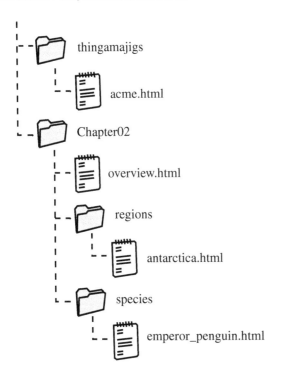

As you might have noticed from the screenshot on page 27, our example HTML is named *overview.html* and located in the directory *Chapter02*, which makes that directory the current one. In the same directory, there are also two subdirectories named *regions* and *species*. The above code fragment demonstrates how to access the files in these two directories from within the *overview.html* file.

Maria: Is our last code fragment located inside *overview.html*?

Professor: It is indeed.

Suppose now, we want to establish a hyperlink from the same *overview.html* document to point to the *acme.html* document, which resides in the *thingamajigs* directory. We can arrive at that directory only by first going to its parent, which is at the same time the parent of *Chapter02*. You move to the parent directory by using two dots:

```
Read about our <a href="../thingamajigs/acme.html">partners</a>
```

Mike: Is this parent directory named *Project42*? At least the address bar on the screenshot on page 27 says it is.

Professor: I didn't include that information into the above directory tree because the name is not important. But yes, *Project42* is the parent directory of *thingamajigs* and *Chapter02*. You can also observe the URL in the lower left corner of the screenshot on page 27 that is displayed when the mouse cursor hovers over the hyperlink. The browser expands the relative URL *species/emperor_penguin.html* to its absolute form *file:///F:/Project42/species/emperor_penguin.html*.

Any more questions?

Mike: Yes. I sometimes notice that certain links open a document not at the beginning but rather at some specific location. How do you do that?

Professor: This is accomplished by so-called *URL fragments*. Here's how it works: you mark a location in your document to which you want to be able to point a link. Usually you do this with the `id` attribute on an `<a>` element:

```
<a id="something-special">Very important indeed.</a>
```

When linking to this location, you use the same id value preceded by the hash sign (#) as the value of the `href` attribute. So if you want to link to the above very important location from within the same file, then you simply say:

```
<a href="#something-special">Read even more.</a>
```

If you want to try this out, you should make a document long enough to scroll in the browser window or you won't notice the effect of following the link.

If you would like to establish a link to the above very important location from another file, all you have to do is prepend the additional (relative or absolute) URL of the destination file to which you are linking:

```
<a href="URL_of_the_destination_file#something-special">
   Read even more.</a>
```

Maria: I suppose the destination file is the file containing the id of `something-special`?

Professor: Exactly.

You will occasionally notice a hash sign (#) alone. Authors sometimes use it as a placeholder in web templates where actual destinations of the links are not known at the time of writing code. Such links, when followed, simply reload the current page.

Maria: Could I use the `<a>` element for an image gallery? For example, to make a clickable thumbnail that would behave like a hyperlink to a larger photo?

Professor: Of course. The HTML5 standard poses no limits as to what the content of this element may be, so long as it is not interactive. For example, you cannot nest a button or another link within a hyperlink. This limitation is understandable since placing interactive content inside the element that is itself interactive would present a conflict of interests. Otherwise, the `<a>` element can contain entire paragraphs, tables, and even entire articles.

A question for you. In the following code, which is a thumbnail and which is a full-size photo?

```
<a href="A.jpg"><img src="B.jpg"></a>
```

Maria: The destination of the link is the *A.jpg* file. Therefore, this must be the large photo. *B.jpg* is the source for the image displayed on the page, which must be the thumbnail.

Professor: Great! We are ready to move on.

2.6   Character Entities

Professor: Consider this HTML:

```
The notation x<y means that x is less than
y, while x>y means that x is greater than y.
```

Let's view it in the browser.

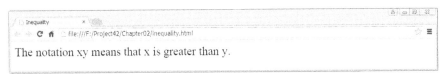

What do you think happened?

Maria: It seems that the browser ignores everything between the angle brackets including the brackets themselves.

Professor: Good observation. What the browser is actually ignoring here is an unknown tag. HTML is a computer language, which follows a certain well defined set of rules that we call *syntax*. Those rules come in the form of combinations of different symbols that define certain commands. They tell a computer what needs to be done. For example, a left angle bracket (<) immediately followed by some text followed by a right angle bracket (>) is an opening tag—a message to the browser that a new element has started. The tag itself is not displayed. The browser does not display the tag even if it does not recognize it. An unrecognized tag is silently ignored.

Let's go back to our example. The problem is not that the tag isn't recognized by the browser, since we didn't use any tag in the first place. The less-than sign has a special meaning for the browser, which interprets it as the beginning of an opening tag rather than displaying it. A character that has a special (reserved) meaning is called a *reserved character*. To be able to display reserved characters, HTML has defined additional syntax rules called *character entities*. Character entities are not limited only to the reserved characters, though. They can be used to display any character. For example, a character that is not present on your keyboard.

The syntax for a character entity is an ampersand (&) followed by an entity name followed by a semicolon (;):

```
&entity_name ;
```

All browsers should support at least these five character entities:

Display Character	Entity Name	Description
<	lt	less than
>	gt	greater than
&	amp	ampersand
"	quot	quotation mark
'	apos	apostrophe

Note that the entity names are case sensitive. There are of course tons of other character entities. If you are interested in experimenting with them, I encourage you to search tables of character entities on the Web.

Now we are able to rewrite our last example:

```
The notation x&lt;y means that x is less than
y, while x&gt;y means that x is greater than y.
```

Instead of entity names, you can also use entity numbers, to which you should prepend an additional hash (#) character. Using an entity number, a character entity can be expressed in the form:

```
&#entity_number ;
```

One advantage of using entity numbers instead of names is that the browser support for numbers is generally better than that for names. I'm talking about more exotic characters, of course, and not the basic five mentioned above. For example, if you would like to display a white chess rook, then you write:

```
&#9814;
```

You can find a list of some common HTML5 entities at *www.tutorialspoint.com/html5/ html5_entities.htm* and some more exotic ones like weather or recycling symbols at *en.wikipedia.org/wiki/Miscellaneous_Symbols_Unicode_block*.

2.7   Homework

Professor: For homework, build a menu system for navigating through pages of a website. Assume the following main menu items: Home, Lighting, Small Home Appliances, Support, and Contact. Let the menu item Lighting include a sub-menu containing items Energy Saving Lamps, LED Lamps, and Halogen Lamps. Under Halogen Lamps add a sub-sub-menu containing the items Halogen Classic and Halogen Spot.

Also write HTML code that displays the following in the browser window.

Maria: Can you give us some more clues as how to build a menu system?

Mike: Likewise, I think we need more detailed instructions.

Professor: OK, I'll give you some hints. A menu system is nothing more than a structured collection of hyperlinks shown on a page. You do not have to point the links to actual documents or you will have to create lots of files. Besides, you don't have to worry about how the menu looks.

Maria: Oh, I see. That shouldn't be too difficult.

Professor: I'm glad to hear that.

As the last thing for today, here is a new list of keywords. Try to connect them with one another to form a mind map, which you can use later as a reference.

In this meeting: structure, presentation, CSS, obsolete elements and attributes, semantics, `<b>`, `<i>`, `<strong>`, `<em>`, `class`, tables, lists, element context, element content, generic elements, `<div>`, `<span>`, HTML5 semantic elements, `<article>`, `<section>`, `<nav>`, `<aside>`, `<header>`, `<hgroup>`, `<footer>`, sectioning element, sectioning root, document outline, `<a>`, `href`, absolute URL, communications protocol, document-relative URL, root-relative URL, URL fragment, unsupported tags and attributes, syntax, character entity, reserved character

Presentation

3.1 Homework Discussion

Professor: How was your homework?

Maria: We did it together and didn't have any serious problems. Maybe because we were able to discuss the questions that came up during the work. Here's the solution of your second question:

```
<p>The character entity for an ampersand is &amp;amp;</p>
```

Professor: Great! I think no discussion is needed here.

Mike: This is the menu part:

```
<ul>
  <li><a href="#">Home</a></li>
  <li><a href="#">Lighting</a>
    <ul>
      <li><a href="#">Energy Saving Lamps</a></li>
      <li><a href="#">LED Lamps</a></li>
      <li><a href="#">Halogen Lamps</a>
        <ul>
          <li><a href="#">Halogen Classic</a></li>
          <li><a href="#">Halogen Spot</a></li>
        </ul>
      </li>
    </ul>
  </li>
  <li><a href="#">Small Home Appliances</a></li>
  <li><a href="#">Support</a></li>
  <li><a href="#">Contact</a></li>
</ul>
```

Since we didn't actually build a whole website, we used hash signs (#) as placeholders for links, like you told us to the last time.

Professor: What pleases me most is that you didn't use a table to build the menu system. It is the look of many menus that fools some people into thinking that they are tables. Logically, a menu is a list of items, but the main menu items are often arranged horizontally, which is a little unusual for a list in general.

Mike: There is, however, one question bothering us. We believe that the same menu will appear at the top of most of the pages of our website. Now, if the pages resided in different directories that are not all at the same level of the directory tree, we would have to use different relative URLs for the same links in the same menu in different documents. That's a lot of work. We believe that absolute URLs are not a solution either. If we had to move the whole website to some other location for one reason or another, then all the absolute URLs would have to change. What shall we do?

Professor: Very good question indeed. There are generally two schools of thought concerning URLs for hyperlinks to the pages within a website. The first teaches that relative URLs should be avoided by all means while the second claims just the opposite. Many arguments have to do with principles well beyond the scope of our course. Personally, I wouldn't use absolute URLs for links within my website. I always test my pages on a local server not open to the public. If I used absolute URLs, then I would be shot off my test server to the already published pages any time I clicked a hyperlink. On the other hand, using relative URLs too clumsily can for example create so-called *spider traps*. Web spiders (also named web crawlers) are programs that systematically browse the World Wide Web, most often for the purpose of indexing. A spider trap is a set of web pages that (intentionally or unintentionally) produce an infinite number of requests from web spiders and may cause a poorly written spider to crash.

In my opinion, if you are consistent and careful with your coding, and you hold to the standard rules, no harm can be done using relative URLs. However, to access files that are not inside the current directory, it is probably the best practice to use root-relative URLs. And by the way, this is also the best solution for a website menu system, which was your original question.

Maria: I see. But then we have to put all our web files and directories into the root directory. I don't think that's a good idea.

Professor: The trick is that the root directory does not have to be unique within a directory tree, which is usually the case under Windows. In fact, each process (program) can have its own idea of what the root directory is. I think that we have arrived at the point where you can learn to set up your first web server.

3.2 Setting up a Web Server

Professor: A *server* is either a physical computer or a computer program to serve the needs of other programs, called *clients*. Since at this stage we will move mostly on the client side, there's really not much you need to know about servers. A server is like a good servant, serving your needs while hiding all the dirty details from you. A web server, for example, helps primarily to deliver web pages when requested from clients by means of Hypertext Transfer Protocol (HTTP). A web server also provides web developers with the means to publish their contents.

Up to this moment, we have been doing just fine without a server because our job has merely been displaying HTML documents inside a browser. Interpreting and rendering HTML content is exclusively the browser's job. *Delivering* the document to the browser is what a web server should worry about. So, for example, if a document is requested by the URL */general/introduction.html*, then, using the Windows local file system, this will translate to *c:/general/introduction.html*, if the request originates from the *C:* drive. However, if the document is requested through a web server, the server could translate the path to some other location, for example *c:/users/meandyou/www/general/introduction.html*.

Mike: What's the point?

Professor: The */general/introduction.html* URL is a root-relative URL. If you are accessing the *introduction.html* file directly through the local file system, then there is only one possible location for the file: in the *general* directory, which in turn should be placed in the root directory of the drive. However, if you are accessing the file through a web server, then you can configure the server to prepend any path you like to the given root-absolute URL.

Maria: That's great! Now we can choose any directory we want to be our site root and use root-relative URLs in navigation menus. Is it difficult to run a server?

Professor: It depends on what type of server you need. Fortunately, you will not have to set up a *production server*, which is a complex task indeed. A production server (also called a live server) serves a website that can be viewed by the public. As for now, you better leave this to one of the many web hosting companies, whose services you can rent quite cheaply.

For developing and testing your website, you need a *development* (also named staging) server, which runs locally on your computer and is not accessible by the public. Setting up this kind of server is fairly trivial and I think you could do it by yourselves. Since you both run Windows, you'll need a so-called WAMP package. WAMP is an acronym formed from the initials of the operating system Microsoft Windows and the three main components of the package: Apache, which is a web server, MySQL, an open-source database, and one of three scripting languages: PHP, Perl, or Python. Don't worry about them, though. We will not use them in our course.

One of the many WAMP packages is *EasyPHP* (*www.easyphp.org*), which is an open source project and therefore freely available. Just go to their website and download and install the EasyPHP DevServer. During the installation, you don't have to do anything except decide to which directory you want the software to install. After the installation has completed, you will find a directory named *data/localweb* just inside the directory you have chosen for the installation. This is your website home, or root directory, and voilà, that's it!

When you build a website, you usually save your main page (the one that a visitor should see first) to a file named *index.html* in the site root directory. This name can be different and is a matter of server settings. The *index.html* file is a page that opens by default when someone visits your website without explicitly writing a file name at the end of a URL.

Now, if you want to test your pages, rename your main page to *index.html* and copy all the files and directories to the aforementioned *localweb* directory. Don't forget to start EasyPHP. When you do, you will find the EasyPHP icon on the Windows taskbar, beside the clock and date in the lower right corner of your screen. If you hover the mouse pointer over the icon, the tooltip should say "EasyPHP (Started)". If you right-click the icon, you will get a menu of options. It is very important that you do not open your web pages by simply opening the files in a web browser. Instead, you right-click the EasyPHP icon and select Local Web from the menu. In place of *file:///C:/...*, you will now see the *127.0.0.1/...* at the beginning of your web address in the browser's address bar.

Mike: Can we try opening our last homework through a server?

Professor: Why not indeed! Just rename your homework file *index.html*, copy it to *localweb*, select Local Web from the EasyPHP menu, and here we go.

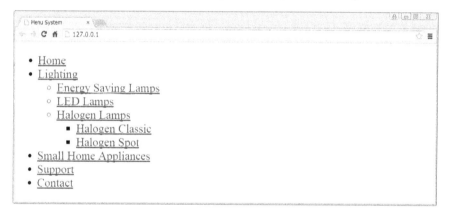

You will notice that *index.html* does not appear in the browser's address bar. By default, the Apache server is configured to hide names of files that it recognizes as default index files.

The menu looks quite ugly, doesn't it? You cannot expect it to look any different because we're looking at the browser's default rendering of a <ul> element. I think it's time now to do some styling.

3.3   Introducing CSS

Professor: Our second language in this course is CSS, which stands for *Cascading Style Sheets*. The language belongs to the group of so-called *style sheet languages*, whose job is to take care of the *presentation* of structured documents, such as HTML documents.

In a nutshell, CSS does just one thing: it tells the browser how things should look on a screen, which is called *styling*. Just like HTML, CSS has its own syntax, which is extremely simple. For example, if you want to paint something blue, you simply say:

```
color: blue;
```

Or, if you want something 200 pixels wide:

```
width: 200px;
```

The above are examples of a CSS *declaration*. A declaration is composed of a *property* and a *value*, separated by a colon (:) and terminated by a semicolon (;). A declaration simply gives a certain value to a certain property.

Mike: I assume that there is a list of CSS properties we can use.

Professor: Of course there is. You'll find a short CSS reference at the end of this book where you can find descriptions of some of the CSS properties and the values these properties can assume.

Maria: How do you tell the browser which element it is whose color and width you want to set?

Professor: There's more than one way to skin a cat, actually. The simplest way is to put the declaration in the right place inside the HTML document. Where would *you* put it? Give it a shot.

Maria: Perhaps inside the element I wanted to style?

Professor: Close, but not quite. *Inside* an element is where you put the element's content.

Mike: I recall that the `class` HTML attribute is used for styling, isn't it?

Professor: It is, although you cannot use the `class` attribute for writing CSS declarations directly. We are coming close, though: a logical place to put a CSS declaration is to write it as the value of an HTML attribute. Only the attribute is not `class` but `style`. Both are global attributes and can be used on any HTML element.

For example, if you want the main heading text to look blue, you write:

```
<h1 style="color: blue;">God, I Feel Blue Today!</h1>
```

Don't forget to use quotes. Of course, you can apply more declarations at the same time by simply writing one after the other, separated by spaces:

```
<h1 style="color: blue; width: 200px;">God, I Feel Blue Today!</h1>
```

Mike: Will all main headings in the document now become blue and 200 pixels wide?

Professor: No, they won't. And, for the most part, that's not very convenient.

This kind of CSS declaration is called *inline* because it is written inside the opening tag of an element. Inline CSS isn't very useful. In order to set or change some property of all the elements of the same type on one page, you must write a declarations for each of the elements separately. If you have lots of pages, you will really work yourself up into a lather to style all of the elements.

In practice, you will exclusively use CSS *rules* (also called *styles*). A rule consists of a *selector* and a *declaration block*. A declaration block is simply one or more declarations enclosed within a pair of curly brackets. However, the real secret hides in a selector, which we are going to unravel later. For now, we'll use the simplest of selectors, which is simply the name of the HTML element you want to style. Here's an example of a CSS rule:

Selector		Declaration Block					
		First Declaration		Second Declaration			
		Property	Value	Property	Value		
h1	{	color:	blue;	width:	200px;	}	

With the h1 selector, you select all the main headers in a document and apply the declarations within the declaration block to all of them simultaneously, which saves you a lot of work.

Mike: Where do we put a rule to make it work?

Professor: You have two possibilities. The first one is called an *internal style sheet*, meaning that you place CSS rules in the same document you're styling. CSS rules come inside a <style> HTML element placed inside document's <head>. Most designers place <style> elements just before the closing </head> tag. For example:

```
<head>
  <!-- ...
    Here come other metadata elements not shown
    in this example.                    ... -->
  <style>
    h1 {                /* Selector and opening brace */
      color: blue;      /* First declaration          */
      width: 200px;     /* Second declaration         */
    }                   /* Closing brace              */
  </style>
</head>
```

In the above example, I used comments to explain different parts of the rule. Note that, inside a <style> element, you should write CSS and not HTML code. Differently from HTML, a CSS comment starts with a slash-asterisk pair (/*) and ends with an asterisk-slash pair (*/). A comment can, of course, span over multiple lines. Same as in HTML, comments are completely ignored by the browser. Remember that you can use comments—apart from clarifying parts of complicated CSS code or making general remarks about the code—also for temporarily switching off certain declarations or rules you don't want to delete completely because you may need them back in the future.

Mike: OK, fine. If I understood this correctly, then I still have to manually apply a style to each and every page of my complete website. That's still a lot of work, isn't it?

Professor: You're perfectly right. Even though a rule will apply to many elements on the page, we still haven't achieved much if there are lots of pages. That's one of the reasons why internal style sheets aren't very popular either.

The second possibility to include CSS rules is an *external style sheet*. This is a separate plain text file containing one or many CSS rules, which you can include in any HTML document using a `<link>` HTML element. An external style sheet should not contain any HTML code, so don't include the opening and closing `<style>` tags. You can name the file whatever you want, yet it is a good idea to use descriptive names. Also, be sure to use the *.css* file extension. For example, you can create a plain text file named *beauty.css* containing the next CSS code:

```
h1 {
    color: blue;
    width: 200px;
}
```

You include this file in your HTML document using a `<link>` HTML element, normally as the last child of the `<head>` element:

```
<head>
  <!-- ...
    Here come other metadata elements not shown
    in this example.            ... -->
  <link href="/css/beauty.css" rel="stylesheet">
</head>
```

The `href` attribute specifies the URL of the desired external style sheet and can accept absolute as well as document-relative or root-relative URLs. The `rel` attribute names the type of the linked resource and must be set to `stylesheet` if you're linking to an external style sheet. If you want, you can include multiple external style sheets simply by using multiple `<link>` elements one after the other.

Mike: You saved the *beauty.css* inside the *css* directory. Is that necessary?

Professor: Actually no. Yet, as your website grows, you'll find it great if you can group files into different directories so you can manage them more easily. I created the *css* directory inside the website root to hold all my styling files.

Consider now that you have tens or even hundreds of pages styled by the above rule. If you change your mind about the width of the main headings, all you have to do is change the value of the `width` property inside one single rule contained in the *beauty.css* file. The rule will automatically come into effect for each and every HTML document that includes it.

Let's briefly summarize what we have just learned. There are three types of CSS: inline, internal, and external. Because of their nature, the only really useful one is external CSS. Inline and internal CSS may be useful for quick experimenting to see how the design might look.

There's another good reason why you should stick to external CSS. It is considered good design if you physically separate structure and content from presentation, which keeps your code cleaner and easier to maintain. So, even if there's only a single element instance you need to style, it is still a good idea to put styling in an external style sheet.

If you don't have any questions, then I think we are ready to move ahead.

Maria: I have a question. How does CSS treat spaces? Are they important?

Professor: In CSS, spaces have pretty much the same role as in HTML. Spaces and newlines are important to make code more readable to humans. For example, I like to put each declaration on a separate line, slightly indented to the right, and with a space after the colon between a property and its value. It is of course up to you how you design your writing. The important thing is that you feel comfortable with the layout of the code.

There is, however, one tiny restriction concerning spaces: you shouldn't put a space between a number and the corresponding unit. For example, if you need to write a value of 200 pixels, the correct form is 200px, but not 200 px.

If you are in doubt, it is always a good idea to validate your code. Just because it works in one browser, it doesn't necessarily mean it is correct and will work in other browsers as well. The same as for HTML, there exist online validators for CSS. For example, you can use the one at *jigsaw.w3.org/css-validator*.

Before you ask: CSS is case insensitive in all aspects under its control. Unfortunately, everything is not under its control and there are situations where case matters. So, same as with HTML, I suggest that you stick to lower-case letters and you'll be fine.

3.4   CSS Values

Professor: We learned that the most basic building block of CSS is a declaration, which assigns a value to a certain property. A question arises: which values are allowed for which properties? To answer it, let's take a closer look at different types of values defined by CSS.

CSS values are categorized into different *data type* groups, and only values belonging to certain data types are allowed to be used for certain properties. There are, for example, data types like <color>, which is used whenever you want to set any color property, or <length>, used for setting different sizes and dimensions. Don't get confused by the angle bracket notation, which resembles an HTML tag. In CSS, angle brackets are used to denote a data type.

Let's start with the most basic <number> and <integer> data types. The first represents a number, either integer or fractional, while the latter represents only an integer. Examples of <number> values would be 42, 1.6, or -12.5, while <integer> values are, for example 7, 99, or -10.

Two more data types that will be useful for us are <length> and <percentage>, used for specifying dimensions and relative values, respectively. The <length> data type is composed of a <number> immediately followed by the length unit, like px or em. Similarly, the <percentage> data type is composed of a <number> immediately followed by the percentage sign (%). I already mentioned that there should be no space between a number and the unit literal. Nor should there be any space between a number and the percentage sign.

Mike: How does that look in practice?

Professor: OK, here's an example. If you look up the font-size property in the CSS reference on page 344, you'll see that a possible value for that property is a positive <percentage> value. Because you know that the <percentage> data type is defined as a <number> immediately followed by the percentage sign, you know exactly that the following are all valid declarations giving a value to the font-size property:

```
font-size: 75%;
font-size: 12.5%;
font-size: 120%;
```

Maria: And what exactly is the effect of setting a font size to 120 percent? 120 percent of what?

Professor: Of the parent's element font size. I'll explain more about that next week when we talk about relative sizes in general. Let's return now to CSS data types.

An interesting one is the <url> data type, which lets you specify the URL of a file. Most usually, that will be a file containing graphics of some kind. For example, if you want a nice picture in the background of an element, you declare:

```
background-image: url(/backgrounds/nicepicture.jpg);
```

You must use the url() function and put the desired URL in parentheses. You specify a URL in the same way as you specify a URL of the href HTML attribute: you can use either an absolute or document- or root-relative URL. Note that relative URLs are relative to the URL of the style sheet and not to the URL of the web page.

In the above example, there are no quotation marks around the URL. However, if there are any spaces within a URL, then quotes must be used. You can either use single or double quotes:

```
background-image: url('/backgrounds/nice picture.jpg');
background-image: url("/backgrounds/even nicer picture.jpg");
```

It is a good practice that you choose one type of quotes and stick to them consistently.

Apart from data types like <url> or <percentage>, there is also a special type of values, called *keywords*. Keywords are used for values that fall under none of the standard CSS data types, such as, for example, `left`, `right`, `center`, and `justify`, used for horizontal text alignment. For example:

```
text-align: right;
```

Other examples of using keywords would be specifying colors (for example, `blue` or `silver`) or even sizes (for example, `large` for the `font-size` property).

Speaking of colors, there exists a <color> CSS data type, which is a little more complex than those we mentioned so far. The most-often used method for specifying color is the RGB model in which red, green, and blue light are added together in various intensities to produce the desired color. The name RGB comes from the initials of the three primary colors: red, green, and blue. There are three ways to specify RGB color. The oldest in use is hexadecimal, where each of the three primary colors is given an intensity as a hexadecimal value from 00 to FF, which correspond to decimal values from 0 to 255.

Mike: Excuse me, what's hexadecimal?

Professor: Hexadecimal (also base 16, or hex) is a positional numeral system with a base of 16. It uses sixteen separate symbols, which consist of nine digits 0–9 to represent values zero to nine, and six (lower-case or upper-case) letters A–F to represent values 10 to 15. Depending on the position of a symbol, its value can be multiplied by different powers of 16, in the same way as the digits in a base-10 system are multiplied by the powers of 10. For example, AB7 in hexadecimal equals $10 \times 16^2 + 11 \times 16^1 + 7 \times 16^0 = 2743$ in decimal. A is 10 and B is 11, of course.

Now, let's try to mix some outer space blue:

```
color: #414A4C;
```

You need to prepend a hash sign (#) before the three two-digit hex values to announce that we are dealing with hexadecimal values. In the above example, we added together red, green, and blue with the hex intensities of 41, 4A, and 4C, respectively.

Maria: It seems that you need a lot of experience to be able to produce the color that you want. Isn't there an easier way?

Professor: Every serious graphics program will let you select a color using your mouse on a visual palette. You simply select the color you want and copy the RGB values to your style. There are also many free color-picking websites available, for example, *www.colorpicker.com*. You can also pick a color from a list like the one at *www.colorhexa.com/color-names*.

If you are not comfortable with hex numbers, you can still use the RGB model using either three <integer> values between 0 and 255 or three <percentage> values from

0 to 100. The integer number 255 corresponds to 100%. Instead of a hash sign, you must use the `rgb()` function and list the color intensities inside parentheses, separated by commas. That way, the same outer space blue can be defined as:

```
color: rgb(65, 74, 76);    /* Integers   */
```

or, using percentages:

```
color: rgb(25%, 29%, 30%); /* Percentages */
```

Note that you cannot mix <integer> and <percentage> values within a single `rgb()` function.

Mike: You mentioned that keywords are also used for colors. Which are they?

Professor: There are many of them. Apart from the 17 keywords defined by CSS2, most browsers now support all of the colors defined by the SVG standard. The 17 CSS2 colors are: aqua, black, blue, fuchsia, gray, green, lime, maroon, navy, olive, orange, purple, red, silver, teal, white, and yellow. If you would like to know more about colors in CSS, then you may want to read about CSS <color> data type at *developer.mozilla.org/en-US/docs/CSS/color_value*.

As you see, there are many ways to specify colors and they all work fine. For consistency, I suggest that you decide on one of them and stick to it. For example, I prefer using hexadecimal notation with an occasional use of keywords only for basic colors like white, black, red, or yellow, for example. It is more convenient and self-explaining to write `color: yellow;` than `color: #FFFF00;`

Maria: I have noticed that some pages use a special effect so you can see through colors. Is that difficult to produce?

Professor: Not at all. You simply add transparency to the RGB model using a so-called *alpha channel*, which specifies the amount of transparency as a <number> between 0 (completely transparent and hence invisible) and 1 (fully opaque). You cannot express this model in hex notation. Instead, you use the `rgba()` function with either three <integer> or three <percentage> values for RGB values. As the fourth value, you specify a <number> from 0 to 1 for the alpha channel. For example, a semitransparent green color will be declared as:

```
color: rgba(0, 255, 0, 0.5);   /* Green, half transparent */
```

All right, enough colors. Let's move on.

Maria: Just a question. I understand that CSS selectors allow you to select an element to which you want to apply the color. But how do you control which part of an element will get the color? I mean, in all your examples you only used the `color` property. What if you want to color just text, or just background?

3.5 CSS Properties

Professor: Oh yes, thanks for asking. I completely forgot. That information is implied in the name of the property. For example, the `color` property is defined to specify the color of text. It should in fact be called `text-color` but, for historical reasons, it is not. You can find some information about different color properties in the CSS reference at the end of this book, where you'll discover properties like `border-color` or `background-color`. Those, I believe, are more self-explanatory.

For example, if you want to annoy your visitors by making all text inside paragraphs white on a yellow background, you declare:

```
p {
    color: white;
    background-color: yellow;
}
```

Note that, because of the p selector, this only changes the text and background color of paragraphs. Headings, for example, retain black text on a white background.

Maria: What if you put another element within a paragraph? For example `<strong>`. Will the text inside it also become yellow?

Professor: As a matter of fact yes, because the `color` property is *inherited*. We will talk about inheritance next time, though.

To further whet your appetite, here's one more example. Later, you can experiment with different CSS properties at home. Say you don't like the default list numbering using decimal numbers and want to change it to use lower-case roman numerals instead. If you look up list properties in the CSS reference on page 349, you will find a property named `list-style-type`. This property allows you to select different bullet types for unordered lists as well as numbering types for ordered lists. Set the `list-style-type` property to the `lower-roman` value in order to get lower-case roman numbering of list items:

```
ol {
    list-style-type: lower-roman;
}
```

Sometimes there exist more properties that style the same aspect of an element. For example, you can style a font by setting things like a font size using the `font-weight` property, a font weight using the `font-size` property, or a font face using the `font-family` property:

```
font-weight: bold;
font-size: 20px;
font-family: sans-serif;
```

You can set all these properties simultaneously using the `font` *shorthand property*:

```
font: bold 20px sans-serif;
```

There exist shorthands for many groups of CSS properties and you can find their descriptions in the CSS reference at the end of the book. Although shorthand properties can come in handy as they save you time and make style sheets more concise and readable, there are some caveats to keep in mind when using them.

First of all, they set *all* the properties they are designed to set, even those that you do not specify explicitly. OK, it's true that they set them to default values, which can't hurt. Except if you have set those properties to different values before—default values provided by a shorthand property will override them. For example, if you declare:

```
font-style: italic;
font: bold 20px sans-serif;
```

then your font won't be set to italic. Namely, the `font` shorthand property sets `font-style` as well, whose default setting is `normal`.

The second thing you should be careful about is the order of the values of the properties that a shorthand property replaces. Generally, shorthand properties do not require a specific order of the supplied values. If the values are of different types, then there is no doubt which is which and the order is really not important. But as soon as there are more values of identical types, their order becomes important. The order is even important in certain cases where the values are of different types. For example, the `font` shorthand property demands that values for the `font-size` and `font-family` properties are supplied second to last and last, respectively. What's more, they are even required or the whole declaration is ignored. A best practice is that, before using a shorthand property, you carefully read what and how does it set and that you always stick to the same order of values even when the order is not important.

3.6   CSS Pixel Unit

Professor: Finally, I would like to say a few words about units, and then I will give you homework. Because CSS units in general need a little more discussion than time allows today, I will just explain what you need to know to complete your homework.

You have learned that for specifying sizes and dimensions, you use the <length> data type, which is a <number> immediately followed by a length unit. The only unit I want you to use right now is *pixel*. Historically, the name is derived from the smallest dot a computer monitor could display, which was called a pixel. Today, devices can display much smaller dots, which could not even be visible to the naked eye but only through a magnifying glass. Nevertheless, pixels survive in the same way as they appeared in older documents.

While the definition of a pixel is highly context-sensitive, a CSS pixel (px) could be defined as the smallest visible length that can be sharply displayed. What is small, visible, and sharp depends on the type of device and its normal use. The important thing is that the pixel always has the same visual appearance, no matter on which device it is shown. Many designers prefer pixels because it is easy to align objects whose lengths are expressed in pixels and because everything that is a pixel or a multiple of a pixel wide is guaranteed to look sharp. Apart from that, a pixel in CSS is considered by some to be an absolute value measuring approximately 1/100 of an inch (0.25 mm).

For example, to specify the font size of 18 pixels for paragraphs, you declare:

```
p {
    font-size: 18px;
}
```

Incidentally, if you do not specify a font size, the default base font size in most browsers is 16 px. The base font size is the one used for normal text, like paragraphs. Don't rely on a browser's default settings like that, though. To stay on the safe side, it's always a good idea to set all sizes by yourself.

3.7   Homework

Professor: For homework, I have prepared for you the following HTML document:

```
<!DOCTYPE html>
<html lang="en">
<head>
  <meta charset="utf-8">
  <title>English Fairy Tales by Joseph Jacobs</title>
  <link href="/styles/fairytales.css" rel="stylesheet">
</head>
<body>
  <article>
    <h1>English Fairy Tales by Joseph Jacobs</h1>
    <h2>The Rose-Tree</h2>
    <p>
      There was once upon a time a good man who had two children: a
      girl by a first wife, and a boy by the second. The girl was
      as white as milk, and her lips were like cherries. Her hair
      was like golden silk, and it hung to the ground. Her brother
      loved her dearly, but her wicked stepmother hated her.
      <q>Child,</q> said the stepmother one day, <q>go to the
      grocer's shop and buy me a pound of candles.</q> She gave her
      the money; and the little girl went, bought the candles, and
      started on her return. There was a stile to cross. She put
      down the candles whilst she got over the stile. Up came a dog
      and ran off with the candles.
    </p>
    <p>
      She went back to the grocer's, and she got a second bunch.
      She came to the stile, set down the candles, and proceeded to
      climb over. Up came the dog and ran off with the candles.
    </p>
```

```
<p>
    She went again to the grocer's, and she got a third bunch;
    and just the same happened. Then she came to her stepmother
    crying, for she had spent all the money and had lost three
    bunches of candles.
</p>
<p><a href="#">Continue...</a></p>
<footer>
    From the e-book English Fairy Tales, collected and
    edited by Joseph Jacobs. Belongs to the public domain.
    Source: www.authorama.com
</footer>
    </article>
</body>
</html>
```

Your job is to produce the following output, but you are only allowed to use CSS rules and put them into a file named *fairytales.css*.

I suggest that you browse through the CSS reference at the end of this book to find suitable properties and values to complete the job.

Mike: I have a question. The fairy tale doesn't end on the page but you already closed the `<article>` element. How does that influence the document outline?

Professor: You're right, that is a problem. However, there is presently no better solution. The current general agreement in the web community is that, when an article is split over many web pages, each part should be contained in its own `<article>` element, even if it is not complete and self-contained. This compromise illustrates how in practice semantics and presentational aspects of the Web don't always quite go hand in hand.

Before we leave, I shouldn't forget the list of today's keywords:

In this meeting: server, client, development server, production server, WAMP, *index.html*, CSS, style sheet language, declaration, property, value, rule, style, in-line CSS, selector, internal CSS, embedded CSS, `<style>`, external CSS, `<link>`, spaces, CSS validator, keywords, data types, <integer>, <number>, <length>, <percentage>, <url>, <color>, RGB, hexadecimal, RGBA, alpha channel, transparency, shorthand properties, units, pixel

More Control over Style

4.1 Homework Discussion

Professor: I'm anxious to see what you have produced for your homework. Please, I'm all ears.

Maria: The homework wasn't very difficult, in fact. What occupied us most was studying the CSS reference to find the appropriate properties to use. First, we focused on the article and we came up with this solution:

```
article {
  width: 500px;
  padding: 30px;
  border: 1px solid;
  font-family: sans-serif;
}
```

There's only one thing we are not certain about. We wanted to include some spacing around the article text and padding did the trick. But we are confused about the exact differences between margin, border, and padding.

Professor: You used the padding and border properties correctly. Namely, the padding specifies the space between the element's border and its content. The border can be made visible, but that is not necessary. The margin is the space outside of the element's border. We'll discuss these three measures in detail in connection with one very important concept, the so-called *CSS box model*, which is on our next meeting's agenda.

Maria: I see. Next, we attacked the two headings. We weren't quite sure how to produce the parentheses around the main heading. Then it occurred to us that we could probably use borders with rounded corners. Again, we added some padding to move the heading text away from the parentheses:

```
h1 {
  border-left: 2px solid;
  border-right: 2px solid;
  padding-left: 10px;
  padding-right: 10px;
  border-radius: 10px;
  font-size: 32px;
}

h2 {
  background-color: black;
  color: white;
  font-size: 22px;
}
```

The styling of the footer is quite straightforward:

```
footer {
  border-top: 1px solid;
  font-size: 12px;
}
```

We also removed the underlining from the link:

```
a {
  text-decoration: none;
}
```

What we haven't solved is how to right align the text of the link. Applying the text-align property to the <a> element didn't work. So we applied the text-align property to paragraphs. Like this:

```
p {
  text-align: right;
}
```

Of course, with this we right aligned the other two paragraphs as well. Still, that's the closest we could come. Why doesn't aligning work on the <a> element?

Professor: Because the text-align property aligns text (as well as other inline content) relative to the borders of the element that contains that text. It does not align the *element*. Hence, if you apply text-align to <a>, the text inside <a> will be aligned along the borders of <a>. Since <a> is not a block element, it does not extend over the whole width of the article but rather squeezes tightly around its content (i.e., the text "Continue..."). That's why the text cannot move any more to the right of the <a> element than it already is. Just for fun, you try to add the declaration

```
border-style: solid;
```

to <a> and <p> elements to be able to observe the position of the borders of elements.

By the way, there's a very handy tool for observing element borders already integrated inside Google Chrome. It is very useful and I will demonstrate how to use it in our next meeting.

Actually, without changing the HTML code that I gave you, there's no way you could right align the link without right aligning the two paragraphs as well. Although you can try the following stunt:

```
a {
  display: block;
  text-align: right;
}
```

One of the problems with that is that now every <a> element becomes a block element. It's true that there aren't any other <a> elements in our example, but you should generally avoid solutions like that.

Maria: If it's working, why?

Professor: It is considered a bad style to change something as basic as is the element's display on well-defined HTML elements in order to get some results that can be achieved more elegantly by other means.

Mike: Another reason is that block elements are not permitted inside paragraphs, are they?

Professor: Right, *block* elements. <a> is not a block element even if you change its display. Remember that the block-inline HTML categorization refers to the default display value of an element. If you change it using CSS, that doesn't change the element's categorization as viewed by HTML. Yet you don't usually want to change an element's display if you have other options.

Mike: I see. And what would be the right solution then?

Professor: You can solve the problem by using the right selector. Today, you will learn how to use different selectors, which is one of the most important skills for efficient CSS coding.

Before we continue, I would like to ask you one last thing about your homework. I've noticed that you've defined font sizes for the headings and footer but you didn't specify the font size for the article. Was that on purpose?

Mike: Yes, we wanted to use the default font size for the ordinary text. You said the other day it was 16 px, didn't you?

Professor: If you use pixels for font sizes, it is not a good idea to rely on default sizes. That is yet another topic that awaits us today.

4.2 Class Selectors

Professor: Let me briefly summarize what we have learned about CSS so far. You can style HTML elements using CSS rules. A CSS rule consists of two basic parts: a selector and a declaration block. Although the declaration block takes care of all the formatting, the real power of CSS hides in the selector. With the selector, you take charge of your page's appearance, telling CSS *what* it is you want to style.

The type of selector that we learned about in our last meeting is the most general one. At the same time it is also the simplest to use and understand. Because it lets you select elements only according to their names, it is called an *element selector* (sometimes also a *tag* or *type* selector). It is not hard to identify element selectors in a CSS rule, for they are named after the elements they style. For example, body, p, h1, a, and so forth. Although they are efficient and easy to use, element selectors have their drawbacks. In your homework, for example, you stumbled upon a problem of how to right align a hyperlink within just one paragraph while keeping text inside other paragraphs left aligned. That is something that you cannot do with an element selector.

Mike: What about an inline CSS declaration? We can use it to style a single element instance, can't we?

Professor: Technically, yes. But even if you want to style a single element, it is not recommended to use inline declaration. For one thing, using inline CSS means mixing structure (HTML) and presentation (CSS), which is considered bad style. Apart from that, you will often want to repeat the same single element instance styling in more than a single file, as you can imagine could happen with the *Continue...* hyperlink from your homework. A more elegant way of tackling the problem is using a so-called *class selector*. You create a class selector simply by making up an appropriate name for it, preceding it with a period (.). For example, let's name our class continue and construct the following rule with the continue class as a selector:

```
.continue {
  text-align: right;
}
```

Notice that a declaration block is no different than that used with element selectors. There is, however, an additional step required if you want that to work. The browser won't know what you want to style unless you explicitly mark an HTML element you wish to style. For that purpose you use the class global attribute on the element. Like this:

```
<p class="continue"><a href="#">Continue...</a></p>
```

Be careful not to include a preceding period with the value of the class attribute. A dot is not a part of a class name, but it is used in a CSS rule to signal that a class name follows rather than an element name.

Mike: If I understood correctly, with a class selector you can actually format just one of the many <p> elements on the page.

Professor: That's right. And not just one, as a matter of fact. You can select one or any number you want, and what's more, the type of an element isn't a limitation, either. You can use the same class selector to match elements of different types. For example, if you want some of your text to bear some special meaning, then you can define the following rule:

```
.something-special {
    ...
}
```

Next, you set the `class` attribute to `something-special` on any element that you want to mark (and design) as special:

```
<h1 class="something-special">Main Special Heading</h1>
<h2 class="something-special">Not Main but Still Special</h2>
<h2>I'm Quite Normal</h2>
<p class="something-special">I'm as special as it gets.</p>
```

Maria: How do I know what names I can give to my class selectors. Is there a list like there is for HTML elements or CSS properties?

Professor: No, there isn't. You can invent any name you like, but you must follow certain rules. Although the CSS specification allows a little more, I suggest you stick to the following:

- Only letters (upper case or lower case), numbers, hyphens (-), and underscores (_) are allowed in class names.
- A name must always start with a letter. For example, `50cent` isn't a valid class name, but `b52s` is.
- Spaces aren't allowed in a class name. You can use classes like `next_exit` or `next-exit`, but not `next exit`.
- CSS class names—like everything else in CSS—are not case sensitive but the `class` HTML attribute is. Effectively, that makes CSS class names case sensitive as well. For example, if a `class` HTML attribute is named `thingamabob`, then CSS class names `Thingamabob` or `THINGAMABOB` won't match that HTML attribute. However, don't use two identically spelled and just differently cased names for two different classes because CSS won't be able to tell the difference. As I already said, sticking to lower-case letters will make your life easier.

Again, don't forget to put a period before a class name in a CSS rule, which must not, however, appear in the `class` HTML attribute's value.

Although names like `h1` or `strong` are perfectly legal class names, try to avoid them if you want to stay out of trouble. Also, try to use descriptive names like `copyright` or `postal-address` that express the semantic purpose of the element. Avoid names that describe the presentation of the element such as `yellow` or `italic`. Semantic names retain their logical meaning even if the page presentation changes. Also don't use class names like ddd or qqq out of laziness, because you'll later regret it trying to figure out what they do and finding yourself constantly browsing your old CSS style sheets for how they are defined.

There's one more possibility for using a class selector. If you combine a class selector with an element selector, then you limit the selected elements only to those that match both element and class selector at the same time. For example, if you want only those `<p>` elements that also have a class named `continue` to be selected, you write this more specific CSS rule:

```
p.continue {
   text-align: right;
}
```

Be careful not to insert any spaces before or after a period. Using this rule, the following element will be selected and styled:

```
<p class="continue"><a href="#">Continue...</a></p>
```

The `<div>` element defined by the next HTML code, however, won't be affected by the above CSS rule, even though it is marked by a class named `continue`:

```
<div class="continue"><a href="#">Continue...</a></div>
```

4.3   ID Selectors

Professor: Another way of selecting and styling a specific element is using the ID selector. The principle is very similar to that of the class selector except that instead of a dot you put a hash symbol (#) before a selector name in a CSS rule. For example:

```
#continue {
   text-align: right;
}
```

If you want an HTML element to get selected and styled by the above rule, you use an `id` HTML attribute:

```
<p id="continue"><a href="#">Continue...</a></p>
```

According to the specification, the only limitations concerning a name you can choose for an ID selector is that it must contain at least one character and must not contain any space characters. For practical reasons, however, I advise you to respect the same rules for naming ID selectors as I gave you for class selectors, including the matter of case sensitivity.

Maria: Excuse me, but I can't see any difference between a class and an ID selector.

Professor: There's actually only a tiny difference, which is conceptual rather than functional. You know, the `class` HTML attribute is used to characterize a group of elements that share some common qualities. On the other hand, the `id` attribute is used to identify an element within an HTML document *uniquely*. Consequently, you cannot have more than one element with the same ID in a document. You will use classes on more general elements like sidenotes or captions. IDs, on the other hand, are more appropriate for things like a page header, footer, or main menu, which are supposed to be unique on a page. If it's too much for you, you can just as well forget IDs for now and only use classes. Anyhow, later in our course we will return to IDs.

Maria: It makes sense.

4.4   Grouping Selectors

Professor: Good. We can continue, then. Using classes is a practical way to style more elements at the same time. Yet, there are examples where this approach would be somehow awkward. Consider, for example, that you want all six heading elements to appear in italics and in blue. Constructing a class for that purpose would mean that you would have to apply a `class` attribute to every heading element on your website. Instead, you can simply group more selectors in a CSS rule by separating them with a comma:

```
h1, h2, h3, h4, h5, h6 {
    font-style: italic;
    color: blue;
}
```

Mike: That's nice. What if I wanted some of them a little different? For example, if I wanted `<h1>` and `<h2>` to be italic and blue and also underlined, and the other four just italic and blue? I guess in that case I would still have to write a separate rule for the `<h1>` and `<h2>` headings.

Professor: That's right. But only for the part that's different. In addition to the above rule you would just have to write this short rule:

```
h1, h2 {
    text-decoration: underline;
}
```

And you can even override some properties already declared in a group. For example, you can change the color of the `<h6>` element from blue to gray:

```
h6 {
  color: gray;
}
```

The dangerous thing here is that you may quickly produce a mess of rules so tangled up that you lose control. My piece of advice is not to try to be too clever and to keep things simple. Another benefit of keeping things simple is that a visitor to your page will be thankful as well. It is much easier to find information on a simple, cleanly designed and well-structured site than on the one cluttered with a potpourri of fonts, colors, and other unearthly creations.

Mike: I will remember that. But as you already mentioned that the property can be overridden—how do you know which property value will win if the property is defined twice?

Professor: Such conflicts are resolved by a CSS mechanism called *cascade*, which we will cover in more detail later. In the example above we have two declarations of the font color for the <h6> element, which are both equally important. It is somehow intuitive to expect that the most recent one will win, which is gray in our case. If you try to move the rule for the <h6> element above the first group rule for the elements <h1> to <h6>, then your <h6> element will of course become blue again.

4.5   Nesting Selectors

Professor: There is a danger that you confuse grouping of selectors with nesting. Consider, for example, the following HTML code fragment:

```
<a>anchor</a>
<strong>strong</strong>
<a><strong>strong anchor</strong></a>
```

Let us style this with the following CSS rule:

```
a, strong {
  border-style: solid;
}
```

Because we have grouped our selectors, we get a border around both elements as you can see on this screenshot.

We also see a little irregular composition of two borders one over the other around the words "strong anchor". This occurred because the words are placed within an element that is placed into yet another element, each of them having its own border.

Imagine now a little different style:

```
a strong {
  border-style: solid;
}
```

Mike: I see no difference.

Professor: Look carefully.

Mike: Oh, I see. The comma is missing.

Professor: Exactly! And that comma makes a huge difference. This time we don't group selectors any more but *nest* them. By the way, such selectors are also called *descendant selectors*. The term is connected with the concept of family relations of HTML elements, which I already mentioned some time ago.

The above CSS rule now reads: apply the styling to every <strong> element that is placed *inside* an <a> element. As a result, only the last two words have a single border placed around them, as you can see on the screenshot.

The border belongs to the <strong> element inside an <a> element, of course.

A question for you: what do you think is the difference between the following two rules?

```
p.important {
  border-style: inset;
  border-color: red;
}

p .important {
  border-style: inset;
  border-color: red;
}
```

Maria: Wait a minute. I see no comma this time, but there's a space in the second rule. The first rule applies to all the <p> elements with the class attribute set to important. And the second one…. I don't know if the space before the period matters.

Professor: Yes, it's tricky. You were right about the first one: we already learned today that the name of the element (in our case p) placed *directly* before a period limits the scope of the class selector to that specific element only. So, as you said, the first rule applies to any <p> element having the class attribute set to important. For example, the first rule (but not the second one) will produce a border around this paragraph:

```
<p class="important">
  This is utterly important!
</p>
```

In CSS—just like in HTML—spaces usually don't have a special meaning when there's already some other symbol separating two identifiers. For example, the following lines all bear exactly the same meaning:

```
border-color:red;
border-color: red;
border-color : red;
```

Recall, however, that when you nest selectors, spaces are used as separators between them. And because .important is already a valid selector by itself, the second of the above two rules is an example of *nesting* selectors. It applies to any element *contained* in the <p> element, and having the class attribute set to important. For example, the second rule (but not the first one) will produce a border around this paragraph:

```
<p>
  <em class="important">This is utterly important!<em>
</p>
```

Mike: Does that mean that I can group and nest any selectors and not just element selectors as you did in your first examples?

Professor: You most certainly can.

Before we continue, I want to mention an *asterisk selector* (*), which is a universal selector shorthand for selecting all HTML elements on a page. For example, you can set a font for each and every element on a page using this rule:

```
* {
  font-style: italic;
  font-size: 16px;
}
```

You have to be careful, though, when using the asterisk selector, since it does not apply to any specific element. Because of that, you may experience unpredictable effects with certain properties for which you may not know just how they behave with different types of elements. Most likely, you will use the asterisk selector nested inside another selector, limiting the asterisk selector's scope to the descendants of the selected element only.

Maria: As we already discovered, you can also set a font for a whole page by applying the font styling to the <body> element. Font CSS properties are inherited, didn't you say that?

Professor: They are indeed. You can use the next rule as well for almost the same effect:

```
body {
  font-style: italic;
  font-size: 16px;
}
```

I said "almost," because there is a fundamental difference between both rules. To explain it, I first need to clarify in more detail how inheritance and cascade work, and tell you about relative units.

4.6   The HTML Ancestry Tree

Professor: As it turned out, you already know by intuition how to use inheritance. To use it properly, however, the intuition is not enough, so we must have a closer look at it.

To explain how inheritance works, I must first tell you about the family relationships between HTML elements. You'll recall that I've already used terms like *descendant* and *child* elements. Every HTML document consists of several elements that are related to one another in the same way as the family members are. The only difference is that an HTML element can only have one parent. We can even draw a family tree for an HTML document. For example, the document on page 48 that you had for your last homework has the family tree shown below.

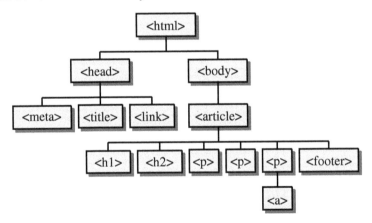

In computer terminology, each value inside a tree (represented by a rectangle in the above picture) is called a *node*, and the top node is called the *root node*. Note that only element nodes are shown in the above tree while an HTML document also contains the document node and text nodes, which are not interesting for us right now.

Fortunately, there are not as many important relations between HTML elements as there are between human family members. There are no cousins or mothers-in-law, for example. You must only know these:

- *Ancestor*: An element that contains other elements is their ancestor. In the above family tree, the `<html>` element is an ancestor of all other elements, while the `<article>` element is an ancestor of `<h1>`, `<h2>`, `<footer>`, `<a>`, and all three `<p>` elements.

- *Descendant*: An element that is contained inside another element is its descendant. In our HTML family tree above, the `<h1>` element is a direct descendant of the `<article>` element but it is also a (more distant) descendant of the `<body>` and `<html>` elements.

- *Parent*: A parent of an element is its closest ancestor. In our example, the `<body>` element is the parent of the `<article>` element, while the `<article>` element is the parent of both headings, all three paragraphs, and the `<footer>` element.

- *Child*: A child of an element is its closest descendant. In the above family tree, the `<title>` element is a child of the `<head>` element.

- *Sibling*: All the elements that are children of the same parent are called siblings. They are just like brothers and sisters. In the above example, both headings, all three paragraphs, and the `<footer>` element are siblings.

Mike: Are `<title>` and `<article>` also siblings?

Professor: You must be careful. Although they are on the same level in the three, they have different parents. Therefore they are not siblings.

4.7 Inheritance

Professor: One of the more important consequences of the family relations between HTML elements is the ability of elements to *inherit* certain traits from their ancestors. Because elements only have one parent, it is easier to predict which characteristics will be passed on to the next generation than it is with humans.

In practice, inheritance works very simply: property values are propagated from parents to their children. For example, if you set font type for a paragraph, then all the child elements of any paragraph will inherit the same font type. If any of those elements have their own children, the same font type will pass even further down to the children of the children.

Maria: When we set the `font-family` property of the article for our homework, inheritance was indeed helping us. The same font family was automatically applied not only to the article's children but also to the `<a>` element, which in our case was a more distant descendant of the `<article>` element.

Also, because `<body>` is the ancestor of all the elements on the page, setting a font on `<body>` will set the font for the whole page.

Professor: Exactly.

Maria: On the other hand, the `border` property was not inherited. Why is that?

Professor: That's for practical reasons. The main reason for the inheritance mechanism to exist is to save you time and energy. For certain properties it is very useful that they pass down to descendant elements while for others it would be extremely frustrating if they were inherited by descendants.

Say, for example, that you write a paragraph containing many other elements like <em> to emphasize parts of text or <a> to include hyperlinks. Besides this, you select some special font shape for your paragraph. It would be quite ridiculous if all text inside the <em> and <a> elements returned to its default style. You would then have to create additional styles for every element nested within the paragraph. Worse still, you would have to be careful when doing that because there might be some other <em> and <a> elements inside of some other section of your document, which you wanted to style differently. That's why font CSS properties automatically pass onto descendant elements.

Thanks to inheritance, you can style a selected portion of your page by simply applying a style to a generic <div> element, or you can use any of the new HTML5 semantic elements like <article> or <footer>.

Your question was why border wasn't inherited. Many properties don't get inherited, and there's a good reason for that. Imagine that the border property was inherited by all the descendants of, say, an article. Then every single <p>, <em>, <a>, or any other element inside an article would also be garnished by a border. What a mess.

Maria: And how do we know which properties are inherited?

Professor: Basically, all the text- and list-related CSS properties are inherited. When in doubt, you can check the short CSS reference at the end of this book. There's a keyword *Inherited* in parentheses after every property that gets inherited.

Maria: OK, let me check. The reference says that the color property is inherited. I do remember that when we played with text color on <article> in our homework, the color was applied to all the text inside the article. Only the text inside the <a> element was not affected. Does this property pass down only to the children?

Professor: Very good observations indeed. But no, it does not pass down only to the children. There's something else going on here, you know. The inheritance mechanism can easily cause conflicts between styles. Imagine that you define two separate styles for the <strong> and <p> elements. For example:

```
strong {
  color: purple;
}
p {
  color: blue;
}
```

The question about the next HTML code fragment is: what will be the color of the displayed text?

```
<p><strong>I feel so strong.</strong></p>
```

On page 57, we already met a similar conflict when we directly set two different colors (blue and gray) to the same element (`<h6>`), except now the conflict has come about more indirectly. Nevertheless, it falls under the same jurisdiction of the already-mentioned mechanism called *cascade*. Both declarations of the `color` property in the example on page 57 were equally specific, for they were both applied to the `<h6>` element directly through the `h6` element selector. Because the specificity of both declarations was the same, the last of the applied styles won.

This time, however, the two styles that apply to the inner `<strong>` element are not equally specific. The style applied directly to the `<strong>` element is more specific than the one applied to `<p>` and only inherited by `<strong>`. That's why the text "I feel so strong" gets a purple appearance, even if the `color: blue;` declaration is more recent.

Thus we have arrived at the second cascade rule, which says that the most specific style always wins. The most specific style is always the one directly applied to the element, or, if there is no style applied to it, the one applied to its nearest ancestor.

Maria: You still haven't quite answered my question. If the hyperlink text appears in a different color even when we have explicitly styled one of its ancestors, then there should have been some more specific style applied to the `<a>` element. But there's not.

Professor: In fact there is. Behind the scenes, browsers have their own predefined (default) styles for certain elements. So even if it looks like the closest style is that of the parent, this is not necessarily so because the browser may have in mind a more specific style for the element.

Mike: And if I apply my own style directly to an element, my style will win because it is more recent than that of the browser. Right?

Professor: Exactly.

4.8 Determining Style Specificity

Professor: Let me summarize the main two rules of cascade: first, the most specific style wins, and second, when the specificity is the same, the most recently applied style wins. Since most severe headaches are caused by misinterpretation of which style is more specific, let us examine this matter a little more closely.

Fortunately, CSS provides a formula for calculating the specificity of a style. In a nutshell, each selector gets a corresponding number of points that reflect its specificity:

- 1 point for an element selector (e.g., `strong`).
- 10 points for a class selector (e.g., `.important`).
- 100 points for an ID selector (e.g., `#superhero`).

- 1000 points for an inline style.

Incidentally, the universal asterisk selector has a zero specificity value.

The points are simply added up for each of the styles and, if two or more styles collide, the one with the highest score wins. Let me illuminate this with an example with four different styles:

```
p .jazzy {        /* 1 + 10 = 11 points */
  color: blue;
}
p strong {        /* 1 + 1 = 2 points */
  color: red;
}
strong {          /* 1 point */
  color: orange
}
#almighty {       /* 100 points */
  color: green;
}
```

Suppose we apply these styles to the following HTML code fragment:

```
<p>
  <strong class="jazzy">I'm blue.</strong>
  <strong>I'm red.</strong>
  <strong class="jazzy" id="almighty" style="color: black;">
    I'm black.
  </strong>
</p>
```

The styling of the first of the `<strong>` elements looks quite straightforward although it is not. As a matter of fact, three styles collide at that element. One of them is with the `strong` element selector, which only has 1 point. The second one is with the nested selector `p strong`, which receives 2 points since it is composed of two element selectors. The winner actually is the third one, the style with the nested selector `p .jazzy`, which collects 11 points (1 from the element selector p and 10 from the class selector `.jazzy`).

Can you figure out how the styling works on the other two `<strong>` elements?

Mike: That's easy. The story with the second one is the same as with the first one, except that the `p .jazzy` selector falls out of the game. The last one has, if I understood you correctly, five styles applied to it.

Professor: Which are?

Mike: All the above four styles plus the inline style, which is obviously the winner with its 1000 points.

Professor: Indeed!

Maria: What about inherited styles? Are the points inherited with them?

Professor: No. Inherited styles do not bring any points no matter how many points an ancestor's style may bear. So inherited properties will always get overridden by a style that applies to the element directly. Of course, if there's no directly applied style, it still holds that the nearest ancestor wins.

In any case, most of the basic rules of cascading are logical so you won't need to go into much detail unless you are up to building really complex style rules.

4.9 Relative Sizes

Professor: Remember earlier this morning when we briefly touched on the question of font size. You used pixels because I told you to and because we hadn't yet covered the other measurement units. In our last meeting I mentioned that a pixel is an absolute unit measuring approximately 1/100 of an inch. Although pixels are great for controlling graphical object sizing and positioning on a page, there are cases when they become impractical.

In your last homework you set font sizes of the `<h1>`, `<h2>`, and `<footer>` elements. Now let's try to add the next declaration to the style for the `<article>` element:

```
font-size: 10px;
```

The `<h1>`, `<h2>`, and `<footer>` elements retain the font sizes that you have set because the declarations are more specific than that of the `<article>` element, which is their parent. Paragraphs, however, don't have their own style and they inherit font size of 10 pixels. Hence we get the output as seen on this screenshot.

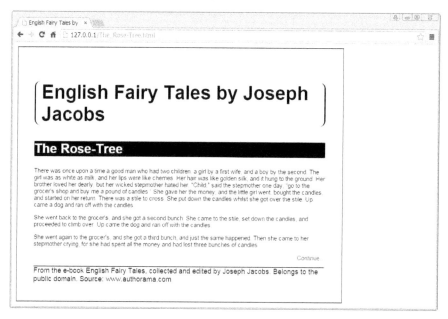

This is not what you would normally expect, though.

Maria: Why are you saying that? Everything looks exactly as you said it would. Paragraph text got smaller and everything else retained its previous size.

Professor: You're right. Yet, that's not the result you would want when you set the font size of an article. If nothing else, the footnote font is now larger than the paragraph font. Ideally, I would want at least the ratios of the font sizes to be preserved when I set the article's base font size, which means that the largest font stays the largest and the smallest stays the smallest. Such behavior is also integrated in the browser's default style sheet.

By default, the browser font sizes are defined *relative* to the basic font size. More specifically, each font size is calculated relative to the font size of its parent. If you remove all the `font-size` declarations from your homework style sheet and only specify the font size for the article, you will notice that no matter how large or small you specify the font, the font size ratios are preserved. And the good news is that you can change these ratios.

There's a measurement unit called *em*. The name comes from typography, where 1 em was traditionally defined as the width of the capital "M" in the current typeface and size. In modern computer typography the term em is preserved but represents the height of the font in question instead. One way of using this measurement unit is for setting the element's `font-size` property relative to the `font-size` property of its parent. For example, if you want to have main headings twice as large as the "normal" text, you write a rule like this one:

```
h1 {
    font-size: 2em;
}
```

If you now put an `<h1>` element inside an element whose `font-size` property is set to, say, 15 px, then this `<h1>` element will have a font size of 30 px.

A short question for you. Consider the following chunk of HTML code:

```
<body>
    <h1>This is so <strong>Important</strong></h1>
</body>
```

If we apply the following styles to the above HTML, what will be the font size, in pixels, of the word "Important"?

```
body {
    font-size: 15px;
}

h1 {
```

```
    font-size: 2em;
}
strong {
    font-size: 1.1em;
}
```

Maria: It is 1.1 times larger than that of the heading, which is two times larger than that of the body. Its size is 33 pixels.

Professor: Well done!

Ems are not used only for font sizes, though. You can size any dimension you like using this unit. That said, the size is calculated a little differently than it is for a font. Instead of the *inherited* value of the font-size property, the basis for calculation is the *computed* value of the font-size property of the element on which ems are used. In CSS terminology, a computed value is a value that is used for inheritance. It is obtained either from directly specified or inherited values.

For example, let's add the following declaration to the above style for the <h1> element:

```
border-bottom: 0.2em solid;
```

What do you think will be the thickness of the border below the heading?

Mike: It will be 20 percent of the heading's font size. The computed value of the heading's font size is two times that of the body—two times 15 gives 30 pixels. The border thickness will be 20 percent of that, which is six pixels.

Professor: Great! I see that you have absorbed the basic philosophy of the em unit, so we are ready to proceed to *percentages*.

Once you understand ems, you will probably have no difficulties understanding percentages as well. For one thing, percentages used with the font-size property work in exactly the same way as ems do. Just keep in mind that a value of 100 percent, for example, means the same thing as a value of 1 em, or, a value of 80 percent means the same thing as a value of 0.8 em. Remember that the format of the <percentage> data type is a <number> immediately followed by the percent sign (%). For example:

```
font-size: 90%;
```

For sizes other than font sizes, however, percentages work a little differently. Ems calculate sizes based on the computed value of the font-size property of the element on which they are used. With percentages things are somewhat more complicated. You must know separately for each value the percentage of what it takes. When applied to the width property, for example, percentages are calculated based on the width of the containing block. The weird thing is that vertical margins, for example, are also calculated based on the width of the containing block and not its height as you might expect. Some properties don't even accept percentages, as, for example, the border property, which can only be sized using a <length> value or keywords `thin`, `medium`, or `thick`. Fortunately, all that information is available in the CSS reference at the end of this book. You simply look up the property you're interested in, and you can also read whether it accepts percentages and how they work for that specific property.

Note that, even though you can use percentages with several length properties, the <percentage> data type is not the <length> data type. The latter is composed of a <number> and a length unit like px, or em.

Maria: You mentioned a containing block. Somehow I've got the impression that it is not the same as a parent.

Professor: You're perfectly right. A containing block is just what it says it is: a block that contains an element. It can either be the element's parent or some more distant ancestor. I will tell you how you can figure out which element represents the element's containing block next time, when I will tell you about positioning.

I still owe you an explanation about the differences between the two styles on page 60 that set `font-size` and `font-style` properties for a whole page. For seemingly the same purpose, we first used the asterisk and then the body element selector.

If you use a single element selector, then the style is applied only to elements of a single type. So, using the body selector, we set the font only on the body. Because the body is an ancestor of all the other elements on a page, the set font is propagated throughout the page via inheritance. Some font properties, however, are modified on certain elements by the web browser's default style sheet. For example, the rule for the <h1> element is defined in some browsers as follows:

```
h1 {
    display: block;
    font-size: 2em;
    font-weight: bold;
    margin: .67em 0;
}
```

This means that you get different font sizes—and sometimes weights, or colors—for different parts of your page. But they are all based on the same font applied to the body, which gives a page a uniform appearance.

The asterisk selector, however, selects *all* the elements on a page. The rule is applied directly to each and every element on a page, overriding any other styles set by the browser. Consequently, with the asterisk selector we sized all the text on the page to 16 px, even headings.

Maria: You said that the asterisk selector had a zero specificity value. How can it override a rule that uses an element selector, which has a higher specificity?

Professor: The default browser's style sheet is less important than author style sheets, so that even an asterisk selector will override the default browser rules.

Any more questions? Otherwise we're done for today.

Mike: Are there any guidelines on when to use ems or percentages?

Professor: Many web designers, including me, prefer using ems for font sizes because of their roots in typography. Also, to keep the so-called "vertical rhythm of the page" it is advisable that the CSS `line-height` property, and vertical margins and paddings (top and bottom) also have values expressed in ems. Namely, if the visitor resizes the font and vertical dimensions do not resize accordingly, the page design may break.

Percentages are more obvious, I guess. For example, you may want some element to occupy the whole width of a page, so you set its width to 100 percent.

4.10   Homework

Professor: For homework, design the following Lorem Ipsum page. Lorem Ipsum is nonsense text derived from a 1st century BC Latin text written by Cicero. It is frequently used as a placeholder text to demonstrate graphic elements of a document presentation. If you need to design your web page but don't have any content yet, you can use any one of the numerous online Lorem Ipsum generators (for example, *www.lipsum.com*).

I think that the document is self-explanatory, I only need to remark that all the gray text is actually hyperlinks, and the list of links at the top of the page is meant as a navigation bar.

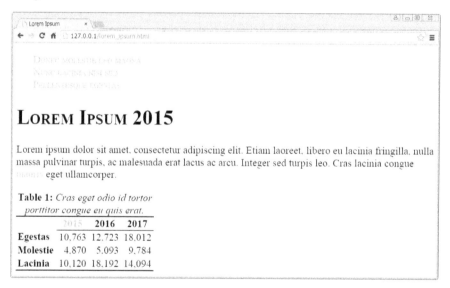

And before going home, a list of today's important keywords for you to ponder upon:

> **In this meeting**: padding, border, margin, selector, element selector, class selector, class names, id selector, id names, grouping selectors, nesting selectors, descendant selectors, universal asterisk selector, cascade, HTML family tree, node, root node, ancestor, descendant, parent, child, sibling, inheritance, style specificity, CSS point system, relative sizes, ems, percentages

Understanding CSS Boxes

5.1 Homework Discussion

Maria: Let me show you our homework. This is the HTML portion of the code:

```
<body>
<ul>
  <li><a href="#">Donec molestie leo magna</a></li>
  <li><a href="#">Nunc lacinia nisi sed</a></li>
  <li><a href="#">Pellentesque egestas</a></li>
</ul>
<h1>Lorem Ipsum 2015</h1>
<p>
  Lorem ipsum dolor sit amet, ...
  ... lacinia congue <a href="#">mauris</a> eget ullamcorper.
</p>
<table>
  <tr>
    <th></th>
    <th class="underline"><a href="#">2015</a></th>
    <th class="underline">2016</th>
    <th class="underline">2017</th>
  </tr>
  <tr>
    <th>Egestas</th>
    <td>10,763</td>
    <td>12,723</td>
    <td>18,012</td>
  </tr>
  ...
  <caption>
    <span class="tablenum">Table 1:</span>
    Cras eget odio id tortor porttitor congue eu quis erat.
  </caption>
</table>
</body>
```

And the rules that we used:

```css
table {
  border-top: 2px solid;
  border-bottom: 2px solid;
  border-collapse: collapse;
}
th, td {
  padding-left: 5px;
  padding-right: 5px;
}
th {
  text-align: left;
}
td {
  text-align: right;
}
th.underline {
  border-bottom: 1px solid;
  text-align: center;
}
caption {
  font-style: italic;
}
.tablenum {
  font-style: normal;
  font-weight: bold;
}
a {
  text-decoration: none;
  color: gray;
}
ul {
  list-style-type: none;
}
h1, li a {
  font-variant: small-caps;
}
```

We had no trouble with the homework, but when we wanted to carry out some ideas of our own we weren't that successful.

Professor: Very good indeed. I'm also glad to hear that there is material left to discuss, but first things first. I'm quite interested in your reasoning about why you did things the way you did. I think that's important.

Mike: The first thing that we weren't able to produce with the sole use of element selectors was the line under the years in the table. So we devised a class for underlining the `<td>` elements, simply by producing a bottom border. We had to use the `border-collapse` property in order to get rid of the spaces between the cells. With those spaces, the line under the years was interrupted.

Professor: I'm sorry to interrupt you, but the name of your `underline` class could have been better chosen. It does not reflect the semantic meaning but it speaks of presentation. You heard about that the last time. A better name would be, for example, `horizontal` to denote that this is a horizontal header. Please continue.

Mike: Then there is a table number, for which there's no special HTML element. So we simply used the generic `<span>` element and applied a class to it, which we named `tablenum`.

Maria: We're particularly proud of the last of the rules. We noticed that the heading as well as the list of the links above are all presented in small caps. So we constructed a group selector with two selectors: an ordinary element selector (`h1`) and a descendant selector (`li a`) specifying to style only the `<a>` elements within the `<li>` elements.

Professor: Splendid! There is, however, one tiny detail that you overlooked. There could be other lists in a document that weren't navigation bars but normal lists. If they contained links, those links would appear in small caps as well.

You could rewrite your last rule to read something like this:

```
h1, .navbar a {
  font-variant: small-caps;
}
```

And the corresponding HTML portion as:

```
<ul class="navbar">
  <li><a href="#">Donec molestie leo magna</a></li>
  <li><a href="#">Nunc lacinia nisi sed</a></li>
  <li><a href="#">Pellentesque egestas</a></li>
</ul>
```

Note that it doesn't really matter whether you start your descendant selector with an `<li>` or `<ul>` element because an `<ul>` element can contain only `<li>` elements anyway.

What's more, you could even use the `<nav>` HTML5 element, which is used for creating a navigation section on a page, as a wrapper around the `<ul>` element. Like this:

```
<nav>
  <ul>
    <li><a href="#">Donec molestie leo magna</a></li>
    <li><a href="#">Nunc lacinia nisi sed</a></li>
    <li><a href="#">Pellentesque egestas</a></li>
  </ul>
</nav>
```

And the CSS portion:

```
h1, nav a {
  font-variant: small-caps;
}
```

Incidentally, there is one little error in your HTML code. The HTML reference on page 327 says that a `<caption>` element can only appear as the first child of a table. So you must put it immediately after the `<table>` start tag in your code to satisfy the specification.

I have one last thing to note. Today you'll learn that there exist special pseudo-classes like `:link` or `:visited`, which give you more control over styling hyperlinks.

Mike: We still have two questions that bother us. First, we wanted to wrap text around the table. I recall that we've already stumbled upon the problem of text wrapping but we haven't solved it yet. Second, we wanted to arrange the navigational links in a line instead of a column. How do we do that?

Professor: The good news is that both your questions fall in the same category, which we shall talk about today. Let me begin with the CSS box model.

5.2   CSS Box Model

Professor: In web design, one very important concept is the so-called *CSS box model*. You know, every element you place on a web page is treated as a little box by a browser. Your life will be easier if you start thinking of boxes as well. The figure below illustrates what the CSS box looks like.

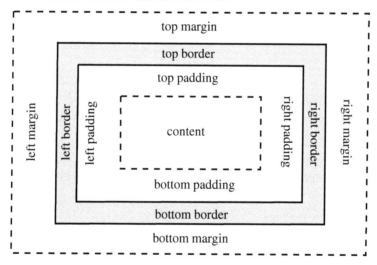

At the core of the CSS box is the *content* box. That's where all text and images appear, while the other three boxes (that is the padding, border, and margin box) are used exclusively to separate the content from neighboring elements. The *padding* clears space immediately around the content. Although no text or images intrude into the padding area, the padding is affected by the background color or image. The *border* represents a border around the padding, which is usually made visible. Since the background color and image expand under the border as well, the background may protrude through the border if it is dotted or dashed. The *margin* represents white space outside of the border and it separates one element from the other. The background color and image don't go under the margin, so the margin is completely transparent.

Mike: You mean white?

Professor: No, transparent. If an element is drawn inside another element, then that other element's color is seen through the contained element's margin.

The perimeter of each of the four boxes of the CSS box is called an *edge*. For example, the rectangle marking off the padding box is called a padding edge. As you can see in the above figure, the margin, border, and padding can be broken down into four segments: top, right, bottom, and left. You can set each of them individually using corresponding CSS properties.

I would like you to pay special attention to the difference between the margin and padding since the two are often confused. Both add space around the content and you use them to separate one element from another. Unless you apply a border or a background color or image, it is actually hard to tell whether space around the content is caused by the padding or the margin. With a border or background, however, the difference becomes significant. Since the padding separates the content from the border, it keeps the content from appearing jammed inside the box. Margins, on the other hand, provide white space between elements and relax the tension between them, which gives a page a lighter appearance.

Maria: So far it looks quite straightforward, but I'm afraid things will become complicated in practice.

Professor: I'm afraid that you're right. Some of the concepts you're going to learn today are pretty difficult to grasp. So, let's dive into some examples, so you can get a clearer picture of the idea.

Consider, for example, the next fragments of CSS and HTML code:

```
.example-box {
  width: 60px;
  height: 60px;
  padding: 10px;
  border: 5px solid;
  margin: 20px;
}


<body>
  <div class="example-box">I am a box.</div>
</body>
```

We will investigate the behavior of the code in Google Chrome. Having loaded the code into the browser, you press Ctrl+Shift+I, or select Tools→Developer Tools from the Chrome menu (≡) at the top-right of the browser window to open the Developer Tools (DevTools for short). On the following screenshot the DevTools window is located in the lower part of the browser window.

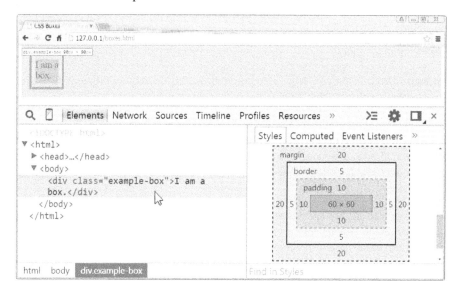

If you don't see the same picture, select the Elements pane from the top menu of the DevTools and the Styles pane from the right part of the DevTools window. In the HTML tree in the left half of the DevTools, select the <div> element (expand the tree if necessary by clicking a corresponding gray triangle) and in the right part of the DevTools you will see the CSS box view of that element. If you don't see it, then scroll down that part of the DevTools window till you do. Notice that the measurements are the same as we defined them by the example-box class. Now let the mouse pointer hover over the same <div> element in the document tree on the left. Notice that hovering the mouse over the element colors the areas over the space occupied by the element in the document window above the DevTools. The exact areas covered by different boxes of the element are colored. Notice also that in the tooltip above the element in the document window you can see dimensions of the element's border box, 90 × 90 px in our case.

Maria: Why is the margin so huge at the right? The large, middle gray area is the margin, isn't it?

Professor: That's true, but let me explain the whole box model. The width and height properties actually control the size of the content box. The content box is a little darker gray box in the center of the box model with the measurements 60 × 60 written inside it as you can see on the screenshot. You can also see the padding, border, and margin, all with the exact dimensions written inside them. These all match the dimensions specified by our CSS rule.

You already noticed that the margin is somewhat special. The margin represents empty space *around* an element and is more flexible. <div> elements, as you may recall, have a block display and they want to occupy the whole available width, which is the body's content box in our case. Because we explicitly made the <div> element narrower by setting its width to 60 pixels, the empty space is filled by automatically extending its right margin, which you can see at the right side of the element.

Let's now select the `<body>` element in the HTML tree and hover the mouse pointer over it.

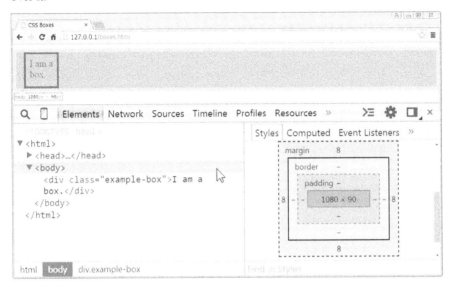

On the right side of the DevTools window you now see the CSS box of the `<body>` element, while at the top, in the document part of the window, you see the actual areas covered by the content and margin boxes of the body.

Maria: Is the margin of the body defined in the default style sheet of the browser? Because we haven't defined it.

Professor: That's right.

If you compare the last two screenshots, you will notice that the width of the body's content box is made just big enough to accommodate the whole `<div>` element including its horizontal (left and right) paddings, borders, and margins. Things, however, are different in the vertical direction. Vertical (top and bottom) margins are ignored and top and bottom border edges of the `<div>` element sit directly on the content edges of the body. In other words, the body content box is just high enough to accommodate the `<div>` element *without* its vertical margins.

Let's now add some padding to the `<body>` element:

```
body {
    padding: 10px;
}
```

The situation is quite different now, as you can see on the following screenshot.

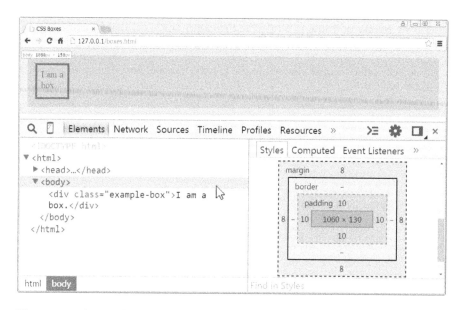

The <div> element is now completely inside the body's content box, even with its vertical margins. You can see that in the document part of the window above the DevTools and because the height of the body's content box is now the same as is the height of the <div> element's margin box, which is 130 px, as you can see in the center of the CSS box on the right side of the DevTools window.

This is probably the time when you can start pulling your hair out, trying to figure out what is happening with the margins. To get some feeling for this, you may want to experiment with the DevTools. If you again select the <div> element from the HTML tree on the left and scroll the right part of the DevTools window to the top, you will see our CSS rule for the <div> element. Hovering the mouse pointer above the rule will produce check boxes, which you can use to switch individual declarations on and off in situ. You can also change the values of the properties if you want, or even add new declarations to the rule.

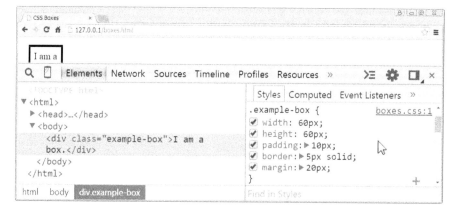

Note, however, that any changes you made to the rules in the DevTools will not affect your original CSS file.

Mike: What are those triangles after some of the property names...? Oh, I get it! Those are shorthand properties and you can probably expand them to see their individual properties.

Professor: Exactly! It's great that you discover things like that without my help.

So, what is happening with the margins? The W3C specification defines the term *collapsing margins*, which means, in a nutshell, that when the *vertical margins* of two elements touch, they merge to form a single margin, whose width is equal to the larger of both. There are of course exceptions to this rule and I will give you a heads-up when we come to them.

Mike: We observed before that when we added some padding to the body, the margins didn't collapse any more. That was of course to be expected, as the body padding separated the margins of both elements.

Professor: Precisely.

Maria: I noticed that the margin of the `<div>` element in our first example touched the top of the window while the margin of the body didn't. I can now explain why that happened. The margins have collapsed and the resulting vertical margin width is 20 px, which is the larger of both widths. Since the body's margin is 12 px narrower than that of the `<div>` element, the body's top margin edge is positioned 12 px below the window top.

Professor: That's right. By the way, the body is contained in the content box of the `<html>` element. You can examine it in the DevTools if you want to.

5.3 Element Display

Professor: We'll now examine how the `display` CSS property can influence the behavior of the CSS box model. Controlling the element's display is one of the more important aspects of web design. By changing the type of display you can considerably alter the way the element finds its position on a page. Let us set the `display` property of the `<div>` element of our last example to `inline`:

```
.example-box {
  width: 60px;
  height: 60px;
  padding: 10px;
  border: 5px solid;
  margin: 20px;
  display: inline;
}
```

We leave the body's padding size at 10 px.

The most obvious effect of setting the element's display to inline is the ignorance of its `width` and `height` properties, which are set to `auto` as you can see at the right side of the DevTools window.

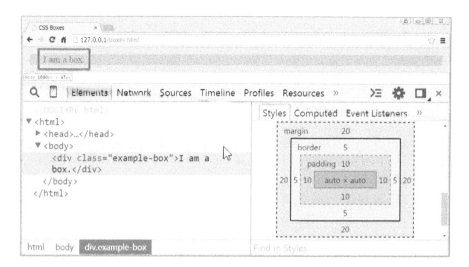

You will also notice that vertical paddings, borders, and margins of the `<div>` element, which now has inline display, are totally disregarded by the surrounding elements. The content box of the body is just high enough to contain the content box of the `<div>` element, which is placed tightly around the text "I am a box." If we removed the body's padding, then the top border of the `<div>` element might even move off the document window. However, horizontal paddings, borders, and margins of an inline element are still respected by neighboring elements. It is important that you understand this, as it influences the positioning of the elements.

5.4 Positioning and Element Flow

Professor: The next important concept connected with the placing of elements on a page is the *element flow*. Normally, elements appear on a page in the same order as they do in the HTML file, which means that they are *in the document flow* or just *in the flow*. If an element is in the flow, then other elements respect its presence and place themselves accordingly. For example, if there's a paragraph after a table in the HTML file, then the paragraph will move further down if the table is made larger, and will move up if the table shortens.

Technically, we control the element's flow with the `position` property. This property can take a couple of keyword values, of which we'll study the values `static`, `absolute`, and `relative`. The default value is `static`, which puts the element in the normal flow.

Depending on the value of the `position` property, we usually say that an element is *absolutely positioned*, *relatively positioned*, or *not positioned*, if the `position` property is set to `static`. Absolutely and relatively positioned elements are often referred to as *positioned*.

Let's carry out a little experiment. Consider, for example, the following piece of HTML code:

```
<body>
  <div id="one">1</div>
  <div id="two">2</div>
  <div id="three">3</div>
</body>
```

Consider also the following CSS style:

```
div {
  width: 40px;
  height: 40px;
  border: 1px solid;
  padding: 5px;
  margin: 5px;
}
```

Because the `<div>` element has block display, the document window shows the three boxes one above the other, as you can see on the screenshot.

Let's set the `position` property of the second element to `relative`. Doing so, the element stays in the flow but can be displaced by any desired distance without disturbing the other elements' positions. The other elements behave exactly as though the relatively positioned element was in the normal document flow. Consider, for example, the following rule:

```
#two {
  position: relative;
  top: 10px;
  left: 25px;
}
```

You set the exact position of an element using the properties `top`, `right`, `bottom`, and `left`. Contrary to what you might think, these properties do not denote the direction of movement, but rather specify the space that is to be inserted at the corresponding side of the element. For example, the declaration `left: 25px;` means that 25 px space will be inserted at the left side of the element, which effectively moves the element to the right.

Applying the above rule to our HTML, the second element moves 10 px down and 25 px to the right while the other two elements stay in place as though nothing happened. You can see this on the following screenshot.

Do you follow me?

Mike: Yes, no problem.

Maria: It's lots of information all right, but nothing that I couldn't understand.

Professor: All right then. Because the absolute positioning is a bit more tricky. We now replace our last style with the next one, which takes the second element out of the flow:

```
#two {
  position: absolute;
}
```

With the second element taken out of the flow, the third element takes the position where the second element was before. Namely, the third element is no longer aware of the presence of the second one. Here's how it looks.

Maria: The second element has also moved. How do we know where an element will move when we take it out of the flow?

Professor: The element in fact didn't move. Basically, it stayed where it had been before we took it out of the flow.

But you're right, the element has slightly changed its position. The reason is that vertical margins of the absolutely positioned elements don't collapse. When we absolutely positioned the second element, it moved down by the amount of its top margin width.

5.5 Containing Block

Professor:We've arrived at one very important concept in CSS layout called a *containing block*. The containing block of an element is basically the rectangular area in which the element is placed. It can be the content box of its parent but that is not necessary. The containing block is so important because every time we set the position, or use percentages to set the dimensions of an element box, we do that relative to the element's containing block.

The containing block for the root (that is `<html>`) element is simply the document window, also called a *viewport*.

Other elements get their containing block according to the value of their `position` property. If `position` is set to either `static` or `relative`, then the element's containing block is the content box of its nearest ancestor with block display. If, however, the element is absolutely positioned, then its containing block is defined by the *padding* edge of its nearest *positioned* ancestor.

Mike: I assume that the containing block does not constrain the contained element.

Professor: Why do you think so?

Mike: When we set the `display` property of the `<div>` element of the `example-box` class to `inline` on page 79, for example, the element was only partly contained in its containing block. The containing block of the `<div>` element was the body's content box, wasn't it?

Professor: Precisely. That is because `<body>` has block display and was of course the nearest ancestor with block display of the `<div>` element in that example. Indeed, an element is not constrained by its containing block and may overflow.

And now, the tricky part. Let's position the second element by setting its `top` and `left` properties to 0:

```
#two {
  position: absolute;
  top: 0;
  left: 0;
}
```

Where do you think the element will appear?

Maria: In the top left corner of the body.

Professor: Yes? Could you be more precise, please? How far, for example, from the edge of the viewport do you think it will be placed?

Maria: Wait a minute! `<body>` is not positioned. Because there are no other positioned ancestors, the containing block of the second element is probably the viewport. The second element will move all the way to the document window's edge. Because

---

of the 5 px margin that we set to our `<div>` elements, the visible border will be 5 px away from the viewport's edge.

Professor: Splendid! Here's the screenshot.

What will happen if I now set the body's position to relative? Like this:

```
body {
    position: relative;
}
```

Mike: The `<body>` element now defines the second element's containing block. So the second element will move further away from the viewport's edge and will align with the first element because both elements live in the same containing block.

Professor: Not quite. What about vertical margins?

Mike: Oh, I forgot that. The second element will be placed 5 px lower because it is absolutely positioned and its vertical margins don't collapse.

Professor: Exactly. Let's check how it looks.

Of course, if I set the body's position to absolute, the body's top margin won't collapse with the first element's top margin any more, and the first and third elements will move down by 5 px, which makes the first and second elements overlap perfectly.

5.6 Hiding Elements

Professor: You can also take an element out of the flow with the `display` property set to `none`. This makes an element completely disappear from the page, while other elements reposition themselves accordingly. Hiding part of the content is useful in many ways. You can, for example, hide a portion of some long text and make it accessible only on demand, or you can hide a sub-menu of a navigation system and make it visible only after the visitor has selected the corresponding top menu item.

Mike: How do you do that?

Professor: You put the removed element back in the flow by setting its `display` property back to `block` or `inline`. You can allow the visitor to trigger that change either through some JavaScript code or the `:hover` pseudo-class. We will cover pseudo-classes later today.

Another way of hiding an element is by means of the `visibility` property, which can be set to `hidden` or `visible`. Unlike the `none` value of the `display` property, hiding an element through the `hidden` value of the `visibility` property leaves the element in the flow and it only leaves an empty space where the element would have been if it had not been hidden.

5.7 Floated Elements

Professor: Two more properties that control the element's position are `float` and `clear`. I guess that the concept behind these two properties is one of the least intuitive and hence the most misunderstood concepts in CSS. I'll try to illuminate the basics so you have a firm foundation from which to study and explore the subject on your own.

You can float an element either to the left or right by declaring `float: left;` or `float: right;` respectively. You cannot float an element to the center, though. Because the concept is basically the same for both floats, left and right, we'll only focus on left floats. So what happens if you float an element to the left?

- The floated element is first positioned according to its normal document flow. Then it is shifted horizontally all the way to the left until its left margin edge touches the left edge of its containing block or the right margin edge of another floated element. Technically, that takes the element out of the flow, although not completely. Text and inline elements still respect its presence.

- All the content that appears above the floated element in the browser window is unaffected.

- All text and inline elements that appear at the same height or below the floated element flow alongside its right margin edge and wrap below it.

- Because the floated element is not in the flow, all the block elements position themselves vertically as though the float didn't exist. However, text and inline content of those block elements will reflow to make room for the floated element.

- Because the floated element is not in the flow, its containing block does not respect its height.

- If there is not enough room for the floated element horizontally, the element will move downward, until it finds enough horizontal space for itself.

- Floating an element implies the block layout and hence changes the computed value of the inline element's `display` property to `block`. One practical implication of this is that you can set a width and height of floated elements.

- You should always set a width of floated elements, otherwise the results can be unpredictable. That is not necessary for images, however, which get their width implicitly from the physical width of the actual image.

- Vertical margins of floated elements do not collapse.

Maria: That's a long list.

Professor: No one expects you to learn it by heart. Instead, you'll use it as a reference when needed.

Let's start out with the same HTML code as in our last example:

```
<body>
  <div id="one">1</div>
  <div id="two">2</div>
  <div id="three">3</div>
</body>
```

We also use the same style:

```
div {
  width: 40px;
  height: 40px;
  border: 1px solid;
  padding: 5px;
  margin: 5px;
}
```

The next rule floats the first two `<div>` elements to the left:

```
#one, #two {
  float: left;
}
```

Do you see what happened?

What a mess! Yet I think you'll be able to explain it.

Mike: Let me try....

The first two boxes are stacked against the left side of the window. More precisely, the left margin edge of the first box touches the left content edge of the document, which is the containing element of all three boxes. The left margin edge of the second box touches the right margin edge of the first box. Both boxes are placed lower by the width of their top margin because margins of floated elements don't collapse. The third box is placed as though the first two didn't exist because they have been taken out of the flow.

Professor: Excellent! Any idea what happened to the number three?

Maria: You said that text always leaves space for floated elements. I suppose the third `<div>` is too narrow for its text to be able to flow at the right side of the second box. So it was forced to overflow its box downwards.

Professor: Exactly! Are you saying that the number three will move up and to the outside of the right edge of the second block if you make the third `<div>` wider?

Maria: I think so.

Professor: OK, let's try and see if that is really so:

```
#three {
    width: auto;
}
```

Indeed, we get exactly the result that you predicted.

It is possible to move the third box down by using the `clear` property. Basically, if you clear an element, that element will move downwards past any floated elements that appear *before* it in the HTML document and are floated to the same side as is the value of the `clear` property of the cleared element. Note that you can clear floated as well as non-floated elements. If you want to know more precisely how the `clear` property works, then you can check the CSS reference on page 362.

So let's clear the third box:

```
#three {
  width: auto;
  clear: left;
}
```

Here's how it looks.

Notice that this move collapses margins, although in a somewhat different way than we're used to. Again, see the description of the `clear` property in the CSS reference if you want to know the details.

Mike: What is the practical use of floated elements? I suppose we can use it for wrapping text around images and tables.

Professor: Indeed, and it's really time that we finally do it. As a matter of fact, the ability to flow text around images really goes back a very long time. So it's no big surprise that it is also quite straightforward to use. Take, for example, the following HTML code with some text and an image:

```
<h1>The Windmills of Europe</h1>
<p>
  Praesent augue leo, blandit dapibus nulla in, vulputate volutpat
  leo. Etiam luctus augue eu est accumsan, sed suscipit eros
  dapibus. <img src="/images/windmill.jpg"> Nam varius, nulla eget
  imperdiet vulputate, risus lacus vehicula nisi, eu dictum purus
  mauris sodales est. Ut consectetur tortor neque, sed rutrum...
</p>
```

The CSS portion of the example is extremely simple:

```
img {
  float: left;
  height: 6.5em;
  margin: .45em   .4em .25em 0;
}
p {
  line-height: 1.2;
}
```

What makes text flow around the image is the `float: left;` declaration. The measures used for the vertical margins and image height were selected so that the top and bottom edge of the image align with the x and base line of the text, respectively.

Maria: What are these lines?

Professor: The base line marks the line on which characters sit, while the x line marks the top of lower-case letters like, for example, a letter x. Hence the name.

You can set the height of an image and its top margin to match these lines by trial and error. First, you set the top margin so that the top edge of the image approximately aligns with the x line of the text and set the image height to align its bottom edge with the base line of the text. You may want to zoom in on the page while doing that, which you can do by pressing `Ctrl+(+)` on the keyboard. To zoom back out, you press `Ctrl+(-)`. After you have successfully aligned both edges, you set the bottom margin so that the sum of the image height and both vertical margins is a multiple of 1.2 em, which is the normal line spacing. Once you have the vertical dimensions set, you can alter the image height by simply adding to or subtracting from it multiples of 1.2 em. Recall that it is important that you use ems in order to keep the "vertical rhythm of the page," even if the viewer resizes the font.

I specified the `line-height` property to make sure that text spacing will always be 120 percent of the font size. The next screenshot shows the final result.

Also observe the vertical position of the floated image. The top of the image is positioned at the same height where the word "dapibus" occurs, which corresponds to the position of the image in the HTML code. Indeed, everything that is above that position stays unaffected by floating the image to the left.

If you put more pictures in your article, you may have some trouble if they get too close to each other. You can see this happening on the next screenshot.

To solve the problem, we clear the `<img>` elements:

```
img {
  ...
  clear: left;
}
```

As a result, both pictures are nicely floated to the left.

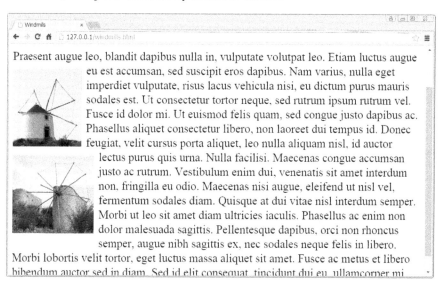

Damn! The last row of text didn't wrap under the second image.

Mike: Why is that?

Professor: The problem is that when using relative lengths, you don't usually get a whole number of pixels. So browsers are forced to round the numbers, and you've just seen a result of such rounding. Until there was only one image, the rounding error wasn't that serious. With two images one on top of the other, however, the compound error has become large enough to prevent the two images from fitting between the lines of text as planned. Interestingly, Firefox doesn't suffer from that problem because of an innovative solution called *sub-pixel rendering*. The technique allows Firefox browsers to render relative sizes exactly, even when they don't produce a whole number of pixels. The result is quite pleasing, as a matter of fact.

Mike: Is there a generic solution to the problem?

Professor: One possible solution is to flow images alternately left and right in order to always allow some lines of text between them. That way, individual errors won't accumulate. Besides, a page could look more balanced if you placed images at alternate sides of the text. If, however, you decide to flow images to the right, you may also consider justifying the text.

Maria: What if we set the paragraph font size to a value that will produce a whole number of pixels when multiplied by 1.2?

Professor: That could work. But you'd also have to be careful that the vertical dimensions don't produce any sub-pixel dimensions. Still, there remains the possibility that your visitor resizes the font. You cannot control that. You can try to experiment at home and see what happens.

Before we continue, I want to show you one more example. The example shows how you can lay out a two-column page using floats. What we're going to do is design a page that has a header, footer, sidebar, and main article. The main article and sidebar will be laid out as two columns between the header above and the footer beneath. This is the HTML portion of the example:

```
<body>
  <header>
    <h1>Lorem Ipsum Integer Fringilla</h1>
    <p>Vivamus dui... leo.</p>
  </header>
  <aside>
    <h1>Consequat</h1>
    <p>Class aptent... fringilla.</p>
  </aside>
  <article class="main">
    <h1>Quisque Eget Quam</h1>
    <p>Lorem ipsum... orci.</p>
    <p>In eu blandit... aliquam.</p>
  </article>
  <footer>
    <p>Cras dignissim... tincidunt.</p>
  </footer>
</body>
```

And this is the CSS portion:

```
.main {
  width:66%;
  padding: 0 2%;
}
aside {
  float: right;
  width: 26%;
  padding: 0 2%;
}
header, footer {
  padding: 0 2%;
  border-bottom: 1px solid;
  border-top: 1px solid;
}
body {
  width: 90%;
}
```

The example does not need many comments, I guess. The `<aside>` element is floated to the right and because the article and sidebar together should occupy 100 percent of their containing block's width, the sum of both widths and corresponding horizontal paddings should amount to 100 percent. I reduced the body width to 90 percent in order to leave some space at the right of the window, which I think makes the page look more relaxed.

Feeding the code to the browser, we get the following output.

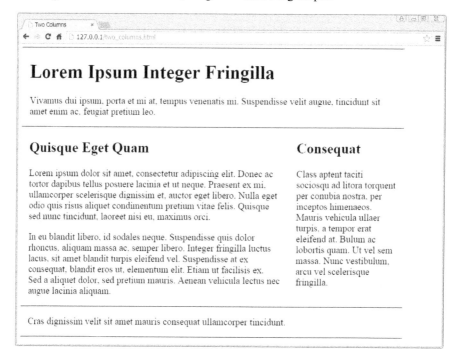

Notice how the `<h1>` headings are rendered smaller inside the elements `<atricle>` and `<aside>`. Recall that this—besides the block display inherent to all the HTML5 semantic elements—is the only specific of the default browser rendering of the four HTML5 sectioning elements.

Maria: Is it possible to see the default browser style settings?

Professor: You can search the Web for default browser style sheets or you can simply inspect the styles in the DevTools, just like we did with our own styles on page 78.

Mike: This example also suffers from the problem of rounding sub-pixel dimensions, doesn't it? If the resulting width of the two columns got bigger than 100 percent, one of them would move downwards.

Professor: You don't have to worry about that. We floated the `<aside>` element but we didn't float the article. So both columns are completely unaware of each other's existence. You can try to increase the widths a little if you want to observe what happens.

Mike: I see.

Professor: Remember how we once tried to flow text around a decorative capital. Because this is no different from flowing text around any other image, you will now be able to do it by yourselves. However, what if you don't want to use images of letters? What if you want to use the capital letters that are part of the text instead? You can accomplish that using the so-called *special selectors*.

5.8   Special Selectors

Professor: Sometimes you want to style some portion of your document, but for one reason or another, you don't like the idea of adding additional `class` or `id` attributes, or even new elements to your HTML code. There are techniques that come in handy in such cases.

You can choose from among several predefined *pseudo-classes* and *pseudo-elements*. These come in handy when you need to select parts of a document that are not furnished with tags per se but are still easy to identify, like the first letter of a paragraph, or a link that the viewer has already visited. All pseudo-class names begin with a single colon (:), while pseudo-element names begin with a double colon (::).

For example, if you want to show the links that have already been visited differently from the ones that have not been, you can use the `:visited` pseudo-class. Can you guys write a rule for me that will color all the visited links gray?

Mike: Is a pseudo-class a selector?

Professor: Correct. Pseudo-classes and elements are actually selectors, and they are used like any other selector.

Mike: Then this will be the rule:

```
:visited {
  color: gray;
}
```

Maria: Will that work? I remember seeing `a:visited` in some code I studied.

Professor: Both are correct. Because `:visited` is a class name, the same rules apply as with the ordinary class names except that you don't use a period before a name. Remember that you can make a class selector more specific by prepending an element name before it. In that sense, `a:visited` explicitly tells that `:visited` only applies to `<a>` elements. This is not necessary because only links can be visited anyhow. So you can just as well drop the `a` and simply write `:visited`.

Note that you are only allowed to style a very limited set of color-related properties of a visited link. The reason for that lies in privacy concerns. People have found out that by using certain JavaScript functions, you can traverse the pages and extract the information about the links that users have visited. This permits quite an exhaustive collecting of personal data, even about the person's identity. Hence the imposed restrictions on styling with this pseudo-class.

Similarly, you can style an unvisited link using the `:link` pseudo-class. There are two more pseudo-classes that you can use to style links, which are the `:hover` and `:active` pseudo-classes. Their use is not limited to links, however. If you're interested in using them, you'll find their descriptions in the reference at the end of this book, of course. Especially interesting is the `:hover` pseudo-class, for it allows you to respond to a mouse pointer moving over an element.

You should be careful when using link-related pseudo-classes because they may override each other if they are not applied in the correct order. You must put them in the so-called *LVHA order*, which is short for `:link`, `:visited`, `:hover`, and `:active`. For example, the `:link` rule, if used, must always be put before any other link-related rules.

Pseudo-elements work much like pseudo-classes do. As an example, let's take the `::first-letter` pseudo-element, which is used to select the first letter of text of any block element. For example, if you want to make the first letter of every paragraph bold, you will write:

```
p::first-letter {
  font-weight: bold;
}
```

If you wish, you can also select the first line with the `::first-line` pseudo-element.

Maria: If the pseudo-class and pseudo-element selectors are used in the same way, what's the difference?

Professor: There's a subtle difference in their interpretation. A pseudo-element often refers to the content of an element that is matched by the selector placed immediately before it. For example, p::first-letter selects the first letter within a paragraph. Pseudo-classes, however, refer to the element itself when placed immediately after another selector. For example, p:first-child selects a paragraph that itself is a first child. You can, of course, also use pseudo-classes as descendant selectors. For example, p :first-child selects the first child of a paragraph, which is quite different.

Apart from that, both selectors have different specificity. Pseudo-elements have the same specificity as element selectors while pseudo-classes have the same specificity as class selectors.

I will just skim briefly over other types of special selectors because I think they are not so difficult to grasp and you can experiment later with them if you wish. Another type of selectors are *attribute selectors*. With these, you can single out elements that have specified a particular attribute. For example, if you want to select all <img> elements with an alt attribute, you write img[alt].

Next, there's a group of so-called *child selectors*, which let you format the children of an element. This is kind of similar to using nested selectors, only nested selectors apply to *any* descendant of an element and not just its children. The child selector makes use of an angle bracket (>) to announce that the second element should be the child of the first one in order for the rule to apply. For example, article > h2 matches all <h2> elements that are children of an <article> element.

One more way of selecting child elements is with any of the predefined child-selecting pseudo-classes like :first-child, :last-child, and :nth-child(). The :nth-child() selector can be quite useful if you want, for example, to color alternating table rows to make large tables more readable.

The last special selector I would like to mention is the *adjacent sibling selector*. You already know that siblings are elements that share a common parent. The adjacent sibling of an element is the element that is placed *directly after* that element in the HTML code. The adjacent sibling selector uses a plus sign (+) to connect two elements. For example, h1 + p will select any paragraph that comes directly after an <h1> element.

I think we are now ready to flow text around an initial. Consider the following HTML:

```
<body>
  <h1>Lorem Ipsum</h1>
  <p>
    Lorem ipsum dolor sit amet, consectetur...
  </p>
  <p>
    Sed elit augue, vehicula sed ante...
  </p>
</body>
```

Can you produce a rule that will make the first letter of the first paragraph bigger, and let the rest of the text flow around it? You're not allowed to change the HTML.

Maria: Let me think.... I can select the first paragraph with an adjacent sibling selector, or maybe with the :nth-child() selector. You didn't explain the latter, so I'm just guessing.

Professor: I didn't explain it because there is too much explaining to do, and you can read all about it in the CSS reference on page 370. In any case, as the name suggests, you *can* select the *n*th child using this selector. The only problem is that you cannot be sure that the first paragraph after the main heading will always be the second child. It's much better to use an adjacent selector. Please continue.

Maria: With the help of an adjacent sibling selector, I first select the first paragraph after the <h1> element, and then apply the ::first-letter pseudo-element to that. Like this:

```
h1 + p::first-letter {
  font-size: 3.9em;
  margin-right: .2em;
  float: left;
}
```

I added the right margin so to create some space between the first letter and the rest of the text.

Professor: Let's see the result.

Looks nice. Now I want to experiment a little. Suppose you want to have an initial at the beginning of every paragraph. You can accomplish that easily:

```
p::first-letter {
  font-size: 3.9em;
  margin-right: .2em;
  float: left;
}
```

Another thing that we'll do is make paragraphs shorter.

Look what happened.

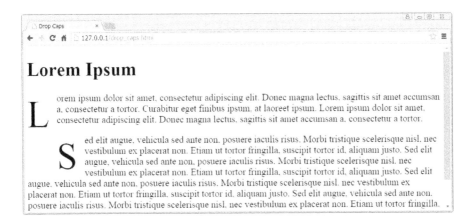

This isn't good.

We floated both initials to the left but the S wasn't placed low enough to float past the L and as a result it floated only far enough to touch the right edge of the L.

Mike: What if we `clear: left`; the initials?

Professor: Unfortunately, you cannot do that because it is the whole paragraph that is placed too high and not just the initial. The trick is to create a separate *block formatting context* for each of the initials. According to W3C, a block formatting context is a region in which block boxes are laid out and in which interactions between floated elements occur. A block formatting context is generated by either the root element, a floated element, an absolutely positioned element, a table cell, or a table caption. Because paragraphs do not create their own block formatting context, both our initials live in the same formatting context created by the root element and hence interact.

Maria: Should we float the paragraphs, then?

Professor: It looks like an obvious solution. By floating the paragraphs, we create a separate block formatting context for each of the initials, which won't interact anymore. Don't forget that you should set the width of floated elements, and that the vertical margins of floated elements don't collapse. The latter means that we have to set one of the vertical margins of paragraphs to zero if we want to keep the same spacing between paragraphs. We'll remove the top margin.

Mike: Do paragraphs have top and bottom margins by default?

Professor: Yes, that's true. However, if you remove the top margin from paragraphs, then you must be careful to check what that means for interactions with other elements that should appear directly before paragraphs. That's why we'll float only paragraphs that come after other paragraphs:

```
p + p {
  width: 100%;
  float: left;
  margin-top: 0;
}
```

Professor: This puts the second paragraph in the same block formatting context as the first initial, which means that you can clear it past the initial. However, that will not be necessary because the width of the paragraph is 100 percent of its containing block width, which in any case moves the paragraph downwards to its own line.

Floated paragraphs will of course cause you new problems. For example, any heading that will follow floated paragraphs will ignore them. To solve that, you can use sections but you'll have to float them as well. Recall that a containing block does not respect the heights of floated elements. However, if you float a containing element, then its height automatically resizes so as to also accommodate all the floats it contains. If you float sections, they will adjust their heights to hold all the paragraphs.

Mike: Things are starting to get complicated.

5.9   Homework

Professor: Floats are not the easiest thing to comprehend, but when you know how to use them they will become a powerful tool for laying out your pages. I suggest that you experiment with them some more at home. Do not forget to inspect content, padding, and other areas in the DevTools, and think about why things turn out as they do.

Besides, I have a very intriguing and interesting project for you to do at home. I would like you to build an attractive menu navigation bar for a web page. For starters, use the following unordered list of links:

```
<header>
  <nav  id="top-navigation">
    <ul>
      <li><a href="#">Home</a></li>
      <li><a href="#">Music</a></li>
      <li><a href="#">Movies & TV</a></li>
      <li><a href="#">Books</a></li>
    </ul>
  </nav>
</header>
```

Your job is to arrange the list items horizontally and design them as you like. You may also use the :hover pseudo-class to highlight the menu item over which the mouse cursor moves. That's the firs part of the homework.

For the second part, you must augment the above HTML list with an additional item (named "More"), into which you insert another unordered list to function as a drop-down sub-menu:

```
...
<li><a href="#">Books</a></li>
<li><a>More</a>
  <ul>
    <li><a href="#">E-Book Readers</a></li>
```

```
    <li><a href="#">Portable Media Players</a></li>
  </ul>
</li>
...
```

Your task is to keep this sub-menu hidden until the visitor hovers the mouse pointer over the "More" menu item. When that happens, the sub-menu will show up.

Finally, for your further contemplation, a list of today's topics:

> **In this meeting**: CSS box model, content, padding, border, margin, edge, collapsing margins, DevTools, element display, element flow, `position`, `visibility`, floated elements, `float`, `clear`, LVHA order, attribute selector, pseudo-class, pseudo-element, child selector, adjacent sibling selector, block formatting context

Behavior

6.1 Homework Discussion

Mike: The homework was quite a challenge, but unfortunately we didn't succeed in finishing it. Here's what we did.

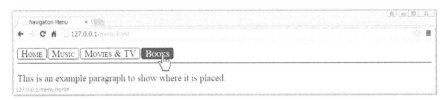

If you hover the mouse cursor over a menu item, the item changes to dark gray and its text to white.

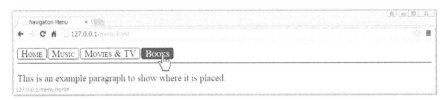

Professor: Nice job! I can't wait to see the code! Show me how you managed to line up the items horizontally.

Maria: After a lot of thought, we were surprised to find out how little effort it took to complete the task. We remembered the example from our last meeting where we arranged two blocks in a line by simply floating them. We used the same trick here and applied it to the list items:

```
#top-navigation li {
    float: left;
}
```

This already placed the links horizontally but the list collapsed and items were not contained in the list any more. What's more, the paragraph, which we added for orientation—and made it float as well—was put to the right of the menu instead of below it.

So we floated the unordered list as well to make it grow vertically and embrace all its floated children. We also made the list 100 percent wide in order to occupy the whole width of the document, which forced the paragraph following the menu to move under it. We also removed the annoying bullets by setting `list-style` to none:

```
#top-navigation ul {
    float: left;
    width: 100%;
    list-style: none;
}
```

Professor: That's it! How did you fine-tune the design?

Mike: First we took care of the list. We figured out that the items were moved to the right because of the extra left padding that lists use in order to indent their items. So we set the `padding-left` property to zero. We added some vertical margins as well:

```
#top-navigation ul {
    float: left;
    width: 100%;
    list-style: none;

    padding-left: 0;
    margin: .4em 0;
}
```

We also floated the header and specified its bottom border:

```
header {
    float: left;
    width: 100%;
    border-bottom: 1px solid black;
}
```

Next, we designed the links within the navigation menu. I don't think this needs any comments:

```
#top-navigation a {
    background-color: #DEF;
    color: black;
    text-decoration: none;
    font-variant: small-caps;
    padding: 0 10px;
    border: 1px solid black;
    border-radius: 5px;
```

```
  margin-right: 3px;
}
```

At the end we added the hovering effect. A menu item and its text change color whenever you hover the mouse cursor over it because of the next code fragment:

```
#top-navigation a:hover {
  color: white;
  background-color: blue;
}
```

Professor: Well done! What about the sub-menu?

Maria: We tried it but weren't successful.

Professor: Please, show me what you have.

Maria: All right. We switched off floating of the inner list's items in order to stack them vertically:

```
#top-navigation li li {
  float: none;
}
```

Then we set the inner list's top margin to zero in order to remove spacing between the sub-menu and the "More" item:

```
#top-navigation li ul {
  margin-top: 0;
}
```

We also managed to hide the sub-menu by setting its display to none:

```
#top-navigation li ul {
  margin-top: 0;

  display: none;
}
```

And that was about it. From then on, we were left without any clue on how to proceed.

Professor: You won't believe how close to the solution you were. All there's left to do is to select a list item over which the mouse pointer is hovering and change the display property of the unordered list within it back to block:

```
#top-navigation li:hover ul {
  display: block;
}
```

There's only a tiny but irritating cosmetic issue left. Note that the inner unordered list is fully contained within the outer list. As a consequence, the outer list expands as the inner list gets block display and shrinks as the inner list hides. Because of that, the rest of the page content moves down when a sub-menu appears, and moves up when it disappears.

If you change `position` of the inner list to `absolute`, then the outer list will no longer contain the inner list, which solves the problem. Besides, it's a good idea to set the list's `width` back to `auto`. This is the complete rule:

```
#top-navigation li ul {
  margin-top: 0;
  display: none;

  position: absolute;
  width: auto;
}
```

Here we are—a really neat two-level navigation menu.

Mike: I have one more question. If I wanted to include the same menu to several pages of my website, is there a way to include it from an external file so that I don't have to copy the code to each of the pages? I mean, just like we included external CSS using the `<link>` element.

6.2   Server Side Includes

Professor: One of the most straightforward approaches is to use so-called *Server Side Includes* or SSI. SSI is a simple server-side scripting language used mainly for the Web. Here's how it works.

First, you must configure the Apache web server so it will interpret your SSI within *.html* files. If you don't want to mess with the server's configuration file, you can create a plain text file named *.htaccess* and save it to the root directory of your website to enable SSI throughout your site. Put the following two lines into the *.htaccess* file:

```
AddHandler server-parsed .html
Options Indexes FollowSymLinks Includes
```

Now you create a file containing HTML code for a menu, or a header, or whatever it is that you want to include in your pages. Note that this file must only contain the code that you want to include in your pages. It is only a fragment of a web page that does not contain any DOCTYPE declaration or elements like <html> or <body>, which are already part of the main, container page. An included file is therefore not a valid HTML document. For example, you may create a file named *my_header.html* and put the following HTML into it:

```
<header>
  <h1>This is my Header</h1>
</header>
```

You can now include this file to whatever page you wish using the virtual argument of the include SSI directive. The directive looks like an HTML comment so as not to intrude upon your HTML code if SSI doesn't work for some reason. The effect of including the file will be the same as if you copied and pasted the content of the included file to the exact location of the include directive. For example, you can include the above *my_header.html* file like this:

```
<!DOCTYPE html>
<html lang="en">
  <head>
    <meta charset="utf-8">
    <title>Including External HTML</title>
  </head>
  <body>
    <!--#include virtual="/includes/my_header.html" -->
    ...
  </body>
</html>
```

Notice that you can use the same root-relative convention for accessing the files with the include SSI directive as you would for accessing external resources with HTML or CSS. The file *my_header.html* is placed in the *includes* subdirectory of the website root.

6.3 Introducing JavaScript

Professor: And now for something completely different.

HTML and CSS are only meant to show your visitor static material to read. While HTML organizes content and gives it a structure, CSS makes things look attractive by taking care of presentation. We even saw an example of how CSS is able to make things happen by using :hover pseudo-class, but this has quite a limited application.

Regardless of how charming and elegant it may look when just a silent pass of a mouse pointer turns up a sub-menu, there is a danger in doing that. Consider, for example, that you hide some vital content and make it accessible to your visitor only through hovering—if the visitor uses a touch screen, he/she will not be able to access it.

Fortunately, there's a third language called JavaScript, which is responsible for making things happen. This makes it a real programming language as opposed to HTML and CSS, which are formatting and style-sheet languages, respectively. We say that JavaScript defines the *behavior* of a web page. Pages equipped with appropriate JavaScript code can become responsive and intelligent. They can, for example, check if someone has filled out a form correctly and let visitors know if they left out necessary information or if they provided information in the wrong form. For example, if they entered a non-numeric character for the day of the month, or a date in the future or too-distant past for the date of birth, which obviously cannot pass as a birthday of a living person.

Although JavaScript is a full-fledged programming language and you will have no choice but to learn some basic programming skills, that doesn't mean it is impossible to use. All in all, programs written in JavaScript are usually much shorter than many applications you are using in your everyday work, such as text processors or digital photo-processing software.

By the way, do you know how to put a chicken in a fridge in three steps?

Mike: First, open the fridge, second, put the chicken inside, and third, close the fridge.

Professor: Right. What about a fish in four steps? How do you do that?

Mike: Open the fridge, take the chicken out, put the fish in, close the fridge.

Professor: Brilliant! That's a joke, all right, but it conveys an extremely important concept. Let me write the instructions down for you:

> Open fridge.
> Take chicken out.
> Put fish in.
> Close fridge.

In computer science, such a list of *instructions* that need to be followed in order to carry out a job is called an *algorithm*. An algorithm is usually written either in plain English or in a so-called *pseudo language*, which a computer does not understand. If you want a computer to understand and *execute* your instructions, you must *code* these instructions in a specific computer language. Such a set of executable instructions is called a *computer program*.

Although the computer will always obey the order in which your instructions are listed, the order itself is not always important. If only the final state is important, you may want, for example, to first put the fish in the fridge (if the fridge is big enough to hold both animals for some time) and only then take the chicken out. But you cannot, by any means, take the chicken out of the fridge *before* opening the fridge doors. That is physically impossible.

One thing you should understand about programming languages is that they are *languages*. Just as you would explain to somebody how to get somewhere or cook some favorite dish of yours in plain English, you also explain to a computer how to do something in a specific programming language. You must never forget that most things we are going to do in our course you already know how to do. All you'll have to learn is to tell a computer how to do it, so the computer will understand you and do exactly what you want it to do.

For example, if you want a computer to add 19 and 23 and display the result in the browser window, then you write the following program:

```
var sum;
sum = 19 + 23;
document.write(sum);
```

Maria: Looks weird.

Professor: But you understand *some* of the code, don't you?

Maria: Yes I do. The sum of the numbers 19 and 23 is written in the same way as we people write it. The last line probably writes the sum to the window, I guess.

Professor: Correct. It's a simple program, but it is important that you understand everything in it. The first line of the above program is a *variable declaration*. A declaration always starts with the var keyword, which is short for a *variable*. The word sum that follows is a user-defined *variable name*. Variables are central to any computer program as they allow the computer to *remember* things that happened in the past. Although you may not always be specifically aware of it, humans also need to remember certain information in order to complete even the simplest tasks. For example, when you count you need to remember the new number each time you increment it. Most of the time you do that unconsciously and only become aware of it when something disturbs you. If you're not concentrating you quickly lose count.

You can think of a variable as a box for storing information that is written on a piece of paper. Once you've put a paper in a box, it is there for you to read any time you want. If you have more boxes, then you need to name them in order to know later on which box to look in for a specific piece of information. That's why you give names to variables.

For example, in the second line of the above program, a value of 42 is first obtained as a sum of 19 and 23 and then stored into a variable named sum. This is accomplished using the equals sign. The final state after these operations can be depicted as a number 42 stored in a box named sum.

Mike: That's like folders in my computer.

Professor: Not quite. The primitive variables, which we are going to meet first, can store only a single value and that's an important difference. For example, consider the following two lines:

```
x = 10;
x = 20;
```

Eventually, x ends up holding only the value 20. The value 10 is simply deleted from memory in the second line to give room to 20.

Maria: How do we choose a name of a variable?

Professor: A name can be chosen arbitrarily subject to certain rules. Note that the official recommendation provides wider allowances, but I think you will be better off by sticking to the following rules when selecting a variable name:

- JavaScript variable names can contain (upper-case or lower-case) letters of the English alphabet. For example, you can name your variable x or give it a more descriptive name like `counter`.

- Names can contain numbers and underscores (_) as well. For example, your variable can be named B42s. If a variable name is composed of more than one word, you can use underscores to separate the words. You must do that because spaces are not allowed. For example, `number_of_lives` is a valid variable name. Many programmers—including me—use a so-called *camelCase* form of writing variable names that consist of more than one word. camelCase is a practice where each new word begins with a capital letter. For example, `numberOfLives` is a camelCase variable name.

- Names cannot start with a number. You cannot use 42 as a variable name while _42 is a perfectly legal name.

- Names can start with a dollar sign ($) as, for example, $awesome. Personally, I don't see any need for using a dollar sign. You will, however, find it used in the popular jQuery library, about which you can read in section B.4 at the end of this book.

- JavaScript is case sensitive, which means that `counter` is considered a different name from `Counter`.

- Reserved words must not be used as variable names. Reserved words are the keywords that have a special meaning in the language. For example, `var` is a reserved word that helps you declare a variable and cannot be used as a variable name. This is a list of the words that JavaScript reserves:

arguments	break	case	catch	class
const	continue	debugger	default	delete
do	else	enum	eval	export
extends	false	finally	for	function
if	implements	import	in	instanceof
interface	let	new	null	package

private	protected	public	return	static
super	switch	this	throw	true
try	typeof	var	void	var
with	yield			

Mike: You mentioned the jQuery library. What's a library?

Professor: In a nutshell, a library is a collection of different solutions that you can readily use in your own code. Libraries exist to spare you the tedious work of writing and testing code for common tasks that have already been written and tested by others.

Back to our program. As we've already figured out, the sum of two numbers is calculated in the second line. The line wouldn't have had any effect were it not for the *assignment operator* (=), which stores the calculated sum in the variable named sum. In general, the assignment operator always takes whatever value is produced on its right-hand side and stores it to the variable on its left-hand side. Because a variable can be placed to the left side of the assignment operator to receive the new value, we sometimes call it *lvalue*, which is short for a left value. Note that a constant number cannot be lvalue. You cannot write 42 = x because a value cannot be stored into a constant number.

Obviously, the last line of the program is used to display the sum to the screen. Don't worry about the whole syntax, just remember for now that a value of anything you put inside the parentheses of the document.write() function will be displayed in the browser window. In our case the function will display the value stored in sum.

Let me mention that the document.write() function is actually almost never used in today's JavaScript programming. In the old days of JavaScript, though, it used to be the only way to display text in a document dynamically and you are likely to see it in existing code. Before we learn other methods, we are going to use it anyway because it is simple and useful for quick experimenting with code.

Maria: And how do we run the code?

Professor: There are many ways to do it. As for now, we'll just be happy with putting the code inside the <script> HTML element, inside the body of the document:

```
...
<body>
  <script>
    var sum;
    sum = 19 + 23;
    document.write(sum);
  </script>
</body>
</html>
```

Now all you have to do is load the file into the browser and you shall see the 42 in the browser's window.

In the last example, there's one more detail that I want you to pay attention to. You instinctively know that 19 has the value 19 and that 23 has the value 23. Such values that appear directly in a program as fixed values are called *literals*. Literal is also any text between (double or single) quotes. For example, `"Hello there!"` is a text literal. Beside literals, programs also contain various names used to refer to different entities such as variables that store values. Such names are called *identifiers*. It is crucial that you understand the difference between the two.

For example, we can put `sum` from our last example in quotes:

```
document.write("sum");
```

Now `sum` is no longer an identifier but a text literal. If you reload the page, you'll see "sum" instead of "42" in the browser window.

Maria: Have you told us how to write comments in JavaScript?

Professor: Oh, yes. The language supports two styles of comments. A single line comment begins with a double slash and ends at the end of the line:

```
//This is only a humble remark.
```

A multiple line comment can span over several lines and goes between two pairs of characters /* and */:

```
/* This is an elaborate discussion about the meaning of life,
which will be completely ignored by the browser. */
```

6.4   Values and Types

Professor: You just learned that a computer program is a list of tasks to be carried out by the computer. Most of the time these tasks involve manipulating *values*. For example, computing the velocity of a bouncing ball in a simulation software or computer game, or calculating interest from the principal investment amount, the annual interest rate, and the number of years the money is invested. You also learned that a value can be represented either as a literal, such as a fixed number or text in quotes, or stored in a named variable. Regardless of a form that a value takes, you shall always ask yourselves about the *type* of that value.

JavaScript distinguishes between four different *primitive* data types, which are summarized in the following table together with some example values:

Type	Example Values
number	`42  2.71828  0xf8  6.022e23  Infinity  NaN`
boolean	`true  false`
string	`"something"  "42"`
undefined	`undefined`

Mike: Why are those types called primitive?

Professor: Simply because they only carry a single value and nothing else. This is in contrast to objects, which are much more complex, as you will see.

Let me ask you something before we continue. What will be the values of x and y after the following code executes?

```
var x;
var y;
y = 10;
x = y + 5;
y = 20;
```

Mike: 25 and 20?

Professor: Think again.

Mike: You said that primitive values only held a single value and that the last value always overrode the previously assigned values. So y should be 20.

Professor: That's true. y has the value 20. But do not overlook the fact that at the time of computing the value of x in the second to last line of the code, the value of y is 10. Therefore, the value of x is 15 and not 25. Note that, just as in the real world you cannot change what happened in the past, the last line of the code has no way of affecting the value of x, which was determined in the previous line.

Let's return to types. The first of the types is number, which represents all kinds of numbers, integers as well as real numbers. In computer terminology we often call real numbers *floating-point* numbers because of the format in which they are stored in memory. If you are interested, you can check the IEEE 754 standard for details of how this format is defined. JavaScript uses the 64-bit floating-point format for both real and integer numbers. That means it can represent real numbers as huge as $\pm 1.7976931348623157 \times 10^{308}$ and as tiny as $\pm 5 \times 10^{-324}$. I think those bounds are not something you should worry about right now, I just give them to you as a curiosity. However, the bounds can become problematic when it comes to integers, which are bounded to the values between -9007199254740992 and 9007199254740992. If you study the IEEE floating-point standard, you might discover that these weird numbers are in fact (plus and minus) two raised to the power of 53.

Mike: Where do these limitations come from?

Professor: You know, numbers have limited space in memory, like a mileage counter in a car. A mileage counter usually has six digits, so it can count miles up to 999,999. Something similar happens with numbers stored in computer memory. However, they are stored as binary numbers, and thus the highest and lowest values seem unusual when expressed in decimal form.

What is probably much more important for you right now is that numbers can be represented by numeric literals in forms that you can find in the above table. The first two, (42 and 2.71828) don't need any explanation. The literal 0xf8 may also be familiar to you.

Maria: Is that a hexadecimal? In CSS we put the hash sign (#) before hexadecimal numbers.

Professor: That's right. Except that in JavaScript, you put either 0x or 0X before a hexadecimal literal.

Mike: Is 6.022e23 the Avogadro constant? It's familiar from school. So maybe e means "times ten raised to the power of"?

Professor: Exactly. That's the so-called *e notation*, which is used in place of scientific (exponential) notation if superscripted exponents like 10^{23} cannot be displayed as in most calculators or computer programs. Instead, the letter e (or E) is used to mean "times ten raised to the power of." Therefore, 6.022e23 reads as 6.022×10^{23}.

Maria: Something still doesn't fit just right. If you can write real numbers up to the order of 10^{308}, why can't you write integers quite so large as well?

Professor: That's a good question. Let's try the following example:

```
var biggest = 9007199254740992;  //Take the biggest possible
biggest += 2;                    //integer and increase it
document.write(biggest);         //by two.
```

In the browser window you see the expected result.

So where's the catch? The fact is that it is not the range that is the problem but the *precision*. A 64-bit floating point value is guaranteed to be precise up to 15 significant digits. That is counted from the first digit that is different from zero, no matter how big or how small the value may be. The numbers with a larger number of significant digits may or may not be exact. For example, if you increment the biggest possible integer by just one, you get 9007199254740992, and if you increment it by three, you get 9007199254740994, which are both wrong. The correct result, which we get by adding two, is just a coincidence. You can also try incrementing the value by one many times and you'll never get anywhere.

Enough of numbers for now. The second of the primitive data types is Boolean. It is used in logical calculus of truth values, and variables of the Boolean type may assume only the values true and false. Next, there is a data type called string, which is nothing more than plain text. In JavaScript, everything you put between (single or double) quotes is considered a string literal.

The last of the primitive types is undefined. This is a special type with only a single value undefined, which is assigned to a variable that has been declared but not given

---

any value yet. For example, if you run the following code, you will see `undefined` written in the browser window:

```
var x;
document.write(x); //Writes undefined
```

Closely related to the `undefined` value is `null`. This is a special value that belongs to the Object type but in practice, `null` can be seen as an exclusive member of its own type. It is usually used to indicate the absence of a value, just like `undefined` is. There is a slight difference in the meaning of both, though. `undefined` is considered as an unexpected, or erroneous, absence of value, and `null` as a normal, or expected, absence of a value.

Maria: Normal in what sense?

Professor: For example, if a program needs a value to be entered from the user, then it is normal to expect that the variable used to store that value doesn't have a value before the user enters it. On the other hand, the programmer might have forgotten to initialize a variable that should have been initialized, which results in an unexpected absence of value.

By the way, `null` is a language keyword while `undefined` is a variable whose value is set to `undefined`. But since `undefined` is read-only according to the specification, that fact should be no more than a technical curiosity to you. And the same goes for `NaN` and `Infinity`—they are read-only variables too.

Mike: I noticed these two between the numbers in our last table on page 110. What are they?

Professor: Oh, thanks, I forgot to explain. `Infinity` is infinity, which means something without any limit. I don't want to talk too much about it, for some great minds have already gone nuts because of infinity. For us, let it be that infinity is, for example, something divided by zero:

```
document.write(1 / 0);  //Writes Infinity
```

In JavaScript, infinity is in fact anything larger than the largest 64-bit floating point number. For example:

```
document.write(1.797693134862316e308);  //Writes Infinity
```

The identifier `NaN` stands for "Not a Number." `NaN` arises if you attempt to carry out an operation that cannot return a well defined numeric value, such as divide zero by zero or multiply a number by `undefined` (declared but uninitialized variable). For example:

```
document.write(0 / 0);         //Writes NaN
document.write(3 * undefined); //Writes NaN
```

Mike: Is there a way to get and write the type of a value?

Professor: There is, actually. You can get the type of a value using the `typeof` operator. You simply put it before the value you want to test:

```
var txt = "I'm plain text";
var x = 4;
document.write(typeof txt);       //Writes string
document.write(typeof NaN);       //Writes number
document.write(typeof x);         //Writes number
document.write(typeof 1.2e-14);   //Writes number
document.write(typeof "3.14");    //Writes string
document.write(typeof true);      //Writes boolean
document.write(typeof "NaN");     //Writes string
document.write(typeof undefined); //Writes undefined
```

Because the `typeof` operator returns a description of a type in a text form, the type of the value returned by the `typeof` operator is `string`:

```
document.write(typeof typeof 42); //Writes string
```

Maria: I wonder if there is any practical use for knowing the types of values. I mean, what's the difference between 3.14 and "3.14"?

Professor: The behavior of both can be quite different. To find out how, we first need to know about operators in JavaScript.

6.5 Operators and Expressions

Professor: What a computer program does is make operations on values to produce new values. The operations are executed by means of *operators*, and the values on which they operate are called *operands*. However, before we plunge into the world of JavaScript operators let's define an *expression*. An expression can be viewed as a phrase that a JavaScript interpreter can *evaluate* in order to produce a value. We say that an expression *returns* a value.

According to this definition, an expression can be as simple as a single constant or literal value, a certain language keyword, or a variable reference. These are called *primary* expressions. For example, the following are all primary expressions:

```
3.14             //Returns 3.14
x                //Returns the value stored in x
true             //Returns true
"something blue" //Returns "something blue"
```

Primary expressions can be used to build more complex expressions by means of appropriate operators. Some of them we have already met. For example:

```
19 + 23     //The + operator for adding two operands.
            //The expression returns 42.
typeof true //The typeof operator for determining the type of
            //an operand. It returns "boolean".
```

Consider, for example, the following expression. What value does it return?

```
2 * 4 + 3 * 3
```

Maria: 17.

Professor: And how do you know that?

Maria: Because two times four equals eight, three times three equals nine, and, finally, eight plus nine equals 17.

Professor: I see you are already familiar with operator *precedence*. Just as you learned in primary school that multiplication is done before addition, JavaScript precisely defines which operator has precedence over which.

What about the next expression? What value does it return?

```
5 - 4 - 3
```

Mike: -2.

Professor: Because you computed it right, I assume you already know about operator *associativity*. Associativity defines whether operators with the same precedence are executed from left to right or from right to left. Subtraction has left-to-right associativity, as do most of the JavaScript operators that take two operands. You will find JavaScript operators organized by precedence, with additional information about their associativity in the JavaScript Mini Reference on page 373.

If you don't like precedence and associativity of an operator, you can always change the order of execution using parentheses—the operators within parentheses will be evaluated first. For example, if you want the subexpression 4 - 3 in the above expression to be evaluated first, then you simply enclose it in parentheses:

```
5 - (4 - 3)  //Returns 4
```

Arithmetic Operators

Professor: I would now like to go through some of the JavaScript operators that you will most likely use in your programs. First, there are arithmetic operators, and you won't be surprised that they include operators for addition (+), subtraction (-), multiplication (*), and division (/). As you already know from primary school mathematics, the second two have precedence over the first two, and all have left-to-right associativity. The subtraction operator can be used as a negative sign operator as well, in which case it has the highest precedence. For example, the following expression returns -12:

```
2 * -6 //Returns -12
```

It is somehow self-evident that you should evaluate the subexpression −6 first, and then multiply it by two. There's simply no other way of seeing it. The good news is that many rules of precedence and associativity are somehow logical and need not be learned explicitly.

The only arithmetic operator that you may not have heard of before is the *remainder* operator, which returns the remainder after a division. The operator's symbol is a percent sign (%). Although in JavaScript this operator operates on integers as well as on non-integers, it is primarily used with integers (the so-called *Euclidean division*). Consider, for example, that you have to divide 41 candies fairly among six children. Each child gets six candies and five will be left over as a remainder. This is summarized in the next expression:

```
41 % 6  //Returns 5
```

You should be careful, though, when you use the remainder operator with negative numbers. In number theory, this operator is called the *modulo* operator where the remainder is always positive. That said, in JavaScript the sign of the result is the same as the sign of the dividend and can be negative. Things become even more confusing because the sign can be computed differently in other languages. In Python, for example, the same operator returns a remainder whose sign is the same as the sign of the divisor.

Relational Operators

Professor: The next group of operators test for a relationship between two values and return the Boolean value `true` if the relationship holds and `false` if the relationship doesn't hold. Basically, you can view the relational operations as asking questions like: "Is the left-hand side value greater than the right-hand side value?" or "Are both sides equal?" If the answer is "yes," then the return value is `true`, or else, if the answer is "no," the return value is `false`. Here's a table of all the relational operators:

Operator	Test of Relationship
<	Less than
<=	Less than or equal to
>	Greater than
>=	Greater than or equal to
==	Equal to
!=	Not equal to
===	Strictly equal to
!==	Strictly not equal to

Notice that some operators are composed of two, or even three symbols. You should never put a space between them or you will break an operator into two or three operators.

The common use of relational operators may be trivial, but still, let me give you some examples to warm you up. Can you tell me what values the following three expressions return?

```
10 < 2
6 >= 6
6 != 4
```

Maria: They return `false`, `true`, and `true`.

Professor: Precisely. What about the next one? Suppose that x has been given the numeric value 4.

```
10 < x < 20
```

Maria: That's `false`.

Professor: Unfortunately it isn't. But let's save the explanation for the next meeting.[1]

Did you notice in the above table that there are two different operators testing whether two values are the same? They use, however, two slightly different definitions of what "the same" means. The first one, the equality operator (==), represents a more relaxed test of sameness. It compares values and doesn't require the types to be the same. For example, the following two expressions both return `true` even though the types, or even values, are different:

```
42 == "42"        //Returns true
undefined == null //Returns true
```

The strict equality operator (===), however, returns `false` if the types don't match:

```
42 === "42"        //Returns false
undefined === null //Returns false
```

The inequality (!=) and strict inequality (!==) operators work in just the opposite way:

```
42 != "42" //Returns false
42 !== "42" //Returns true
```

There's one very special case where equal is not equal, which involves the NaN value. NaN will not compare equal to any other value, including itself. Consider the next two expressions:

---

[1]If you're curious, you can find the explanation on page 145.

```
NaN === NaN //Returns false
NaN !== NaN //Returns true
```

Thut's bizarre. As a consequence, if you want to check whether a variable has received the NaN value, you must compare a variable to itself:

```
x !== x //Returns true if and only if x is NaN.
```

Mike: Excuse me, but why are `null` and `undefined` equal? Is there a rule or must you learn it by heart?

Professor: Both. It's a rule that you must learn by heart. You are going to hear about this the next time when we talk about type conversions.[2]

Logical Operators

Professor: There exist three logical operators to perform Boolean algebra. Most often you will use them to combine two or more relational expressions to build more complex expressions. You may have already heard of them, but just in case, let me review them briefly. In common for all of the three operators is that they operate on Boolean values `true` and `false`, and they return Boolean values as well.

The first one is logical AND (`&&`), which returns true if and only if both of its operands are `true`. For example, the following cxpression returns true only if both a *and* b are equal to zero:

```
a == 0 && b == 0
```

Mike: Can you write this shorter: `a && b == 0`?

Professor: You can, but it means something completely different. If you check the operator table in the JavaScript reference on page 373, you will find out that the equality operator (`==`) has precedence over logical AND (`&&`). Therefore, the expression `b == 0` is evaluated first. The value of this expression is then combined with the value of the variable a through the logical AND operator. Your expression will thus evaluate to `true` if and only if b is zero and a is `true`. I must warn you that JavaScript operators are very precisely defined and you cannot build expressions relying on your intuition.

The second Boolean operator is logical OR (`||`). It returns `true` if *any* one of the two operands evaluates to `true`.

For example, the following expression will return `true` if at least one of the variables a and b is zero:

---

[2]You can learn why `null` and `undefined` are equal on page 144.

```
a == 0 || b == 0
```

Put differently, it will return `true` if either a is zero *or* b is zero *or* both are zero.

The last of the Boolean operators is logical NOT (!), which *inverts* the value of the expression. That means it makes `false` from `true` and vice versa. You place this operator before a Boolean expression whose value you want to invert. For example, if the value of x is `false`, then the expression `!x` will return `true`.

By the way, because the logical NOT operator operates on a single value, it is called a *unary* operator. As you have probably noticed, most operators combine two values and are hence called *binary*. This property of operators is called *arity*.

Operations with Strings

Professor: Let me now answer your question, Maria. You asked what was the difference between the literals 3.14 and "3.14". You already know that, although both hold the same value, the first one is a number while the second one is a string.

Mike: But are the values really the same? I mean, the second one has quotes.

Professor: Quotes are not part of the value. They are just part of a string literal and they tell the JavaScript interpreter that the value between them is a string type. You can try and write both values to the browser window:

```
document.write(3.14);    //Writes 3.14
document.write("3.14");  //Writes 3.14
```

In both cases the value 3.14 appears in the window without quotes. So, what's the difference, you ask? Let's try the following code:

```
document.write(3.14 + 1);   //Writes 4.14
document.write("3.14" + 1); //Writes 3.141
```

As expected, the first line writes 4.14 to the window. The second one, however, will produce 3.141. That's because the plus sign behaves differently with strings—it becomes the *concatenation operator*, which concatenates strings even if they contain numbers.

Relational operators also behave differently with strings as they do with numbers. Consider, for example, the next two lines of code:

```
document.write("mountain" > "molehill"); //Writes true
document.write("2" < "14");              //Writes false
```

The first line writes `true` not because a mountain is bigger than a molehill, of course. It writes `true` because, alphabetically, a mountain comes after a molehill. Relational operators compare strings alphabetically.

Similarly, the second line writes `false` because, alphabetically, 2 comes after 14.

Maria: Why is that so?

Professor: When comparing strings alphabetically, only the first characters of both strings are used in comparison. If they are equal, then the second characters of both strings are compared, and so forth. This continues until two different characters are found or strings are equal. In that sense 2 comes after 14 just like "n" comes after "me" in a dictionary.

Maria: Yes, but how do I know that 2 comes after 1 in the alphabet? And how are numbers related to letters?

Professor: Now you know that 2 comes after 1.

OK, seriously. Every character is stored in a computer memory as an integer, also called a *character code*. When JavaScript compares characters, it actually compares their character codes. Which character is represented by which code is defined by the *Unicode encoding*. If you are interested, you can check the table at *unicode-table.com*. For example, the character 1 has the hexadecimal code 31 (decimal 49) and 2 has the hexadecimal code 32 (decimal 50). Letters have bigger codes so any number will generally compare as being smaller than any letter.

Although spaces seldom make a difference in JavaScript code, it is very important that you be aware of spaces within strings, which are characters in their own right. The next expression, for example, evaluates to false because of an extra space at the end of the second string:

```
"I'm different" == "I'm different " //Returns false
```

Likewise, two spaces are different from one:

```
" " != "  " //Returns true
```

Assignment Operator

Professor: It's time to draw your attention to the fact that none of the operators we've discussed so far has any side effects. All variables involved in a computation retain their previous values intact. The question arises: how can a computer program work if nothing it does leaves any traces behind? That's why we need the *assignment operator* (=), which can copy the value returned by an expression on its right-hand side to a variable on its left-hand side. Perhaps you will recall that what we can put to the left of an assignment operator is called lvalue. For now, only variables can be lvalues, but you'll see later that there are some other expressions that can act as lvalues as well.

This is the syntax of using the assignment operator:

```
lvalue = expression
```

Maria: What about `document.write()`? It has a side effect, doesn't it?

Professor: That's not a type of side effect that I had in mind. The `document.write()` function writes a value of an expression to the browser window, which only the user can read. The code cannot read this value back, so the value is lost as concerns the JavaScript program. Besides, `document.write()` is not an operator.

Back to the assignment operator. Because the value of *expression* has to be evaluated *before* it can be copied to *lvalue*, it comes as no surprise that the assignment operator has almost the lowest precedence of all the operators. Besides, the = operator has right-to-left associativity, which enables you to set several lvalues to the same value in a single expression. For example, you can set variables x, y, and z to zero this way:

```
x = y = z = 0
```

Mike: Does the assignment operator also return a value as other operators do?

Professor: Of course it does. Otherwise the above multiple assignment expression wouldn't work. The assignment operator returns the value that is being assigned. For example, the expression z = 0 returns zero. In the above example, this expression is evaluated first because of the right-to-left associativity of the assignment operator. Next, the returned zero is assigned to y. This assignment also returns zero, which is finally assigned to x.

Shorthand Operators

Professor: Some operations are very common in computer programs, and to make life easier for programmers there exist shortened notations for them. For example, if you want to multiply x by 2 so that x will actually receive the new, multiplied value, you should not only multiply x by 2 but also store the result back to x:

```
x = x * 2
```

You could, however, use the shorthand operator *= to do the same:

```
x *= 2
```

In fact, all five arithmetic operators have their corresponding shorthand form:

```
lvalue op= expression
```

where *op* can be replaced by any one of the operators +, -, /, *, or %. The above syntax actually means this:

```
lvalue = lvalue op expression
```

You will often increase or decrease a value by one:

```
counter = counter + 1 //Increases counter by one
counter = counter - 1 //Decreases counter by one
```

Using a shorthand notation, you can rewrite the last two lines as:

```
counter += 1 //Increases counter by one
counter -= 1 //Decreases counter by one
```

You can make them even shorter:

```
counter++ //Increases counter by one
counter-- //Decreases counter by one
```

Operators ++ and -- are used very often to increase or decrease the value of a variable by one. They are called *increment* and *decrement* operators, respectively. As you can see, they are unary operators. Both can be placed either before or after a variable with a slight difference in operation. In both cases the value of a variable is changed but if an operator is placed after a variable, it returns the old value. Only if an operator is placed before a variable, it returns the updated value. Consider, for example, the following example:

```
var x = 10;
var y, z;
y = ++x;
z = x++;
```

What will be the final values of the variables x, y, and z when the code executes?

Maria: Can you declare several variables with a single var keyword? I also noticed that you initialized x while declaring it.

Professor: That's right. You can give a variable an initial value right away, and you can declare many variables using a single var keyword, provided you separate individual declarations by commas.

So what do you think will be the final values in the above example?

Maria: Because x was incremented twice, it should be 12, and because the expression ++x returns the new, incremented value of x, I think y will be set to 11.

Professor: You said the incremented value of x was 12. So why 11?

Maria: Yes, but in the third line x is only incremented for the first time. The fourth line has not even started yet.

Professor: Absolutely. Go on.

Maria: In the fourth line x is incremented for the second time, but the expression x++ returns its old value, the one it had before this second increment. Hence, z will be 11 as well.

Professor: Perfect.

The Conditional Operator

Professor: It's nice when a computer can undertake some tedious calculations for you, but it would be even nicer if it could make *decisions* based on those calculations. For one thing, that is what computers usually do. Even tasks as simple as calculating the absolute value of a number involves decisions whether or not to multiply the number with minus one. The conditional operator (?:) allows a program to make decisions like that. It works with three operands, which makes it a *ternary* operator. Because it is the only one of its kind, it is sometimes called simply *the ternary operator*. Although you will see the conditional operator written as ?:, it actually does not appear quite like that in code because one of the three operands goes between the symbols ? and :.

With the conditional operator, you write expressions using the following syntax:

```
condition ? if_true : if_false
```

The operator works as follows. First, `condition` is evaluated, and if its value is true, then `if_true` is evaluated and its value is returned. However, if `condition` is false, then `if_false` is evaluated and its value is returned. Notice that only one of the expressions `if_true` and `if_false` is actually evaluated as part of a conditional operator evaluation.

Basically, this is it. Could you now write an expression that will replace the value of x with its absolute value?

Mike: `condition` will probably be something like x < 0 because the value should only change if it is negative. Then, in place of `if_true`, I would put x *= -1. And that's it.

Professor: You're close but that won't work, I'm afraid. As for any operator, you must provide *all* the necessary operands for the conditional operator as well. If they are three, then they are three. You should put in something as a third operand even if there's nothing to do when `condition` is `false`. Besides, it is not a common practice to make assignments within the conditional operator although it is not wrong. It's better to let the operator just return the right value and then assign that value to an appropriate variable via the assignment operator. The next expression replaces the value of x with its absolute value:

```
x = x < 0 ? -x : x
```

This works because the conditional operator has precedence over the assignment operator, and the less-than operator has precedence over the conditional operator.

6.6 Concluding Remarks and Homework

Professor: There's a lot of stuff we covered today, so let me just summarize some important things. Being able to evaluate the value of an expression is vital. An expression is composed of one or more values and, possibly, operators to manipulate those values. Several things are important here:

- You should know *what* a certain operator does and *when* it does it.

- To find out *what* a certain operator does, you sometimes need to know the type of operands. The addition operator, for example, becomes the concatenation operator when used on strings.

- The *when* is defined by operator precedence and associativity, which define the order in which operators are evaluated. You already know that multiplication and division have precedence over addition and subtraction, and you will probably remember that the assignment operator has almost the lowest precedence. If you are uncertain, you can always use parentheses to change the order in which operators are evaluated.

- Only the assignment operator has a side effect. Because shorthand operators implicitly include an assignment operator, they have a side effect too.

When in doubt, you can experiment so as to better understand how a certain operator works.

Of course, there arc a lot more operators and values that JavaScript has in store for you, but I think you've learned enough to make a decent start in JavaScript programming.

For your homework, write two short JavaScript programs. Let the first one read an integer and test its parity. That is, write to the browser window whether the number is even or odd. The second program should also read a number and inform the user whether the number is a prime number. You may recall from high school mathematics that a prime number is any natural number that is greater than one and has no divisors other than one and itself.

Maria: How can a program read a number?

Professor: You can use the `prompt()` function. When called, this function opens a dialog that allows a value to be entered. When the user selects OK, the dialog closes and the function returns whatever value the user has entered. Inside the parentheses you write a message to the user, typically in the form of instructions as to what should he/she do. In order to capture the returned value, you must assign it to some variable using the assignment operator. For example:

```
var myValue = prompt("Enter an integer greater than one.");
```

Note, however, that although the `prompt()` function is easy to use, you should avoid using it if you strive for good web design. Just like the `document.write()` function, you will only use `prompt()` until you learn enough to be able to use much better alternatives.

Before you go, these are today's keywords for you to reflect upon:

In this meeting: JavaScript, programming language, behavior, instruction, algorithm, computer program, variable declaration, `var`, variable name, reserved words, `<script>`, `document.write()`, literal, identifier, value, data type, number, floating point, largest number, smallest number, precision, Boolean, string, `NaN`, `undefined`, `null`, primitive types, operator, expression, operator precedence, operator associativity, arithmetic operators, relational operators, logical operators, operators and strings, assignment operator, shorthand operators, conditional operator, `prompt()`

Controlling Program Flow

7.1 Homework Discussion

Professor: Did you have any trouble with your homework?

Mike: The first one wasn't that difficult. We discovered that it is even possible to write it with a single conditional operator, combined with the remainder operator for checking whether the number is divisible by two. We use the concatenation operator to make the message more explanatory. Parentheses are necessary because the concatenation operator has precedence over the conditional one. This is the code:

```
var valueToCheck = prompt("Enter an integer.");
document.write("The entered number is " +
                (valueToCheck % 2 == 0 ? "even." : "odd."));
```

Professor: I like your solution and I see that you understand how JavaScript operators work.

What about the second homework?

Maria: It was much harder. As a mater of fact, we could solve it only for numbers up to 20.

Professor: Why 20?

Maria: OK, we could have written the program for any number but its size would then grow beyond the limits. Maybe we overlooked some important detail? This is our solution:

```
var x = prompt("Enter an integer greater than one.");
var counter = 0;

counter = x % 2 == 0 && x != 2 ? counter + 1 : counter;
counter = x % 3 == 0 && x != 3 ? counter + 1 : counter;
counter = x % 5 == 0 && x != 5 ? counter + 1 : counter;
```

```
counter = x % 7 == 0 && x != 7 ? counter + 1 : counter;
counter = x % 11 == 0 && x != 11 ? counter + 1 : counter;
counter = x % 13 == 0 && x != 13 ? counter + 1 : counter;
counter = x % 17 == 0 && x != 17 ? counter + 1 : counter;
counter = x % 19 == 0 && x != 19 ? counter + 1 : counter;

document.write(counter == 0 ? "Prime number." :
                             "Not a prime number.");
```

The program simply counts the number of times the division leaves a remainder of zero. If there's no such division, the tested number is a prime number. Of course, we don't want to count situations where the number is divided by itself, which also returns a zero remainder. That's why we included the inequality comparisons.

Professor: Good. I see that you tested only divisions by prime numbers. There's really no need for testing with other numbers. For example, if a number is evenly divisible by four, then it must also be evenly divisible by two. Thus you don't need to test the division by four. The same goes for six—if the number is evenly divisible by six, then it must also be evenly divisible by two and three, and there's no need to divide by six.

Your program can in fact test the numbers up to 361 because you must only try the divisions with numbers that are less than or equal to the square root of the tested number. Observe that, if a number n is a product of two positive integers, only one of them can be greater than the square root of n. So, if there existed a divisor greater than the square root of n, then a divisor smaller than the square root of n would also exist and you would already have found it.

Before we start today's work I want you to understand a fundamental difference between the two tasks you had for your homework. The first one could be solved using a single test, while the second one required *repeated* testing. This cannot be done by simply writing many lines of code because you don't know *how many times* the test should be repeated. This information is not available at the time of writing the program. Besides, just copying and pasting the same code is not programming. Fortunately, there are tools that allow for automated repetition of tasks, and that's one of the topics of today's meeting.

Mike: One question: sometimes you write a semicolon at the end of a line and sometimes you don't. Is there a rule so we know exactly where to put a semicolon? We experimented and discovered that semicolons are not necessary.

Professor: A semicolon terminates a *statement*. Which brings us to another important topic in JavaScript.

7.2 Statements

Professor: So far we've only been working with expressions. Recall that an expression is evaluated in order to return a value. However, technically speaking, you need a *statement* in order for that to happen. That is similar to English phrases, which should be put into sentences. And just as sentences are terminated and separated from each other with periods, so are statements terminated with semicolons. For example, you can form a simple statement from an expression by adding a semicolon:

```
expression;
```

This kind of statement is called an *expression statement*. Note that, if `expression` has no side effect, such a statement makes no sense. Of course, you can build more complex statements by combining several subexpressions so that something meaningful happens. For example, to write the value of `expression` to the browser window, you can compose a statement like this:

```
document.write(expression);
```

A semicolon is important because it makes a statement out of an expression and *executes* that statement. It is, however, not necessary to put a semicolon at the end of the statement if it is the only one, or the last one on the line. Yet I strongly recommend that you *always* terminate a statement properly for consistency.

Flowchart

Professor: When you organize expressions into statements and execute them, the order of execution becomes of paramount importance. So far, execution has simply followed a list of instructions in the exact order in which they were written. Still, the conditional operator slightly changed that by evaluating only two of the three expressions. As we proceed, you'll see more and more situations where the order of execution is not quite that linear. Especially at the beginning it is useful to have a graphical tool that will allow you to illuminate what is going on in your program. The tool is called a *flowchart* and it is exactly what it says it is: a chart which graphically represents the flow (order) of the statements' execution.

A flowchart is composed of two different types of boxes depicting two basic types of actions to be carried out:

- A processing step
- A decision

A processing step is denoted by a rectangle having an entrance, an exit, and a description of an activity to be accomplished.

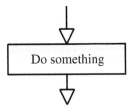

I think the graphics are more or less self-explanatory: you follow the arrow entering the box, do whatever it says should be done inside the box, and, finally, you exit the box.

A decision is depicted as a diamond with one entrance and two exits.

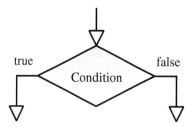

You again start by following the arrow entering the box. Then you have two possibilities: if the condition inside the box evaluates to `true`, you exit the box via the left arrow, and if the condition evaluates to `false`, you exit the box through the right arrow.

With different combinations of processing and decision boxes, you can assemble any algorithm you like.

Maria: We still don't have repetitions.

Professor: You can connect an arrow exiting a box to the entrance of an arbitrary box. If one of the arrows exiting a decision box is directed backwards, you get a repetition, as you will soon see.

Expression Statement

Professor: As already mentioned, the simplest statement in JavaScript is an expression that has a side effect terminated with a semicolon, which is called an expression statement. For example, if you want to increase the value of `counter` by one, the next statement will do it:

```
counter++; //Increments counter
```

We can draw a flowchart for this statement, which looks like this.

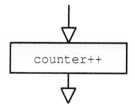

Another example of an expression statement is a function call. For example:

```
document.write(counter);
```

Compound Statement

Professor: There are often situations when you need to write several statements while the rules only allow one. On such occasions you use a *compound statement*, which is no more than a sequence of statements joined into a single one using a pair of curly brackets. For example:

```
{
  counter++;
  x -= counter;
}
```

Notice that a compound statement doesn't end with a semicolon.

Maria: You indented the lines of code inside the curly brackets. Is this necessary?

Professor: No, it isn't. Indentation is completely optional, but every programmer uses it to optically separate statements that are subordinate to another statement. If the hierarchy of statements is clearly visible from the code, the code is much easier to read, understand, and, most importantly, maintain.

The flowchart of the above compound statement looks like this.

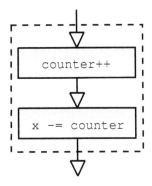

Notice that from the outside, a compound statement looks like a single statement with a single entrance and exit, which is indicated by a dashed rectangle.

if/else Conditional

Professor: Apart from expression statements, there exists another group of so-called *control statements*. These statements allow you to change the JavaScript default order of code execution, which follows the order in which statements are written.

The if/else statement is the essential control statement that empowers JavaScript to make decisions. Because this statement actually represents a point in your code where the flowchart branches into two paths, it is sometimes known as a "branch." The statement has two forms, the simpler of both having the following syntax:

```
if (expression) statement
```

This statement works as follows. First, *expression* is evaluated, and if it returns true, then *statement* is executed. Incidentally, statement is sometimes called a *body* of the if statement. Otherwise, if *expression* is false, then *statement* is not executed. For example, the following fragment of code will assign zero to the variable x if and only if the value of x is null or undefined:

```
if (x == null) {
   x = 0;
}
```

The code can be illustrated by the following flowchart.

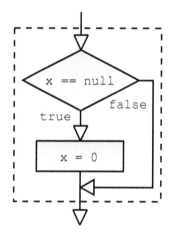

There are two important aspects I would like to point out regarding this flowchart. First, notice how the whole if statement only has a single entrance and exit regardless of the branch inside. This single entrance/exit policy is an important quality of the structured programming languages to which JavaScript belongs. Second, the expression statement x = 0; can be replaced by an arbitrary JavaScript statement, including empty, compound, or even another control statement. As a matter of fact, any rectangle inside any flowchart can represent any JavaScript statement, which is possible precisely because of a single entrance/exit property of JavaScript statements.

Mike: I'm a little confused. Why did you use a compound statement composed of only a single statement?

Professor: I didn't have to. I could just as well write this code:

```
if (x == null) x = 0;
```

You will, however, see many programmers (including me) consistently write curly brackets around statements subordinate to control statements like `if`. It makes your code more transparent and less prone to obscure errors, especially when you nest one control statement into the other.

The above form of the `if` statement does nothing if `expression` evaluates to `false`. If this is not what you want, you can use the augmented form of the statement with the `else` part:

```
if (expression) statement_true
else statement_false
```

This form of the `if` statement will again evaluate *expression* and only execute *statement_true* if *expression* evaluates to true. If *expression* evaluates to false, however, it will execute *statement_false*. For example, if you want to assign a string either "odd" or "even" to the variable `msg` depending on the value of `x`, you can write:

```
if (x % 2 == 1) {
    msg = "odd";
}
else {
    msg = "even";
}
```

The code can be represented by the following flowchart.

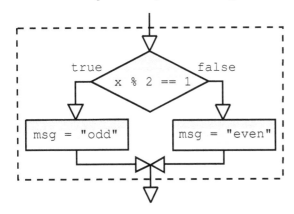

Maria: OK, that's the same as the conditional operator.

Professor: Not quite. The logic is the same but there are differences. First, the conditional operator returns a value while the `if/else` statement doesn't. That's because statements don't return values. As a consequence, *statement_true* and *statement_false* are actually statements that should have side effects. Second, because the conditional operator will operate on expressions and not statements, you cannot conditionally execute control statements with it. In short, the conditional operator can be viewed as a condensed form of the `if/else` statement.

Truthy and Falsy Values

Professor: The first line of our last example could also be written shorter, without a comparison but with the same result:

```
if (x % 2)
```

You see, JavaScript is quite flexible when it comes to data types. If it doesn't like the type, it implicitly converts it so as to match the requirement. For example, the expression x % 2 doesn't return either `true` or `false` so, technically, we cannot use it as a branching condition, which requires a Boolean value. However, knowing that JavaScript implicitly converts zero to `false` and one to `true` when it requires a Boolean value, we can drop the comparison part of the expression in this particular case.

But zero is not the only value that converts to `false` when a Boolean value is expected. Altogether, there are five such values:

```
0
undefined
null
NaN
""    //A string without any character
```

All these five values work as `false`, and are, together with the `false` value itself, called *falsy* values. All the other values are called *truthy* values. You have to be careful, though, when leaving out explicit comparisons, because results are not always identical as was the case in our last example. For instance, this is how you can test whether x has been given a value:

```
if (x != null) ...
```

In this example, the body of the `if` statement will be executed only if x is not `null` or `undefined`. In the following example, however, the body will be executed if x is not `false`, or any falsy value, for that matter:

```
if (x) ...
```

It is really important that you know exactly what types of values you expect in your program and what you want to achieve. For clarity, it is almost always better to use explicit comparisons, although they may not be required.

`while` Loop

Professor: Another type of branching is going back upon the same statement again to repeat portions of your code. A control statement that allows such branching is called a *loop*, and one repetition of a statement or statements within a loop is called an *iteration*. JavaScript's fundamental loop is the while statement, which has the following syntax:

```
while (expression) statement
```

If *expression* is falsy, then *statement* (also called the loop's body) is skipped and the while loop is terminated. The execution continues with the code immediately following the while loop. Alternatively, if *expression* evaluates to a truthy value, then *statement* is executed and *expression* is evaluated again. That repeats until, eventually, *expression* returns a falsy value. Note that *statement* should leave a side effect such that it influences the value of *expression*, or else the loop will never end if *expression* starts off truthy.

As an example, let's multiply two non-negative integer variables x and y by adding x copies of y together. This is the code:

```
while (x-- > 0) {
  product += y;
}
```

The result of the multiplication is stored in product. Note that product should be initially set to zero in order for the code to work properly.

The flowchart of the while loop from the above example looks as follows.

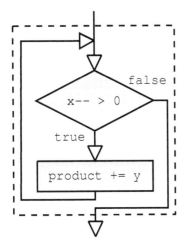

Recall that the decrement operator (--) returns the old value of a variable when placed behind it. Therefore, when x is one, it is decremented to zero, but the old value—which is one—is used in comparison and the body is executed once more.

This example was fairly trivial but I suggest you nevertheless try it at home. You should add variable declarations and initialize them to proper values. You can even let the user enter both multiplicands. Also try to figure out what would happen if product wasn't initialized, and then run the code to see whether your prediction was right. You can also try and rewrite the code so it will work for negative integers as well.

Maria: Can we now write a program that checks for prime numbers using `while`?

Professor: We are running out of time, so you can do that at home. Here is a possible solution, but use it only after you have tried to do it by yourselves:

```
var x = prompt("Enter an integer greater than one.");
var counter = 0;
var n = 2;
while (x > n) {
  if (x % n == 0) {
    counter++;
  }
  n++;
}
document.write(counter == 0 ? "Prime number." :
                            "Not a prime number.");
```

7.3 Design a Simple Program

Professor: Understanding how a simple piece of code works is often easier than writing your own program. Before you start, you shall do some planning, which doesn't require any specific computer language knowledge at early stages. Recall our fridge problem where we wrote a list of instructions to be followed in plain English or *pseudo language*, which we called an algorithm. We are going to write an algorithm in pseudo language for our next problem as well.

The problem is as follows. Write a program that reads some numbers until a zero is entered, and calculates and writes to the browser window the arithmetic mean of the entered numbers. The arithmetic mean is the sum of a sequence of numbers divided by the number of numbers in the sequence. Can you compose an algorithm?

Maria: Perhaps something like this:

> Read numbers until zero is entered.
> Count the numbers.
> Sum the numbers.
> Divide the sum by the count of the numbers.
> Write out the result.

Professor: OK, it's a start. You might have a tiny problem with the implementation, though. You won't be able to store the numbers so you can count and sum them later.

Maria: Why is that? I can use variables, can't I?

Professor: Ask yourself some questions. For example: "How many variables do I need?" Or this one: "How do I tell a program to store each value in its own variable without making the program unreasonably long?"

These are questions indicating problems very similar to those connected with your prime numbers homework. The story is repeating itself: the more numbers you want to include in the calculation the longer your program will be. Unfortunately, you cannot solve this problem using just variables of primitive data types.

Maria: I see. Does that mean we have to wait until we learn more?

Professor: Not necessarily. You can rewrite the first three lines so that individual numbers are not stored. There's really no need for storing them.

Maria: OK...? Oh yes, now I see a solution. I will count and sum the numbers on the fly, which means I don't have to store them individually. Perhaps something like this:

> Set `counter` and `sum` to zero.
> **Repeat**:
>> Read number. If it is zero, stop repeating.
>> Add number to `sum`.
>> Increment `counter` by one.
>
> **End Repeat**
> Divide `sum` by `counter`.
> Write out the result.

Professor: Brilliant! The next step is to code the algorithm in JavaScript. That should be fairly easy now.

Mike: I will try:

```
var counter = 0, sum = 0;
var number, mean;
while (number = prompt("Enter a number:")) {
  sum += number;
  counter++;
}
mean = sum / counter;
document.write(mean);
```

Professor: I like how you used the fact that the assignment operator returns the value assigned to `number`. In your case, when a falsy value is returned by `prompt()`, the `while` loop will stop.

The last step in program development is testing. Basically, testing is executing a program with the purpose of finding errors in it. There's quite some theory behind it, but we'll just use some investigative intuition with a little help of the DevTools.

When you start your program and enter some numbers, for example, four, five, six, and zero, you notice that the loop doesn't stop but the prompt opens once again. Now you hit Cancel and the page loads and you see the number 1140 written on it. It is next to impossible to figure out what happened without a proper tool.

Trying to locate errors in your code is called *debugging* and a tool that helps you to do it is called a *debugger*. The name stems from the ancient era of computing, when occasionally real bugs were found to dwell inside a computer. Computers offered them shelter and plenty of food like crumbs from sandwiches and pizzas, the programmers' basic nutrition sources. Acids in those bugs' feces damaged the sensitive electronic circuits causing hardware failures. Nowadays, a software failure is also called a *bug* to honor those first bugs. But that's just one of the many stories about the etymology of "computer bug."

The DevTools has integrated debugging tools, a tiny portion of which we're going to use now. Press `Ctrl+Shift+I` to open the DevTools and then select Sources from the menu. You should see something like what appears on this screenshot.

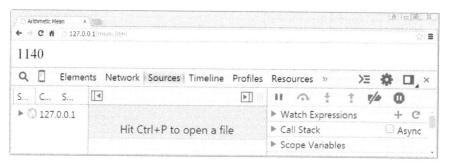

Next, click the disclosure triangle at the left side of the DevTools beside the 127.0.0.1 number. The navigator expands and you select the file in which your code is located. After you do that, you should see the source code of your program inside the DevTools window.

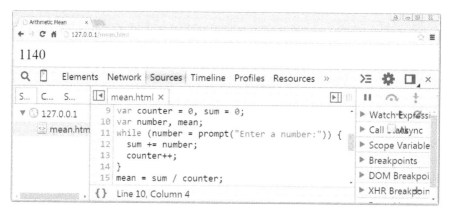

What we are going to do now is to pause the program execution at the point we suspect is critical and observe the values of some variables. To do that, we need to set a *breakpoint*, which is a point in the program where the execution will pause. We can set a breakpoint simply by clicking the number before the line where the program should pause. I think that line 12 is the most interesting for us so we set a breakpoint to this line. A dark gray arrow appears to indicate a breakpoint. To remove a breakpoint, simply click the arrow and it will disappear.

Next, we add variables whose values we want to observe. Click the plus sign next to the Watch Expressions menu item at the right side of the DevTools and enter the name of the number variable so we can watch its value. By the way, you can in fact watch any expression here and not just variables.

Now you run the program again by refreshing the browser window (press F5) and input the number 4 into the prompt. After pressing OK, you should see the following situation.

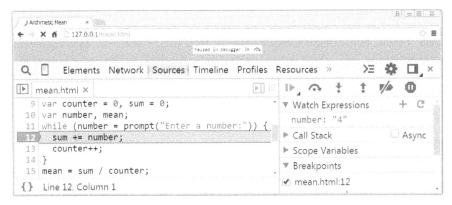

If you watch carefully, you might already see where the problem is.

Mike: Below the Watch Expressions item, I notice that number has the correct value but it is in quotes. Does that mean it is a string?

Professor: Precisely. The type of the value that prompt() returns is a string. You can observe what happens next if you execute just the line of the program where the execution has paused. You do that by selecting Step Over (the curved arrow over a dot above the Watch Expressions menu item). Alternatively, you can press F10. The value of sum will also be of interest to us now, so let's add it to the watch expressions the same way we added number. The situation is as seen on this screenshot.

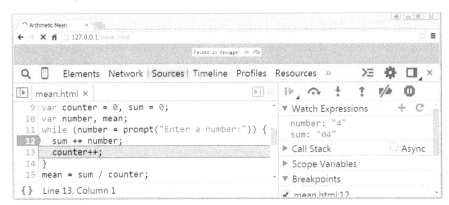

Notice that as soon as line 12 has been executed, the stripe over the code moves to the next line, so you always know which line of code will be executed next.

Now that we discovered that prompt() returns a string, it comes as no surprise to us that number was concatenated to sum. Remember that the + sign works as a concatenation operator when applied to strings.

Mike: But sum isn't a string.

Professor: As you'll learn shortly, it isn't necessary that both operands be strings in order for the concatenation to take place.

However, let me ask you a question. Why do you think the loop didn't stop when you entered zero, and what should you enter if you want it to stop?

Maria: Let me think. Since a return value of prompt() is a string, entering a zero returns the string "0", which is not a falsy value. We need a falsy value to stop the loop. I think that we shouldn't enter anything, just press OK. If we do that, the return value will be an empty string (""), which is a falsy value.

Professor: That's right.

Mike: Why did hitting Cancel stop the loop? And when I hit Cancel right away, before entering any numbers, the result was NaN. Why is that?

Professor: If you press Cancel, then the prompt() function returns null, which is a falsy value. When you entered no numbers, sum and counter both remained zero. Recall that dividing zero by zero is not defined and therefore returns NaN.

Now we need a mechanism that will enable us to manipulate data types explicitly.

7.4 Type Conversions

Professor: We've already seen that JavaScript is very flexible about types of values. When, for example, JavaScript expects a Boolean value, you can supply any type of value and JavaScript will convert it to a Boolean value. Some values, which we named truthy, convert to true, and others, which we named falsy, convert to false. The same goes for other types. If, for example, JavaScript wants a number, then it will convert any value that is not a number to a number.

There are two things that you need to know about type conversions: first, *to what value* does a value convert, and second, *when* does the conversion occur. For the sake of completeness, let us summarize the most basic conversions in the following table.

Value	Converts to:		
	String	Number	Boolean
undefined	"undefined"	NaN	false
null	"null"	0	false
true	"true"	1	
false	"false"	0	
"" (empty string)		0	false
" " (string with one or more spaces)		0	true
" 42 " (numeric string)		42	true

`"three pigs"` (non-numeric string)		NaN	true
0	`"0"`	false
-0	`"0"`	false
NaN	`"NaN"`	false
Infinity	`"Infinity"`	true
-Infinity	`"-Infinity"`	true
42 (any non-zero number)	`"42"`	true

The conversions are pretty straightforward. You are already familiar with the last column, which covers truthy and falsy values. The second column lists conversions to strings, which don't change values at all except for negative zero where the minus sign is lost.

Conversions to numbers, which you find in the third column, are a just a little bit more knotty. I mentioned the other day that `null` represented an expected absence of value. You will therefore not find it surprising that this value is converted to zero. Next, `true` in digital logic usually denotes a switched-on switch and is often represented by a logical one. On the other hand, `false` denotes a switched-off switch and is represented by a logical zero (the absence of electrical current, if you want). If you can accept that, then these two conversions start looking quite obvious, too.

The last category listed in the third column shows conversions from strings. Strings that contain pure numeric data, possibly with leading or trailing spaces, simply convert to the number found in quotes. Next, an empty string converts to zero. This is not hard to remember because an empty string is a falsy value just like a zero is. Finally, any string that cannot be converted to a number becomes `NaN`.

Maria: Now I'm confused. I remember earlier this morning when we talked about truthy and falsy values. You said something to the effect that the expression `x != null` is only false if `x` is `null` or `undefined`. Is that so?

Professor: That's right.

Maria: What if `x` is zero, for example? Zero is a falsy value and `null` is a falsy value. So they should not be different since they both convert to `false`. I therefore think that the expression `0 != null` should return `false`.

Professor: I see your point. It's true that both zero and `null` convert to `false` any time such conversion occurs. However, conversions do not always occur. Relational operators do not require Boolean values as their operands, so the truthy and falsy conversions do not occur in connection with the inequality operator. That's why a comparison of zero and `null` evaluates to not equal.

Let's take a look at the most common situations in which type conversions take place. The basic rule is that JavaScript always converts values to the type it wants in a particular situation. It is straightforward that conditionals and loops require Booleans as their conditional expressions. Logical operators also require Booleans as their operands. In such cases the truthy and falsy conversions take place.

Mike: You mean the values are converted to those in the last column of the above table?

Professor: That's right. For example, consider the next statement:

```
if (cond) { /* Do something */ }
```

The value of `cond` will invariably be converted to a Boolean value `true` or `false`, depending on its original value.

Mike: What if I place some complex expression in place of a variable? Do all the variables inside the expression convert to Booleans?

Professor: Oh no. Variables themselves do not change at all. It's only the return values of expressions and subexpressions that are converted. What's more, they are converted on an as-needed basis, respecting the evaluation order as defined by operator precedence and associativity. If a subexpression can be evaluated without any type conversion, then it is evaluated so. Only the return value is converted, if that is necessary, of course.

Unlike with Booleans, it is not as obvious what JavaScript wants in the case of arithmetic operators. In most cases it wants numbers and converts everything it finds to a number. A few examples:

```
false + 2 * true //Returns 2: Booleans convert to numbers
"15" / "5" //Returns 3: strings convert to numbers
"23" - 1    //Returns 22: the string "23" converts to a number
"x" * 5     //Returns NaN: the string "x" can't convert to a number
13 % "2"    //Returns 1: the string "2" converts to a number
```

When, however, the addition operator is used, it behaves as a concatenation operator as soon as at least one of the operands is a string. The other operand is then converted to a string:

```
var x = "2";
true + " colors"   //Returns "true colors"
7 + " brothers"    //Returns "7 brothers"
"2" + 5            //Returns "25"
x += 1;            //x becomes "21"
x++;               //An exception: x becomes a number (i.e., 3)
```

Notice that the increment operator (++) converts a string to a number even though it is just a shorthand operator for adding one to the value of a variable.

Explicit Conversions

Professor: There are situations when you want to perform certain type conversions explicitly. For instance, in our arithmetic mean example we should have converted the number entered by the user—which was returned as a string—to a number before adding it to the sum. Any idea how to do that?

Maria: Perhaps we could subtract a zero from a string?

Professor: True. As a matter of fact, any basic arithmetic operation other than addition will do the trick:

```
number - 0
number * 1
number / 1
```

And you can use a unary positive sign operator as well:

```
+number
```

However, the easiest and the most obvious method to perform an explicit type conversion is to use any of the functions `Number()`, `String()`, or `Boolean()`.

```
Number("3.1416")  //Returns 3.1426
String(true)      //Returns "true"
Boolean("")       //Returns false
```

Going back to our arithmetic mean program, you can replace the `while` loop condition with either one of the following two expressions:

```
number = +prompt("Enter a number:")
number = Number(prompt("Enter a number:"))
```

Incidentally, sometimes you might encounter the following two conversion expressions:

```
x + ""   //Same as String(x)
!!x      //Same as Boolean(x)
```

Mike: Where does the second one come from?

Professor: Well, remember that the logical NOT operator (`!`) accepts a Boolean value and returns a Boolean value as well. In fact, `!x` returns `true` if x is falsy and returns `false` if x is truthy. If you use the logical NOT once again, you finally get `true` from a truthy value and `false` from a falsy value.

Conversions and Relational Operators

Professor: I will conclude today's meeting with what is probably the most important of all the situations that account for type conversions. These occur in connection with equality (inequality) and comparison operators.

The most straightforward is the strict equality operator (===), which performs no type conversion whatsoever. The values as well as types on both sides of the operator should be equal if you want the operator to return `true`:

```
"42" === 42          //Returns false
true === 1           //Returns false
null === undefined   //Returns false
```

The strict inequality operator (!==) is exactly the opposite of the === operator, and it performs no type conversion either:

```
"42" !== 42          //Returns true
true !== 1           //Returns true
null !== undefined   //Returns true
```

The very special case, as you already know, is the `NaN` value. This value is never equal to any other value, including itself. Therefore, if you want to check whether a value of x is `NaN`, you write:

```
x !== x //Returns true if and only if x equals NaN
```

The equality operator (==) is less strict. Essentially, it works like the strict equality operator except that it attempts certain type conversions if it discovers that the operands are not the same type. The operator uses the following rules:

- Values `null` and `undefined` are considered equal.
- If a number is compared to a string, then a string is converted to a number and the converted number is used in comparison.
- If either value is `true` or `false` it is converted to 1 or 0, respectively, before used in comparison.

For example, the next comparisons all return `true`:

```
null == undefined
5 == "5"
false == "0"
```

In the last of the above lines, `false` is converted to the number zero before the comparison is made. Since the types are still different, the string `"0"` is then converted to the number zero and the comparison is performed again. Because both numbers are now the same, the expression returns `true`.

The inequality operator (!=) is of course the exact opposite of the equality operator.

Comparisons other than equality and inequality can only be carried out on strings and numbers. There are three possible situations:

- If both operands are strings, then they are compared alphabetically.
- If at least one operand is not a string, then both operands are converted to numbers and then compared.
- If either operand is (or converts to) NaN, then a comparison always returns false.

In view of the second rule, the following two expressions will return true:

```
false < "1" //Returns true
2 < "14"    //Returns true
```

The third rule tells us that the following two expressions will return false:

```
"I'm the most epic line of computer code ever." > 0 //Returns false
"I'm the most epic line of computer code ever." < 0 //Returns false
```

Recall our last meeting, when we encountered the following expression where x was set to four:

```
10 < x < 20
```

Now you might be able to explain what the return value of this expression is and why.

Maria: You said the other day it wasn't false so it must be true. OK, I guess that x is first compared to 10 because of the left-to-right associativity of the less-than operator. Because x is 4, this obviously returns false, which is then compared to 20. The second of the above rules says that if at least one operand is not a string, it must convert to a number. The false value is therefore converted to zero. Since zero is less than 20, the whole expression returns true.

Professor: Splendid! As a matter of fact, such an expression is meaningless for it *always* returns true regardless of the type and value of x. If you want to check whether x lies in an interval between 10 and 20, you should write:

```
10 < x && x < 20
```

In that sense, you can look at the logical AND operator as the intersection operator. The above expression returns true for all values of x that belong to the intersection of two sets: the set of values greater than 10, and the set of values smaller than 20.

Mike: Is it also true that logical OR can be seen as the union operator in situations like this?

Professor: Yes. It is indeed.

7.5 Homework

Professor: I guess it's time we call it a day. For homework you write a program that computes and writes out the greatest common divisor (GCD) of two integers. The GCD of two integers, as you might recall from high school mathematics, is the largest positive integer that evenly divides both numbers. You use the Euclidean algorithm, which is fairly simple and straightforward:

> Read two integers.
> **Repeat** as long as they are different:
>> Calculate the absolute difference between both numbers.
>> Replace the larger number by the absolute difference calculated in the previous step.
>
> **End Repeat**
> Here both numbers are equal. They are also equal to the GCD of the two integers entered at the beginning.

Lastly, here is a list of today's keywords:

> **In this meeting**: expression, statement, semicolon, order of execution, flowchart, processing step, decision, expressions with side effect, compound statement, code indentation, conditional, `if/else`, truthy value, falsy value, loop, `while`, debugging, breakpoint, watch, step, type conversions, explicit conversions

Introducing Objects

8.1 Homework Discussion

Maria: We're quite excited about showing you our homework today because we came up with several solutions and we're not sure which is best.

Professor: Let me see.

Maria: Here's the first one:

```
var a = 63, b = 98;
while (a != b) {
  if (a > b) {
    a -= b;
  }
  else {
    b -= a;
  }
}
document.write("GCD is " + a);
```

Professor: That's a pretty forthright solution. I have one question, though. If a and b are equal, then the `else` part is executed and b becomes zero. From here on, neither a nor b can change and you're stuck inside an infinite loop. Have you thought of that?

Mike: This cannot happen. If a and b are equal, then the `while` loop condition is `false` and the `if/else` statement doesn't even start.

Professor: Exactly. And what are your other solutions?

Maria: We experimented a little with the DevTools and discovered a handy tool for printing intermediate results. Just before the closing brace of the `while` loop we inserted a `console.log()` function call, which we used to write the values of a and b to the JavaScript Console within the DevTools:

```
   . . .
  else {
    b -= a;
  }
  console.log("a=" + a + ", b=" + b);
}
document.write("GCD is " + a);
```

Inside the JavaScript Console we got the following values.

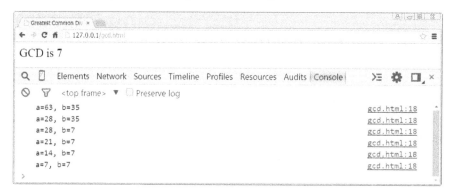

Professor: Excellent. What you did is called *tracing*. Tracing is a way of recording information about how a program executes, and is typically used for debugging purposes. Please, go on.

Mike: Inspecting the values, we discovered that the same subtraction is sometimes repeated several times if the larger variable stays larger after the subtraction. We hence thought the program could execute much faster if we rewrote it as follows:

```
var a = 63, b = 98;
while (a != b) {
  while (a > b) {
    a -= b;
  }
  while (a < b) {
    b -= a;
  }
}
document.write("GCD is " + a);
```

Professor: OK, I see your point. I don't think that there are actually fewer comparisons in this solution, though. You can try at home to count comparisons in both programs, but you'll have to do it for many different value combinations of a and b to get a general picture. Besides, speed is not something I would worry about in this particular example.

What I like about this solution is that it makes what is actually going on more evident. You can see right away that one and the same variable is repeatedly being subtracted until it becomes smaller.

Maria: Exactly. Precisely this observation led us to the conclusion that we were actually computing the remainder of a division. So we tried the following solution:

```
var a = 63, b = 98;
while (a != b) {
  if (a > b) {
    a %= b;
  }
  if (a < b) {
    b %= a;
  }
}
document.write("GCD is " + a);
```

But unfortunately it doesn't work.

Professor: Let's take one more look at the sequence of values produced by your first program. In the third line of the Console, the value of a is exactly four times that of b. The statement a %= b; will therefore make a equal zero. Because a is now smaller than b, the statement b %= a; is executed next. Since the remainder of division by zero is not defined, b becomes NaN. End of story.

If you want to use the remainder operator, this is a possible solution:

```
var a = 63, b = 98;
var tmp;
while (b != 0) {
  tmp = b;
  b = a % b;
  a = tmp;
}
document.write("GCD is " + a);
```

The trick is that you copy the old value of b to a with help from a temporary variable tmp. Meanwhile, you use the old value of a to compute the remainder and assign it to b. If you do that, a is always greater than b at the end of each iteration. Notice also the changed loop condition, which now compares b to zero.

Interesting in this last solution is that the first iteration of while just swaps the values of both variables if the run starts off with b greater than a.

Mike: By the way, we also extended the multiplication program you gave us on page 135 to work with negative integers. First we computed the sign of the product and absolute values of multiplicands, and in the end we multiplied the product with the sign:

```
var product = 0, x = 6, y = 8;
var sign = x * y >= 0 ? 1 : -1;
x = x >= 0 ? x : -x;
y = y >= 0 ? y : -y;
```

```
while (x-- > 0) {
  product += y;
}
product *= sign;
```

Professor: Good work.

8.2 `switch` Conditional

Professor: Today you are going to learn how to use objects, which will allow you to build considerably more elaborate programs. But we'll also cover two more control statements of JavaScript. The first one is the `switch` conditional.

You already know the `if/else` statement, which allows you to choose between two alternatives. Sometimes, however, you have more than two alternatives to select from, in which case it would be wasteful to repeatedly evaluate one and the same expression in multiple `if/else` statements. That's where another control statement called `switch` jumps in. Its syntax is as follows:

```
switch (expression) {
  case expression_1: statements_1
  case expression_2: statements_2
  ...
  case expression_n: statements_n
  default: other_statements
}
```

It looks complicated at first glance but it's really quite simple. First, *expression* is evaluated and compared to the expressions from *expression_1* to *expression_n*, one after the other, using the strict equality operator (`===`). As soon as a comparison evaluates to `true`, no more comparisons are made. Instead, the statements at the right begin to execute, starting with the statement after the `case` keyword where a comparison first returned `true`. For example, if *expression* equals to *expression_2*, then all the statements starting from *statements_2* are executed.

Incidentally, you can use as many `case` statements as you like. However, if none of the comparisons returns `true`, then *other_statements* are executed, which are put after the `default` keyword. This is optional, though, and you don't have to include the `default` keyword.

Mike: I don't understand this. It seems that *other_statements* are always executed, regardless of the actual outcome of the comparisons. Am I missing something?

Professor: Very good observation, indeed. The `switch` statement is usually augmented with `break` statements to reach its full functionality, but more about that later.

As an example, consider a problem where you have to append an appropriate suffix to a number in order to produce an ordinal number. In the English language, there are four different suffixes: "st," "nd," "rd," and "th." From what we have learned about the `switch` statement we can produce the following code:

```
var place = 2;
switch (place) {
  case 1: place += "st";
  case 2: place += "nd";
  case 3: place += "rd";
  default: place += "th";
}
document.write("You have won " + place + " place!");
```

Here's a flowchart of the switch statement used in our program.

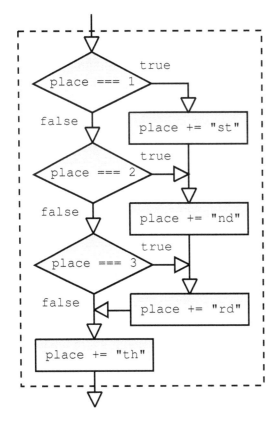

Maria: I see that Mike was right. "th" is always appended, and worse still—in our case "nd," "rd," and "th" are all appended. So we get the message, "You have won 2ndrdth place."

Professor: You're becoming critical, aren't you? But you're right, of course. To solve this problem, we need to use a break statement. With break, you can exit the switch statement whenever you like. It is evident from the above flowchart that we need three extra exit points in order for the program to work correctly. That means we need to add three breaks:

```
var place = 2;
```

```
switch (place) {
  case 1: place += "st"; break;
  case 2: place += "nd"; break;
  case 3: place += "rd"; break;
  default: place += "th";
}
document.write("You have won " + place + " place!");
```

In this version, the `switch` statement terminates correctly as soon as the suffix "nd" has been appended. Of course, `break` doesn't change the flow of the statements outside of `switch`. When the `switch` statement in our example terminates—either regularly or through a premature exit caused by one of the `break` statements—the `document.write()` is always executed after it.

Mike: Don't we need braces to make a compound statement after `case` keywords?

Professor: No, we don't. You can place as many statements following each `case` keyword as you like. They will all be executed until either `switch` ends or `break` is encountered.

Maria: I think our program still doesn't work properly. What about values above 20? The program will produce "21th" instead of the correct "21st," and "22th" instead of "22nd," and so on.

Professor: You're right again. This pattern actually repeats every 10 numbers, except for 11, 12, 13, 111, 112, 113, and so on. There are of course more solutions and this is one of them:

```
switch (place % 100 <= 20 ? place % 20 : place % 10) {
  case 1: ...
```

8.3   Math Object

Professor: I think it's about time we started moving to a higher level of programming. You learned quite a bit about primitive data types and how to manipulate them using operators and some basic control statements. However, as your programs become more complex, you will soon need to organize data and code more efficiently in order for your projects to stay under control. The concept of *objects* offers you one aspect of such data and code organization. Essentially, objects group data and code that logically fit together and hide unnecessary implementation details from other parts of a program. In object-oriented terminology, data and code that are integrated in an object are called *properties* and *methods*, respectively. Properties define the state of an object, while methods define its behavior.

You can think of an object in the same way as you think of a real-world object. For example, your mobile is quite a complex device, yet you don't have to be a rocket scientist to be able to use it because its actual implementation is hidden from you. There's quite some data stored inside your mobile. Apart from the most obvious contacts, there is also information like whether Bluetooth® is on or off, or whether hiding your identity is enabled. These are all properties of your mobile. There are also certain things you can *do* with your mobile. You can, for example, take a photo, listen to your favorite music, or even make a call. These are methods of your mobile.

Mike: Am I thinking right if I see properties as variables and methods as functions?

Professor: You couldn't be more right about that. As a matter of fact, properties are sometimes also referred to as *member variables* or *member data*, and methods are called *member functions*. They are also known under a common name, *members*.

By the way, everything in JavaScript is an object. Even primitive data types have their object counterparts.

It can happen that mathematics isn't precisely your cup of tea, but sooner or later you discover that you can't do without it. That's why JavaScript has a built-in object called Math. In section E.12 on page 409 at the end of this book you will find more details about all that this object has in store for you. When you understand the principles, you will know how to use it.

The Math object has a set of properties. For example:

```
Math.E   //Base of the natural logarithm, approximately 2.71828
         //(Euler's number).
Math.PI  //Ratio of the circumference of a circle to its diameter,
         //approximately 3.14159.
```

Notice a dot (.) following the name of the object. This is an operator used to access properties of an object. You can think of PI as a variable being integrated inside the Math object. When you want to access it, you first name the object followed by a dot access operator, and then PI.

Maria: If PI is a variable, can I change its value?

Professor: No, you can't. That's out of the scope of our course, but properties of JavaScript objects have certain attributes attached to them. Specifically, the PI property has the `writable` attribute set to `false`. Just think of `Math.PI` as a constant, or a variable whose value you cannot change. You can use it freely in all other situations, though. For example, you can compute the circumference of a circle:

```
circ = 2 * Math.PI * radius;
```

Alternatively, you can access a property of an object using square brackets ([ ]) and the property name in string form:

```
Math["PI"]
```

An advantage of using square brackets is that you can use any string as a property name and not just a valid JavaScript identifier.

A much more important advantage is that you can store the name of a property in a variable and use it with a square bracket property access operator, something you cannot do using a dot operator. For example:

```
var what = prompt("What do you want to know?");
document.write(Math[what]);
```

If you enter PI, for example, then you get the number 3.141592653589793 written in the browser window.

The Math object has also many useful methods. These are some examples:

```
Math.abs(x)      //Returns the absolute value of x
Math.ceil(x)     //Returns the smallest integer greater than or equal
                 //to x.
Math.floor(x)    //Returns the largest integer less than or equal
                 //to x.
Math.round(x)    //Returns the value of x rounded to the nearest
                 //integer.
Math.cos(x)      //Returns the cosine of x.
Math.sin(x)      //Returns the sine of x.
Math.pow(x, y)   //Returns x to the power of y.
Math.sqrt(x)     //Returns the square root of x.
Math.random()    //Returns a pseudo-random number between zero
                 //(inclusive) and one (exclusive).
```

Professor: Note that none of the above expressions have any side effects. They just compute and return values.

Mike: Excuse me, but I don't see any advantage in using objects. What's the use of writing Math.PI or Math.abs() when we could simply write shorter PI or abs() and get the same results?

Professor: I see your point. At this stage it really doesn't make much sense. You'll just have to wait until you learn more about objects so you can appreciate them.

If you feel like gambling, maybe now is your chance. We're going to program a lucky lottery number generator.[1] There are a zillion lottery rules across the globe and one of the simplest to program is the one where you must select n different numbers from the range of numbers from one to N. For example, in Lotto Texas®, you select six numbers from one to 54.

So the problem before us is to draw six different random numbers out of the range of numbers from one to 54. Any idea how to do this in a program?

Maria: Basically we repeat something six times. Like this:

---

[1]If you are a gambling addict, then this is not an example for you. You can read further, but you do so at your own risk.

```
var n = 7;
while (n--) {
  //Do something 6 times
}
```

Professor: You're close. But what's wrong?

Mike: If I remember correctly, n is decremented *after* its value is returned. That means that when n is one, it is first evaluated as a truthy value and only then decremented to zero. As a consequence, the loop body is executed once more. I think that n ends up with the value of -1 when the loop finishes. The loop therefore runs seven iterations.

Professor: Not bad at all. You've obviously done your homework. Now we have to find out *what* we need to repeat.

Maria: We need random numbers. The `Math.random()` function could do the trick but the range is not right.

Professor: That means we have some extra work to do.

Mike: We can multiply whatever `Math.random()` returns by 54 and round the result.

Maria: That way you'll get 55 and not 54 possible numbers.

Professor: You are a sharp observer, congratulations! If you multiply the range from zero to one by 54 and then round a number to the nearest integer, then you can get integers from zero to 54 inclusive, which is all together 55 different numbers.

Before we write a final solution, I would like to say some words about the `random()` method. Each time the function is called, it returns a pseudo-random real number greater than or equal to zero and less than one. It is very important that one can never appear as the function's return value.

Maria: Why did you say "pseudo-random"?

Professor: Because the returned numbers only appear to be random. The numbers returned by the `random()` method have statistical properties of random numbers, but they are generated by a deterministic algorithm. They are not truly random as, for example, the roll of a dice, but for many practical applications they are all right. This is an interesting topic but, unfortunately, we do not have time to explore it in more detail.

Putting it all together, we come up with the following possible solution:

```
var x, n = 6;
while (n--) {
  x = Math.floor(Math.random() * 54) + 1;
  document.write(x + " ");
}
```

Mike: You said that multiplying by 54 will give us 55 possible different values.

Professor: If I used `round()`, then yes, I would get 55 possible different values. But I used `floor()` because it is more fair. The problem with rounding is that it rounds the value to its nearest integer, so only numbers from zero to approximately 0.49999 are rounded to zero. This is only half as much as there are values rounded to, say, one. Every value from 0.5 to approximately 1.49999 is rounded to one, which makes the probability of getting one twice of that of getting zero. On the other hand, `floor()` rounds values down to the nearest integer, so all numbers get equally sized ranges. Note that, because `random()` never returns one, the largest possible value returned by the expression `Math.random()*54` is in fact less than 54, which means that `floor()` in our program can never return 54.

Maria: Shouldn't the drawn numbers in our lottery all be different? I suspect that our solution does not guarantee that.

Professor: Exactly. And that's the trouble with this solution. There is absolutely no memory of the numbers that have already been drawn, and it can easily happen that a number is drawn two or even more times.[2]

8.4   `do/while` Loop

Professor: One strategy of securing that two random numbers are different is to draw them again if they are equal. If they're still equal, then draw them again. And again. Eventually, they ought to turn out different. You probably guessed that we need a loop for that, and you might have noticed that the first drawing should be performed *before* we test for equality. There's a slight inconvenience with the `while` loop in that its body is executed *after* the loop condition is checked. While the problem is not fatal, it is still more comfortable to use the `do/while` loop for our particular problem. The `do/while` loop is very much like the `while` loop, with one fundamental difference: the loop condition is tested at the end of the loop, which means that the loop body is executed at least once. This is the syntax of the `do/while` loop:

```
do
   statement
while (expression);
```

This loop isn't used as often as its `while` counterpart but you will find it occasionally in existing code, and you may even want to use it for yourselves, as we're going to do now.

Maria: Why have you put a semicolon at the end of the `do/while` statement? Other control statements don't have it.

Professor: Most statements terminate with a semicolon but there are exceptions. A compound statement and `switch`, for example, do not require a semicolon at the end.

Now, if you want to draw two different numbers, this is the code that does it:

---

[2]We will solve this problem using arrays and continue with the lottery example on page 175.

```
var x1, x2;
do {
  x1 = Math.floor(Math.random() * 54) + 1;
  x2 = Math.floor(Math.random() * 54) + 1;
} while (x1 == x2);
document.write("The winning numbers are " + x1 + " and " + x2 ".");
```

Of course, here's a flowchart as well for more clarity.

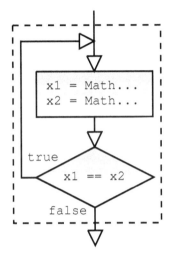

Professor: As you can see, the program flow loops back to selecting another two numbers as long as they are equal. When they are different, the loop terminates.

Mike: The program is not working.

Professor: What do you mean "not working"?

Mike: I tried to open it in the browser but nothing happened.

Professor: Press Ctrl+Shift+I.

Mike: You mean open the DevTools? OK, here we go.

Professor: It says that we have a syntax error in line 13 of our source code. There's an unexpected string in the line where we call document.write(). I see three strings in that line. Which one do you think is making trouble?

Mike: Oh, yes! The last one. There's no operator between x2 and the literal string " . ".

Professor: Precisely. These are two separate expressions and you cannot put one expression after another without an operator in between. I forgot to type + there. But that's just perfect because we discovered that the DevTools also offer a handy tool for finding syntax errors. When things go wrong, opening the DevTools shall become second nature.

Note that in real programs a *syntax error* is normally much easier to catch than a *logic error*. A syntax error can be either a spelling mistake or an error in grammar or context, such as a missing operator or an undeclared variable. The JavaScript Console will usually direct you to the exact location of a syntax error. Logic errors, however, are flaws in the conception of the algorithm and there's no way a tool can tell you automatically what or where an error is. A logic error can also be a consequence of a *semantic error*. This type of error occurs when you misuse syntax rules in a way that they are not violated but the semantics or meaning is not what you had in mind. An example of a semantic error was when we mistakenly used a plus sign as a concatenation instead of an addition operator in our arithmetic mean example. However, you already know how to use various debugging tools to find logic and semantic errors.

Let's return to our lottery numbers. We learned how to produce two different random numbers but in fact we need to find six different numbers. Unfortunately, our last approach isn't a very elegant solution to the problem when we are after more than just a few numbers. For one thing, we would need six different variables to hold six different values. Next, we would have to write 15 different comparison expressions because every selected number must be different from any other selected number. If we wanted to produce more than six different numbers, then things would get significantly worse. Does all this sound familiar to you?

Maria: We have again stumbled upon exactly the same problem as we did with prime numbers, and with the arithmetic mean calculation.

Professor: Indeed we have. Only this time we actually *need* to store individual numbers. We're going to learn how to do that the next time.

8.5 Date Object

Professor: Another handy object built into the JavaScript language is Date, which is used to represent a single moment in time. There is one fundamental difference between Math and Date objects, though. As you don't really need more than a single Math object in a program, it is not hard to imagine the usefulness of having many Date objects, each holding a different moment in time. That's why the Date object is not an "ordinary" object. Rather, it represents a *class of objects* and is hence a data type. You can view Date as a blueprint for making *object instances* of the Date class.

If you therefore want to use a Date object, you first need to construct an object instance by means of the Date() *constructor* and the new operator:

```
var thisMoment = new Date();
```

Notice that the object is declared using the `var` keyword like any other variable. The `Date()` function, which has the same name as the class of objects it represents, is called a constructor because it helps to construct a new object. If you want to create an object, then you must use the constructor together with the `new` operator. It is also possible to call `Date()` as an ordinary function, without the `new` operator, in which case it only returns a string representing current time and date and you don't get an object.

Maria: Are the rules for naming an object the same as they are for a variable?

Professor: Indeed they are.

Mike: You said that it is possible to have many different moments in time. How do you set them?

Professor: You can set a specific moment in time already at the time of the object creation. There are many different ways to call the constructor. For example:

```
//Creates a Date object instance set to the current time and date.
new Date()
//Creates a Date object instance set to milliseconds past midnight
//1/1/1970.
new Date(ms)
//Creates a Date object instance representing a specific moment
//in time.
new Date(year, month, day, hours, minutes, seconds, ms)
```

The name of the constructor is always the same and the JavaScript interpreter decides which one to use depending upon the number and types of the passed arguments. If no arguments are provided, then the current time and date are used.

The second form when one numeric argument is passed is interesting. It is taken as the number of milliseconds past (or before, if it is negative) midnight on January 1, 1970 using Coordinated Universal Time. This is known as the standard UNIX epoch, or simply *the epoch*.

Mike: What's Coordinated Universal Time and the UNIX epoch?

Professor: Coordinated Universal Time, or UTC, is the main time standard used to regulate time all over the world. You can think of it simply as time in London when no daylight savings time is in effect. UTC replaces Greenwich Mean Time, or GMT, which is still occasionally used as the synonym. The gist of the standard is that the same date and time in UTC gives precisely the same time instant no matter where in the world you are located. *Local time* is usually given as an offset from UTC and depends on the time zone you're in and whether daylight savings time is in effect. For example, local time in New York is UTC minus five hours in winter.

An *epoch* is a moment in time selected as the origin of a particular era and it provides a reference point from which time is measured. For example, the epoch of the civil calendar used internationally is the Incarnation of Jesus. UNIX is one of the oldest operating systems and the term UNIX epoch is used because it was first defined and used for the UNIX system. If you are interested, then you may want to check the *www.timeanddate.com* website, which is a great resource of time- and date-related information.

So this number of milliseconds since midnight on January 1, 1970 is also used internally by the Date object to store a single moment in time. For example, if you need a Date object that is set to one hour past midnight on January 1, 1970 (UTC), you write the following statement:

```
var distantPast = new Date(3600000);
```

Maria: This is a huge number already. How do you write modern dates without losing precision?

Professor: You don't lose precision that quickly. Recall that you can write integers as large as 9007199254740992 using the 64-bit floating-point representation. If these are milliseconds, that's more than a hundred million days. To be exact, you can move in the range from minus to plus 273,785 years since the UNIX epoch. That's a lot.

Professor: There's one more constructor in the above example, which allows you to specify time and date in local time. Note that you don't have to provide all the arguments as long as you're omitting them from the right. For example, if you want to supply hours as an argument, then you should at least supply year, month, and day as well, while the rest can be omitted:

```
//March 13, 2017 at 8 p.m. (local time):
var partyTime = new Date(2017, 2, 13, 20);
```

It is important that you specify hours as an integer from 0 to 23, days from 1 to 31, and months from 0 to 11.

Mike: It's weird that days should be counted from one and months from zero.

Professor: As a matter of fact, there's a good reason for that. You'll find out why that is so in the next meeting.

OK, an object is now created but what can you do with it? The Date object has no properties that can be accessed directly. Instead, all reading and writing of time and date values is done through methods, of which they are many, and we will only take a short glance at some of them:

```
getDay()     //Returns the day of the week, an integer from zero
             //(Sunday) to six (Saturday).
getFullYear() //Returns the year, in four-digit format.
```

```
getMonth()     //Returns the month, an integer from zero (January)
               //to 11 (December).
getDate()      //Returns the day of the month, an integer from
               //one to 31.
getHours()     //Returns the hours, an integer from zero (midnight)
               //to 23 (11 p.m.).
getMinutes()   //Returns the minutes, an integer from zero to 59.
getSeconds()   //Returns the seconds, an integer from zero to 59.
getTime()      //Returns the internal, millisecond  representation
               //of the Date object.
toString()     //Returns a string representation of time and date.
```

You can also set all the values, except the day of the week, of course:

```
setFullYear(...) //Sets the year, and optionally the month and day.
setMonth(...) //Sets the month, and optionally the day of the month.
setDate(...)  //Sets the day of the month.
setHours(...) //Sets the hours, and optionally the minutes, seconds
              //and milliseconds.
setMinutes(...) //Sets the minutes, and optionally the seconds and
              //milliseconds.
setSeconds(...) //Sets the seconds, and optionally the milliseconds.
setTime(...)  //Sets time and date using the millisecond format.
```

Note that all these methods except getTime() and setTime() return and set values using local time. If you want to set universal time, you can use any one of their counterparts, which have an additional "UTC" in their names. For example, setUTCHours() or getUTCHours().

Maria: That's a lot of data to remember.

Professor: Don't even think of remembering all this stuff. The important thing is that you understand the principles. If you need to set hours, for example, you'll just look up an appropriate method from the list or, better still, from the language reference such as the one at the end of this book.

Speaking of principles, how do you think you can call, for example, the getDate() method of the Date object?

Mike: I can use the dot access operator:

```
dayOfMonth = Date.getDate(); //Wrong: Date is a data type.
```

Professor: That won't work. Remember, Date is not a "proper" object but a data type and you cannot call methods on data types. You need an object instance for that. For example:

```
var partyTime = new Date(2017, 2, 13, 20); //Creates an object
                                           //instance.
dayOfMonth = partyTime.getDate(); //Correct: partyTime is a Date
                                  //object instance.
```

However, there do exist some methods that can be called directly through the Date object. In contrast to the methods listed above, which are called *instance methods* because you can call them only through object instances, they are called *static methods*. For example:

```
Date.now()  //Static method that returns the millisecond
            //representation of the current time. It returned
            //1422025494222 at the time of testing these lines.
Date.UTC(2015, 0, 1) //Static method that returns the millisecond
            //representation of time and date specified in
            //universal time.
```

We're now ready for a more practical example. Let's make a page that will count down the time to the 2016 Summer Olympics. This is the algorithm:

> Construct two Date objects, one for the present moment and one for the time of the opening ceremony. The ceremony will take place on August 5, 2016 at 8 p.m. Brasília time.
>
> **If** the opening ceremony starts in the future **Then**
>
>> Compute the time difference in milliseconds between both time stamps.
>> Divide the difference by 1000 and round down to the nearest integer to get the number of seconds until the opening ceremony.
>> Divide the seconds by 86,400 (= 60 × 60 × 24) and round down the result to the nearest integer to get the days.
>> Divide the remainder of the previous division by 3600 to get the hours.
>> Divide the remainder of the previous division by 60 to get the minutes.
>> The remainder of the previous division are the seconds.
>
> **End If**

The tricky part is computing the time difference. I believe that you are bewildered by all the available Date methods, some operating in local, others in universal time. In fact, things aren't that complicated once you recognize that internally, time is invariably stored using universal time. Local time is only used occasionally when Date objects interact with the environment. For example, when you call a Date() constructor without arguments, it automatically reads the current time in universal time disregarding time zone and daylight savings time settings. When you want to get specific information from the object, you can do that using local as well as universal time without having to worry about the conversions.

To keep things under control, we'll calculate the time difference using universal time, of course. We just discovered that no extra work is needed on our part to convert the current time from local to universal time. The Date() constructor does it for us.

However, a little more work needs to be done to create the time stamp of the event. We cannot rely on the `Date()` constructor to do the work this time but must do it manually. For that purpose, we use the static `Date.UTC()` method to convert the opening time of the ceremony to the millisecond representation, and pass the obtained milliseconds as an argument to a `Date()` constructor. Brasília time is UTC minus three hours in August. That means the time of the opening ceremony is actually 8 p.m. plus three hours when specified in universal time, which is 23. Finally, the next expression returns the number of milliseconds since the epoch for the start time of the opening ceremony:

```
Date.UTC(2016, 7, 5, 23)
```

The algorithm itself is not very complicated. The only way to compute a difference between two `Date` objects is to subtract their millisecond time representations. After that, necessary divisions are performed to get the days, hours, minutes, and seconds. Here's the code:

```
var msg = "Countdown to Rio 2016: ";
var now = new Date();
var opening = new Date(Date.UTC(2016, 7, 5, 23));
var seconds;
if (opening.getTime() > now.getTime()) {
  seconds = opening.getTime() - now.getTime();
  seconds = Math.floor(seconds / 1000);
  msg += Math.floor(seconds / 86400) + " days, ";
  seconds = seconds % 86400;
  msg += Math.floor(seconds / 3600) + " hours, ";
  seconds = seconds % 3600;
  msg += Math.floor(seconds / 60) + " minutes, ";
  msg += Math.floor(seconds % 60) + " seconds.";
}
document.write(msg);
```

You can see the result on the next screenshot.

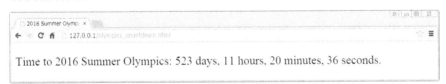

Time to 2016 Summer Olympics: 523 days, 11 hours, 20 minutes, 36 seconds.

Mike: The countdown is not running, is it?

Professor: No, it isn't. The message is produced when the page is loaded and isn't updated until the next page reload. When you learn about the DOM and events, we are going to change that and get a running counter without the need to reload the page.[3]

---

[3]This example continues on page 237.

8.6 Concluding Thoughts and Homework

Professor: I still owe you an answer to your question as to what's the advantage of using objects, which isn't that obvious with the `Math` object. That is because there is never more than a single `Math` object in a program. The story is different with the `Date` object, though. Notice in our last example how the calls to `now.getTime()` and `opening.getTime()` generally return two different values. If `now` and `opening` were not objects, then you would have to use something like `getTime(now)` and `getTime(opening)`, for example. That would still be perfectly all right as long as there are no other arguments. Take, for example, that you want to set the hours of the `partyTime` object, which you can do using the expression `partyTime.set Hours(10)`. Without an object you would have to write a function call like, for example, `setHours(partyTime, 10)`. That wouldn't be quite as evident as is the object notation, which makes the hours of what you're setting more obvious.

Another advantage of using objects is that you don't have to worry about the internal representation of time and date when using `Date` objects. This representation could just as well change in the future but all the code using `Date` objects would still work because it does not depend on the actual form in which time and date are stored.

Still another use of objects is as *namespaces*. A namespace acts as a container for a collection of identifiers, which makes it possible to have more identifiers with the same name (within different namespaces, of course). This is similar to having several different files with the same name stored in different directories. An example of using a namespace would be if you wrote your own mathematical functions using the same names as those of the `Math` object and made them members of your own object.

For today's homework I want you to search the Internet for a formula that, given an arbitrary year, computes the Easter date in the Gregorian calendar, which is not fixed. Then write a program that prints the Easter date for the selected year in a human-readable format.

For the second part of your homework, write a program that, for a given year, computes and writes out the date of Thanksgiving Day in the United States, which falls on the fourth Thursday of November.

Before we leave, I ought not to forget a list of today's keywords:

> **In this meeting**: `console.log()`, tracing, `switch` conditional, `break`, object, property, method, member, dot member access operator, square brackets member access operator, `Math`, pseudo-random, `do/while` loop, syntax error, semantic error, logic error, object data type, class, object instance, constructor, `new`, UNIX epoch, instance method, static method

---

Understanding Arrays and Strings

9.1 Homework Discussion

Professor: I'm curious to find out when next Easter and Thanksgiving come. Did you write your programs?

Mike: There are actually many different algorithms for calculating the date of Easter and we just picked up the first one that looked trustworthy. Basically, we just copied and pasted it but as it was written in some other language (I think it was Python) we were confused by the double slash operators (//), which weren't comments. We discovered that they are actually divisions except that they neglect the remainder. So we replaced them with ordinary divisions and rounded down the quotient to the nearest integer using `Math.floor()`. This is our solution:

```
var a, b, c, d, e, f;
var day, month, easter;
var year = prompt("Please, enter the year.");

a = year % 19;
b = Math.floor(year / 100);
c = year % 100;
d = (19 * a + b - Math.floor(b / 4) -
        Math.floor(
            (b - Math.floor((b + 8) / 25) + 1) / 3) + 15
        ) % 30;
e = (32 + 2 * (b % 4) + 2 * Math.floor(c / 4) - d - (c % 4)) % 7;
f = d + e - 7 * Math.floor((a + 11 * d + 22 * e) / 451) + 114;
day = f % 31 + 1;
month = Math.floor(f / 31);
easter = new Date(year, month - 1, day);
document.write("Easter is on " + easter);
```

Professor: You've done a great job. Did you try to decipher what the calculations are all about?

Mike: Frankly speaking, no. We read that the algorithm holds for any year in the Gregorian calendar, which is any year including and after 1583. The calculations have to do with positions of the Moon and other astronomical stuff.

Professor: Did you test the program or did you just trust the source? All in all, it is also possible that you made a mistake when replacing the division operators, isn't it?

Maria: Yes, we ran the code in a loop, incrementing the year automatically. Then we checked the printed dates and they agreed with those published on several pages in the web.

Professor: Very good. What about Thanksgiving Day?

Maria: That was simpler because all we had to do was to find out which day is the first Thursday of November and simply add 21 to it to get the fourth Thursday. We used `getDay()` to determine the day of the week and `setDate()` to increment the day until it was Thursday. This is the code:

```
var year = prompt("Please, enter the year.");
var day = 1;
var TD = new Date(year, 10, day); //Create the first of November
while (TD.getDay() != 4) {          //While it is not Thursday move
  TD.setDate(++day);                //one day forward.
}
TD.setDate(day + 21);               //Move three weeks forward
document.write("Thanksgiving Day is on " + TD.toString());
```

Professor: Good job. Before we continue, do you have any questions?

Mike: When we tried to construct a `Date` object of the year 2016, we got January 1, 1970. As a matter of fact, we keep getting the same result no matter which year we use. Why is that?

Professor: How did you construct your object?

Maria: Like this:

```
var myYear = Date(2016);
```

Professor: Check the list of constructors on page 159. Which one do you actually use in this example?

Mike: Oh, what a silly mistake! By passing only a single argument we used the second one, which takes milliseconds. So it's 2016 milliseconds past midnight on January 1, 1970. We probably should have passed at least two arguments in order to use the third constructor, shouldn't we?

Professor: That's right.

Maria: One more question. We tried to automatically compute Thanksgiving Day of the following year if Thanksgiving of the given year was already over, but we weren't able to compare dates. At the end we were completely puzzled when we tried this:

```
var day1 = new Date(2016, 0, 1);
var day2 = new Date(2016, 0, 1);
console.log(day1.toString());
console.log(day2.toString());
if (day1 == day2) {
  console.log("The days are equal.");
}
else {
  console.log("The days are not equal.");
}
```

The program produced the message "Days are not equal." Why are two identical `Date` objects compared as not being equal?

Professor: That's a good question. There is a fundamental difference between primitive values like Booleans or numbers, and objects. Primitives are manipulated and compared *by value*, which means that they are equal if they have the same value. Of course, there couldn't be another way of comparing them. Objects are different, though. They are manipulated and compared *by reference*, and even when they have exactly the same properties with exactly the same values, their references are not necessarily equal.

In JavaScript, object values are in fact *references* that only *refer* to the objects in memory. A reference can be seen as a value that represents the *location* of the object and not the object itself. So, if you have two identical objects stored in two different locations in memory, the test of equality of those two objects will return false because it compares the *locations*, which are different.

Mike: How can you then compare two dates?

Professor: You can compare milliseconds returned by `getTime()`, for example. In a similar way we used milliseconds to compute the time difference.

Mike: Oh, yes, I remember.

Professor: There's one more important consequence of objects being references rather than values. Consider the next simple code:

```
var day1, day2 = new Date();
day1 = day2;         //Copy the reference, not the object.
day1.setDate(21);    //Set different days for day1 and day2.
day2.setDate(25);
if (day1.getDate() == day2.getDate()) {
  console.log("The days are equal.");
}
```

In the second line of the program, I assigned the value stored in day2 to the variable day1. Recall that the assignment operator makes a copy of the value on its right and stores it to the variable on its left. What did I actually copy?

Maria: You copied the reference?

Professor: Precisely. Now we have two references that both refer to one and the same object in memory. If we change any part of that object, both references sense that. That's why the program writes out "The days are equal." We are actually comparing the days of one and the same object, although through two distinct references.

Maria: I would like to check whether I understood references correctly. The other day you said that a variable is like a box in which you can put some information written on a piece of paper. That piece of paper represents a value while the name of the variable is written on the box. Correct?

Professor: Correct. Please, go on.

Maria: Now, for an object you actually need two boxes: the name of the object is written on the first box, inside which there's a paper holding the information about where the second box is. The object itself is actually placed in that second box.

Professor: That's exactly how it is. The next drawing is an illustration of the Date object and the two references day1 and day2 from the above example.

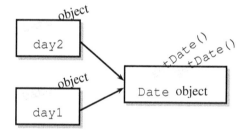

Maria: So, when you assign an object to another variable, you really only make a copy of the piece of paper in the first box and put that copy in yet another box. But how do you make a copy of an actual object?

Professor: That's not exactly a forthright task in JavaScript, I'm afraid. If you want to make a Date object holding the same moment in time as some existent Date object, for example, then you simply construct a new one, passing the milliseconds of the first one to the Date() constructor. Like this:

```
var copy = new Date(original.getTime());
```

Now you have two distinct objects, copy and original, both of which hold exactly the same moment in time. Because they are distinct, any change to either one of them will not affect the other one.

It's time for us to move on.

9.2 `Array` Object

Professor: You often have to handle a set of closely related data, in which case you can store that data into various properties of the same object. Consider, for example, a rectangle, which can be described by the coordinates of its upper left and lower right corners, and perhaps a color. If a rectangle is rotated within a given coordinate system, then an angle of rotation is also necessary. All these data could be stored as values of the properties of a single object called "rectangle." We will see next week how you can devise your own object types or even add properties and methods to existing objects.

There's one object type built into the JavaScript language that already allows you to store and manipulate a collection of data. This is the `Array` object, which is a specialized form of a JavaScript object. Generally, an *array* is an ordered collection of values, which are called *elements*. It is important that each element has a numeric position called an *index* within the array. JavaScript arrays are *zero-based*, which means that the first element in an array has the index value zero.

Just like `Date`, the `Array` object is a data type and we need a constructor in order to create object instances to work with. You can create an `Array` object in any of the following three ways:

```
var a1 = new Array();  //An empty array
var a2 = new Array(5); //An array with room for five elements
var a3 = new Array("start", 6, 2); //An array with three elements
```

The first line constructs an empty `Array` object to which you can add elements later. Let me note that JavaScript arrays are dynamic, which means that they will grow or shrink automatically as you add or remove elements.

In the second line, the elements are not defined but the space in memory is allocated for future needs. I think that you'll rarely use this second constructor, though.

In the last of the above lines, an array with three elements is created and the arguments of the constructor become the elements of the new array. Notice that an element of an array can be of any type. I used a string literal and two numbers but you can use any type you like, even an object. You're also not limited to literals but may use an arbitrary expression for an element.

If you call the `Array()` constructor with a single argument, the second constructor will be called if that argument is a number. Don't forget that and make the same mistake you did when you called the `Date()` constructor with a single year argument.

If, however, you supply a single non-numeric argument, then the third constructor is called:

```
//An array with one element set to a string value "I'm single."
var myArray = new Array("I'm single.");
```

I guess it's a matter of taste but some programmers prefer to create arrays using array literals rather than constructors, which use square brackets and commas. We too are going to use array literals in our course. For example:

```
var a1 = [];                //An empty array
var a2 = [6, 2, "That's it."]; //An array with three elements
```

It is very important to have a firm picture of how an array is organized. You can think of it as a stack of numbered boxes (elements), each holding its own value. You can access individual values with the [] operator, which you put after the name of the object with an appropriate index inside. Here's what the a2 array from the above example looks like:

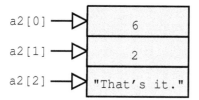

You can access the second element, for example, simply by writing a2[1]. Note that you can put an arbitrary expression inside the brackets as long as it evaluates to a non-negative integer. You'll soon see that this is a powerful tool in the hands of a programmer for accessing array elements.

Finally, you can do with an individual element just about anything you can do with an ordinary variable or object of the same type without affecting other elements of the array whatsoever. For example:

```
a2[0]++;              //Increments the first element
a2[1] = "two";        //Assigns a new value to the second element
console.log(a2[0]);   //Writes 7
console.log(a2[1]);   //Writes two
```

It's time for putting it all together in an example program that actually does something. Imagine you would like to print dates using the full names of the months and the days of week, rather than abbreviations as produced by the toString() method. It's true that the recent implementations of JavaScript allow you to use full names, in English as well as in many other languages, but we'll do it our own way anyhow. If nothing else, it's a good exercise. Besides, you may sometimes want to use some more exotic names like, for example, Middle English names, which standard implementations do not support.

Because we cannot compute names using a mathematical formula, the only possible solution is to define a list of names that we want to use. For example, we can define an array containing the names of days of the week:

```
var dayOfWeek = ["Sunday", "Monday", "Tuesday", "Wednesday",
                 "Thursday", "Friday", "Saturday"];
```

This array will serve us as a *lookup table*. We can look up a name from the above array simply by providing an appropriate index. The following expression, for example, returns the second name from the list:

```
dayOfWeek[1]   //Returns "Monday"
```

The real power of arrays, however, is the ability to use an expression as an index. We can use a value returned by the getDay() method as an index to the dayOfWeek array:

```
var today = new Date();
dayOfWeek[today.getDay()]
```

Maria: Now I see why getDate() and getMonth() return values starting from zero and getDate() returns values starting from one. That's because you may want to look up the names of the months and the days of the week from an array, which is zero based in JavaScript. On the other hand, days of the month don't have names and there's no need for looking anything up from an array. That's why numbers starting from one are returned by getDate().

Professor: That is indeed so.

Now we have everything that we need to write a program that prints today's date in words, in an unabbreviated form:

```
var dayOfWeek = ["Sunday", "Monday", "Tuesday", "Wednesday",
                 "Thursday", "Friday", "Saturday"];
var month = ["January", "February", "March", "April", "May", "June",
             "July", "August", "September", "October", "November",
             "December"];
var today = new Date();
var msg = "Today is ";
msg += dayOfWeek[today.getDay()] + ", ";
msg += month[today.getMonth()] + " ";
msg += today.getDate() + ", ";
msg += today.getFullYear();
document.write(msg); //Writes Today is Sunday, March 1, 2015
```

In addition to the elements themselves, an Array object also holds the information about how many elements it has, which is stored in the length property. Notice that, because numbering begins with zero, the value of length is always one greater than the index of the last element. As an example, examine the following fragments of code:

```
var a = [5, 88, 21]; //Constructs an array with three elements
a[2]               //Returns 21
a.length           //Returns 3
a[a.length - 1]    //Returns 21
```

Observe that the last of the above expressions will always return the last element of the a array regardless of its size.

Incidentally, it is not imperative that an array contains elements with contiguous indexes starting at zero. If some elements are missing in between, then we say that the array is *sparse*. The value of the length property of a sparse array is not equal to the number of elements but simply one greater than the index of the last element:

```
var a = []; //Constructs an empty array
a[5] = 42;  //Adds an element to the array and sets its value to 42
a.length    //Returns 6
```

In the above example, I constructed an empty array and then dynamically added the sixth element to it. The elements a[0] to a[4] do not exist, although some of older browsers wrongly set them to undefined. In practice, however, you will rarely use sparse arrays, so I will not go any deeper into the subject.

Mike: I'm just curious: is there any connection between square brackets used for accessing a property of an object and square brackets used for accessing an element?

Professor: Both work the same. Recall that a property name within brackets should be a string. Of course, JavaScript doesn't mind at all if you provide a numeric index instead of a string. It simply converts the numeric value to a string according to the general type-conversion rules. For example, the index 0 becomes the string "0", which is then used as a property name. In that sense, all indexes are no more than property names.

That said, there are two fundamental differences between an array index and an ordinary object property name. You should, of course, first be able to clearly distinguish an array index from an object property name. Any property name that is an integer between zero and $2^{32} - 2$ is an array index. For example, 3 or 42 are array indexes but -5 or 1.4 are regular object properties with names "-5" and "1.4", respectively.

Maria: Does that mean that any object type can have indexes?

Professor: No, of course not. Indexes as such only apply to array objects. If you use property names 3 or 42 with a date object, for example, they will behave just like ordinary properties, and not like true indexes.

Mike: Are you saying that we can add our own properties to an existing object?

Professor: That's right, but let's leave that topic for our next meeting.

If you therefore use an integer between zero and $2^{32} - 2$ as an array property name, it will behave like a true index. The first difference in using an index rather than a regular property name is that access to array elements is mostly significantly faster than access to regular object properties. That's due to the fact that JavaScript implementations typically optimize arrays for index access.

The second difference is that when you use a numeric index to add a new element, the `length` property of an array is automatically updated for you.

Maria: You said before that arrays were dynamic in the sense that you could add or remove elements. You showed us that an element is added automatically by simply giving a value to a non-existent element. But how do you remove elements?

Professor: One way of doing that is to explicitly change the value of the `length` property, which will automatically truncate or grow the size of the array. If you make `length` smaller, then you effectively delete elements from the end of an array.

There also exist certain array methods like, for example, `pop()` and `push()`, which you can use to explicitly add or remove elements.

I will now show you the `for` loop, which is commonly used in connection with arrays.

9.3 `for` Loop

Professor: The `for` statement is a looping instrument similar to the `while` loop but it is often more convenient when the code has to do with counting of some kind. Counting consists of three elementary tasks: first, setting a counter to an initial value, second, deciding when to stop, and third, deciding upon the size of the increment (or decrement for countdown). A variable used as a counter in a `for` loop is first initialized and then tested before each iteration of the loop is executed. At the end of each iteration the counter is incremented or otherwise updated. All three central operations upon a counting variable, the initialization, the test, and the update, are neatly grouped inside the parentheses at the beginning of the `for` loop. This is the loop's syntax:

```
for (initialize; test; update) do_something
```

The expressions *initialize*, *test*, and *update* are responsible for manipulating a counter variable, and *do_something* is a statement that represents the loop body.

I thought it best to explain how the `for` loop works in a two-stage example of printing the 10 times multiplication table. In the first stage we simply fix one of the multiplicands to, say, three, and print a single row of the table:

```
var x = 3, y;
document.write("<table border=\"1\">");
document.write("<tr>");
for (y = 1; y <= 10; y++) {
//Prints numbers 3, 6, 9, 12, ..., 30 in one table row.
  document.write("<td>" + x * y + "</td>");
}
```

```
document.write("</tr>");
document.write("</table>");
```

Notice how I used strings of HTML code as arguments of `document.write()` to produce markup dynamically. The code is of course interpreted by the browser as any other HTML. Notice also the backslash characters (\) preceding the quotes around the value 1 inside the `<table>` opening tag. The backslash is a so-called *escape character*, which has a special meaning in JavaScript strings. Together with the character that it precedes it forms an *escape sequence*, defining a character that you could otherwise not use within a string. For example, when a string literal starts with a double quote, you cannot use double quotes within that string because the first occurrence of a double quote ends the string. You can use the escape sequence \" instead. Another example of using an escape sequence would be a newline character, which can be represented as \n.

Mike: Is this similar to character entities in HTML?

Professor: It is indeed. But if you don't like to use escape sequences, you can just as well use different types of quotes instead, as single quotes inside double quotes:

```
document.write("<table border='1'>");
```

Here is the flowchart of the for loop from the above example.

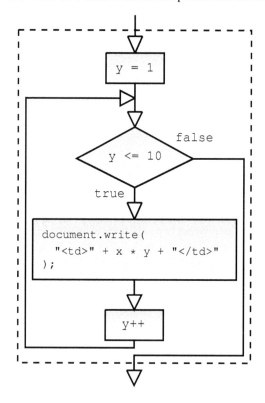

You can see that the initialization of the loop counter, which is the expression y = 1, is performed right at the beginning of the for loop. It is performed only once because no path loops back before that expression. Next, the loop test is evaluated, which is the expression y <= 10. If that expression evaluates to false, then the loop terminates. Otherwise, the loop body is executed, which outputs the start and end tags of a table cell and prints the value of the expression x * y in between. Finally, y is incremented before going back to be tested again, and the story repeats itself.

Although this loop is somehow more complex than the other two loops that we know, I don't think it needs any more explanation.

Mike: You said the other day that any rectangular box in a flowchart can be replaced by an arbitrary statement. Looking at the for loop in the above example, I find it hard to believe that this is really so.

Professor: You're quite right about that. It's true that *initialize*, *test*, and *update* are expressions and not statements. There are always exceptions to the rules, aren't there?

In order to produce the complete multiplication table, we just need to repeat the part of the code responsible for writing a table row. We can do that by simply putting the code into yet another for loop, iterating x from one to 10. Like this:

```
var x, y;
document.write("<table border=\"1\">");
for (x = 1; x <= 10; x++) {
  document.write("<tr>");
  for (y = 1; y <= 10; y++) {
    document.write("<td>" + x * y + "</td>");
  }
}
document.write("</tr>");
document.write("</table>");
```

The time is ripe to finish our lucky lottery number generator. I propose the following algorithm:

> **Repeat*** six times:
>> **Repeat****
>>> Set number to a random value between 1 and 54.
>>> **If** number doesn't exist in drawnNumbersList **Then**
>>>> Add number to drawnNumbersList.
>>>> Stop repeating**.
>>> **End If**
>> **End Repeat****
> **End Repeat***
> Print the six lucky numbers (contents of drawnNumbersList).

I think that we shouldn't have any difficulties implementing the algorithm except for the part that checks whether `number` already exists. This can be done as follows:

> Set an auxiliary variable `exists` to `false`.
> **Repeat** for every element of `drawnNumbersList`:
>> **If** `number` is the same as the value of the element **Then**
>>> Set `exists` to `true`.
>> **End If**
> **End Repeat**

Incidentally, the approach that I have just taken is called *top-down* design. Using this approach you basically break down a problem into a composition of several less-complex subproblems. Then you further break down these subproblems to even simpler ones until they are simple enough for you to implement.

Every time the above loop finishes, the variable `exists` holds the information about whether or not `number` already exists in the `drawnNumbersList` array. It is not hard to see that this is, in fact, a single `for` loop containing a single `if` conditional. The following are the necessary ingredients to build the loop:

- Initialization: `exists = false` and `i = 0` (`i` is chosen to be an index to the `drawnNumbersList` array).
- Loop test: `i < drawnNumbersList.length` (the index of the last element is one less than the length of the array).
- Update: `i++` (go to the next array element).
- Loop's body: if `number` is equal to the `i`th element of `drawnNumbersList`, then set `exists` to `true`.

Notice that `exists` is not a counter but I still chose to initialize it at the same spot as the loop counter. That's an obvious thing to do because `exists` should be initialized to `false` every time the loop starts anew, and it is much easier to understand what the loop is doing if you keep all the necessary loop initialization stuff in one place.

We only need to find an appropriate operator to join the two initializations into a single expression.

Maria: We can use a logical AND operator to do that.

Professor: That won't work because of *short-circuit* evaluation. Namely, when two expressions are joined by a logical operator, it can happen that the value of the expression that is evaluated first already decides the final return value of the whole Boolean expression. In such a case the other expression is short-circuit evaluated. Effectively that means it is not evaluated at all.

Consider, for example, the following Boolean expression:

```
(i = 0) && (exists = false)    //The expression on the right is not
                               //evaluated.
```

Because the expression i = 0 returns a falsy value, there is no way that the whole Boolean expression can evaluate to true, no matter what the value of the expression exists = false is. This expression is therefore never evaluated.

We will use a *comma operator* (,) to join both initialization expressions. This is probably the only situation where you're going to use this operator, but it's still good to know exactly how it works. The comma operator evaluates both of its operands (from left to right) and returns the value of its right operand. Consider, for example, the following expression:

```
i = 0, exists = false    //Returns false
```

Here the variable i is first set to zero and then exists is set to false. The return value of the whole expression is equal to the return value of the second assignment, which is false. We're not going to use this return value, though. The important thing is that both assignments are joined in a single expression and that both actually happen.

Mike: How can we stop repeating a loop from within an if statement? I'm talking about the loop on page 175 marked by two asterisks.

Professor: We can use a break statement. You already know that break terminates a switch statement but you can use it to terminate loops as well. Whenever a break is encountered within a loop, that loop is terminated and the program execution continues immediately after the loop that has just terminated.

This is now a completed lucky lottery number generator:

```
var number, exists;
var drawnNumbersList = [];
var i, n;
for (n = 0; n < 6; n++) {
  while (true) {                                    //An infinite loop
    number = Math.floor(Math.random() * 54) + 1;
    for (exists = false, i = 0; i < drawnNumbersList.length; i++) {
      if (number == drawnNumbersList[i]) {
        exists = true;
      }
    }
    if (!exists) {
      drawnNumbersList[n] = number;
      break;                                        //Exits the infinite loop
    }
  }
}
document.write(drawnNumbersList);
```

Maria: Won't break terminate the program altogether?

Professor: No, it won't, because `break` terminates only the innermost loop in which it is placed. If that loop is nested within another loop, then the outer loop will continue its execution as though nothing happened.

Notice how I produced an infinite loop by simply putting the constant value `true` in place of the loop condition. Since `true` always evaluates to `true`, the loop can only be stopped by means of `break`. By the way, instead of `while (true)` you will sometimes also see `for (;;)` which also makes up an infinite loop.

Mike: I suspect that this program can be made shorter, without `break`, without `number`, and using a `do/while` loop instead of the `while` loop.

Professor: You know, I think you're right. I was planning to revisit the example later today anyhow and we may just as well try to make it shorter using your suggestions.

Speaking of `break`, there's a similar statement called `continue`. The difference is that `continue` does not terminate the loop execution completely. It only terminates the current iteration and jumps directly to the loop condition, which is then tested for the next iteration.

Even though `break` and `continue` can sometimes make for elegant solutions, you should limit yourselves to using them exclusively at the start or end of the loop body, only as additional or alternative checks. Otherwise you may later overlook them, getting yourself in some nasty trouble.

Maria: The drawn numbers produced by your code aren't sorted.

Professor: Of course they aren't because we haven't implemented any sorting yet. But we'll be coming to it shortly.

9.4 Array Methods

Professor: Now, you already know that a JavaScript array has a single `length` property, which is automatically updated to always be one larger than the highest index of the array. While array elements are also properties, they are usually optimized for a significantly faster access than ordinary properties. There are three different constructors for the `Array` object, but they can be replaced by array literals, and we decided to use them as they are shorter and there's no danger of accidentally calling a wrong constructor if you use a single numeric argument.

There is more to arrays, though. You'll discover that the `Array` object implements quite a few array-manipulating methods. Two of them we are going to meet now and you'll find more in the JavaScript reference in section E.3 on page 374. Even what you find there is just a selection of what this object has in store for the programmer.

For our lucky lottery number generator we are going to use two methods, `sort()` and `indexOf()`. The first one is kind of self-explanatory—it simply sorts the array elements. The elements are reordered *in place*, which means that no new array is created. The elements of the array on which you call the `sort()` method are rearranged.

The `indexOf()` property searches for an element and returns its index. If no element is found, then it returns -1. Here are a couple of examples:

```
var myArray = [7, 9, 3];
var x = myArray.indexOf(9);   //x becomes 1
myArray.sort();               //myArray becomes [3, 7, 9]
x = myArray.indexOf(9);       //x becomes 2
```

Maria: Does `sort()` only work for numbers?

Professor: As a matter of fact, elements are sorted alphabetically—or according to the order determined by the character encoding, to be exact. If the elements are not strings, they are converted to strings.

Mike: What abut `indexOf()`? Does it also convert elements to strings before doing comparisons?

Professor: `indexOf()` makes no conversions. It uses the strict equality operator (===) for comparison. The next call, for example, will return -1:

```
x = myArray.indexOf("9");   //x becomes -1
```

Maria: What if there's more than one element equal to the passed argument?

Professor: Then the index of the first element that is found is returned. The search always begins at the beginning of an array. Alternatively, if you want to start a search from some other position, you can pass an additional argument that specifies an array index at which to begin the search. The next example uses this possibility to find the positions of all the elements whose value is 9:

```
var myArray = [1, 6, 2, 9, 2, 5, 9, 9, 3];
var nines = [];
var index = myArray.indexOf(9);
while (index != -1) {
  nines[nines.length] = index;
  index = myArray.indexOf(9, index + 1);
}
console.log(nines);            //Writes 3, 6, 7
```

The third line of the code searches for the first occurrence of the number nine. If there's one found, then the `while` loop starts. Inside the loop the index of the found number is first appended at the end of the `nines` array. Then the next number nine is searched for. Notice that I entered `index + 1` as a start position for the search, which is one element after the last successful find. If I continued the search at the position `index`, then I would be getting the same element over and over again.

The same program can be written a little shorter if slightly more enigmatic:

```
var myArray = [1, 6, 2, 9, 2, 5, 9, 9, 3];
var nines = [];
var index = -1;
while ((index = myArray.indexOf(9, index + 1)) != -1) {
  nines[nines.length] = index;
}
console.log(nines);            //Writes 3, 6, 7
```

We can now rewrite our lucky lottery number generator using the methods sort()
and indexOf(). We will also use your suggestions from page 178, Mike, to reduce
the length of the program. The program is now considerably shorter, and with an
additional sorting feature:

```
var drawnNumbersList = [];
var i, n;
for (n = 0; n < 6; n++) {
  do {
    drawnNumbersList[n] = Math.floor(Math.random() * 54) + 1;
  } while(drawnNumbersList.indexOf(drawnNumbersList[n]) < n);
}
drawnNumbersList.sort();
document.write(drawnNumbersList);
```

What's wrong with the sorting? We got the numbers 10, 25, 3, 31, 4, and 48.

Mike: The elements are arranged alphabetically, aren't they?

Professor: That's right. The sort() method compares strings.

Maria: Is sort() therefore useless for us?

Professor: Not necessarily. We just have to do some more work to make it useful for
our purpose. We need to provide a criterion by which the array elements should be
compared. Next week you're going to learn how to do that.

9.5 String Object

Professor: Next on our agenda is the String object. Strings could be seen as arrays
of characters, which makes String quite similar to the Array object. The main difference between both is that you can change the value of array elements but there's no
way you can alter parts of the text stored in an existing string. We say that strings are
immutable. The following example demonstrates that:

```
//Constructs a String object:
var myString = new String("Testing, testing...");
var letter = myString[3];   //Copies the fourth letter ("t")
myString[0] = "7";          //No-go: strings are immutable
```

The String() constructor is quite straightforward—you simply pass any value to be converted to a string as its argument.

In the second line, I used an index access operator ([]) to access the fourth character, which is the letter "t," and assign its value to letter. The last line of the example, however, doesn't work. It won't change the myString string because strings are immutable. Note that this last line doesn't generate an error, though. It simply fails silently.

Maria: That's odd, I could swear that we've already changed a value of a string, haven't we?

Professor: Technically, yes. You can assign a whole new text to a string:

```
myString = "Something new";
```

But that's not the same string any more. It's not even an object. The above line didn't change the object value but created a whole new data instance, which is now a string primitive. The next example demonstrates that:

```
var myString = new String("Original");
console.log(typeof myString);    //Writes object
myString = "Something new";
console.log(typeof myString);    //Writes string
```

Anyway, from a practical point of view, it is important that you cannot change just a portion of a string in the way you can change individual elements of an array.

Just like arrays, strings also have the length property, which holds the number of string characters.

You can manipulate strings using one of the many methods of the String object. These are some examples:

```
var myString = new String("Still testing...");

myString.indexOf("S")    //Returns 0: upper-case "S"
myString.indexOf("s")    //Returns 8: lower-case "s"
myString.indexOf("til")  //Returns 1
myString.indexOf("TEST") //Returns -1: not found
myString.slice(10)       //Returns "ing..."
myString.slice(2, 4)     //Returns "il"
myString.toUpperCase()   //Returns "STILL TESTING..."
myString.toLowerCase()   //Returns "still testing..."
```

Because strings are immutable, the methods that modify the string only *return* a modified string—the original string is not altered in any way. Do you want any extra explanation of the above examples?

Mike: I see that indexOf() works the same as indexOf() of the Array object except that it can search not only for a single character but also for the whole sequence of characters. toUpperCase() and toLowerCase() are also obvious, but I don't think I can decipher slice(), except that it returns a part of the original string. How does it work?

Professor: slice() is used to extract a substring from the original string. It accepts as an argument the index of the first character to be included in the extracted substring. The returned substring then contains all the characters from the specified character to the end of the original string. With the second argument, which is optional, you determine the last character to be included in the substring. Note that the last character will be the one with the index one smaller than the given second argument.

The fact that many string methods return a string allows for a so-called *method chaining*. You can use a property or method directly on a returned value without using any intermediate variables. Here are two examples of method chaining:

```
myString.toUpperCase().slice(2, 4) //Returns IL
myString.slice(2, 4).length        //Returns 2
```

Let's top off our discussion about strings with an example that rounds a number to two decimal places. If the number is an integer, or has only a single decimal place, then the program should still produce a result with two decimal places. For example, 7.928 should end up as 7.93, 10 as 10.00, and 5.2 as 5.20. Any ideas on how to pull that one off?

Maria: Math.round() comes first to my mind but it rounds to the nearest integer. Besides, I expect that we're going to work with strings, aren't we?

Professor: We'll make use of them, all right. Please, go on.

Maria: I can multiply a number by 100, then Math.round() it, and divide it back by 100. That's OK if a number has at least two decimal places, otherwise.... No, that's not going to work. Instead of dividing, I can simply insert a dot. Like this:

```
var number = 12.4;
var dotPosition;
var result;
number *= 100;
number = Math.round(number);
number = String(number);
dotPosition = number.length - 2;
result = number.slice(0, dotPosition);
result += ".";
result += number.slice(dotPosition);
console.log(result);              //Writes 12.40
```

Mike: I think that you cannot use number.length because number is a primitive string and not a String object. You forgot to use new with the String() constructor.

If I remember correctly, `String()` called without a `new` operator is a type conversion function that returns a primitive string.

Professor: That's true, although it doesn't really matter whether `number` is a string primitive or object. Whenever you use a property or method on a string primitive, a temporary `String` object called a *wrapper object* is created automatically behind the scenes. When you stop using the object's properties and methods, the wrapper object is destroyed. This effectively blurs the distinction between primitive and object string types and you can use a primitive string as though it was an object:

```
var s = "I am primitive";
console.log(s.length);       //Writes 14
console.log(s[3]);           //Writes m
console.log(s.slice(0, 4));  //Writes I am
```

Still, there is an important difference between primitive strings and object strings. Consider the next code fragment:

```
var strObject1 = new String("Something");
var strObject2 = new String("Something");
console.log(strObject1 == strObject2);
console.log(strObject1 == "Something");
```

What do you think the result will be?

Maria: I remember that objects are compared by reference. In the above example we compare objects that have equal values but are distinct. Both comparisons therefore return `false`.

Professor: The first comparison indeed returns `false`. The second comparison, however, has a primitive string on one of its sides. Because of that, the string object `strObject1` on its other side is converted to a primitive string and both primitives are compared by value, which returns `true`.

The same object-to-primitive conversions take place when you compare string objects using greater-than or less-than operators. Let me summarize the rules for you:

- If both operands of the equality operator (==) are string objects, then the operator returns `true` if and only if both operands refer to one and the same object. If they refer to different objects, the return value is `false` even when both objects hold identical strings. No object-to-primitive conversion occurs here.

- If one operand of the equality operator (==) is a primitive string and the other is an object string, then the object is converted to a primitive before comparison. Both primitives are then compared by value.

- If either operand of a comparison operator >, <, >=, or <= is a string object, then that object is converted to a string primitive. The string primitives are then compared alphabetically.

Of course, just the opposite of what was said for the equality operator is true for the inequality operator (!=).

Mike: You didn't mention the strict equality (===) and inequality (!==) operators. I guess that no conversion is made with them.

Professor: That's true.

Note that string primitives might not compare as equal even if they appear equal. They may be encoded using different encoding rules. But that isn't something you should worry about right now.

Mike: How can you compare two string objects by value if you want to find out whether they are equal?

Professor: You can use the toString() method, which returns a primitive string representation of an object. Note that other object types also have this method, which is usually called automatically by the system whenever a conversion to a string is necessary. But you can call toString() explicitly as well. For example, you can compare two string objects like this:

```
var s1 = new String("I'm the same");
var s2 = new String("I'm the same");
console.log(s1 == s2);                    //Writes false
console.log(s1.toString() == s2.toString()); //Writes true
```

Maria: What if I compare a string object to a number?

Professor: What do you think?

Maria: Perhaps the string object should convert to a string primitive first. According to what you have told us so far, this sounds like the most reasonable solution to me. I guess that after that the standard rules for primitive types apply, which means that the string primitive will further convert to a number.

Professor: Exactly.

Mike: I remember seeing a Number object somewhere. Could it be that JavaScript can create a wrapper object for a primitive number as well?

Professor: You're quite right about that.

Mike: If the Number object also has the toString() method, then we could rewrite the sixth line of our rounding-to-two-decimals example on page 182 like this:

```
number = number.toString();
```

Professor: Yes, that is indeed possible. You decide which one you prefer.

9.6 Homework

Professor: By the way, do you like Sudoku?

Maria: Isn't that a number puzzle where you have to fill out the missing numbers in a 9×9 grid so that each column, each row, and each of the nine 3×3 boxes contain all of the numbers from one to nine? Yes, I like to solve that puzzle.

Mike: Me too. Are we going to program a Sudoku generator?

Professor: Not so fast. First we are going to develop a web page that will assist a visitor in solving a predefined Sudoku puzzle, inasmuch as it will warn him/her about a number that he/she has placed in a row, a column, or a 3×3 box that already contains that number. The program will, however, only check the number that has just been placed in relation to the already-placed numbers, not in relation to the final solution of the puzzle.

For homework, you try your best to implement the part of the algorithm that checks whether a number can be placed in a certain cell. You can use the following Sudoku puzzle to work with:

5	7		4			2		
3	8			2	7			
1		7	3					
7		2	8				3	
3		4				1		6
1				6	4		7	
			2	7		1		
		1	9			2	3	
	4			1		5	9	

Since the puzzle has a two-dimensional layout, the most logical data structure to hold the numbers is a two-dimensional array. JavaScript does not support two-dimensional arrays per se but as there's no limitation on type of array elements, you can construct an array of arrays. For example, you can work with the following two-dimensional array, which describes the above Sudoku puzzle:

```
var initial = [
  [null, 5,    7,    null, 4,    null, null, 2,    null],
  [null, 3,    8,    null, null, 2,    7,    null, null],
  [null, 1,    null, 7,    3,    null, null, null, null],
  [null, 7,    null, 2,    8,    null, null, null, 3   ],
  [3,    null, 4,    null, null, null, 1,    null, 6   ],
  [1,    null, null, null, 6,    4,    null, 7,    null],
  [null, null, null, null, 2,    7,    null, 1,    null],
  [null, null, 1,    9,    null, null, 2,    3,    null],
  [null, 4,    null, null, 1,    null, 5,    9,    null]
];
```

Do you perhaps have any ideas as to how to access a value in the above array? Like the second value in the first row, for example.

Mike: Since `initial` is an array of rows, we can select a row simply by using the array access operator (`[]`). For example, this is the first row:

```
initial[0] //The first row
```

Because this is also an array, we can apply another array access operator to select a single element from it:

```
initial[0][1] //The second element in the first row
```

Professor: Splendid! Now you know everything to complete your homework.

Maria: Excuse me, what exactly is it that we have to do?

Professor: Oh, I'm sorry. I really should have given you more specific directions. You first need to obtain a number from one to nine, and then the row and column indexes from zero to eight of the cell in which the player wants to place that number. Once you have these three numbers, you try to check whether the number can be placed in the required cell. You must check whether the same number isn't already in the same row or column. That's the easy part. The tricky part is to check whether the same number isn't already in the same 3×3 box. For example, you cannot put the number five in any of the four empty cells of the upper left 3×3 box because that box already contains that number.

Just a few more keywords to summarize today's meeting and then I guess we're done for this week:

> **In this meeting**: comparing dates, value, reference, `Array` object, index, `length`, array literal, array element, lookup table, sparse array, `for` loop, initialize, test, update, short-circuit evaluation, comma operator, `break`, `continue`, infinite loop, array methods, `sort()`, `indexOf()`, `String` object, immutable object, object-to-primitive conversion, method chaining, wrapper object, `toString()`, two-dimensional array

Understanding Functions

10.1   Homework Discussion

Professor: How was your homework? Are we ready to play Sudoku?

Maria: We can check placed numbers but we didn't implement a user interface.

Professor: Don't worry about that. Soon you will learn how to do a user interface as well. Please, show me what you've got.

Maria: These are the variables that we used:

```
var number = 3, row = 4, col = 3;
var exists = false;
var i, j;
var iStart, jStart;
```

The variables `number`, `row` and `col` define a number that we want to place in a grid and where we want to place it. We used an auxiliary variable `exists` for pretty much the same purpose as we did last time in our lottery program.

In the last line we declared the variables `iStart` and `jStart`, which will hold the indexes of the upper left cell of a 3 × 3 box. We used these indexes as a starting position for searching a box for a number.

Since each row is an array, checking a row was the most straightforward task:

```
//Checks a row:
if (initial[row].indexOf(number) != -1) {
  exists = true;
}
```

Just slightly more complicated was checking the column. We had to check each cell separately:

```
//Checks a column:
for (i = 0; i < 9; i++) {
  if (initial[i][col] == number) {
    exists = true;
  }
}
```

Mike: Checking the 3×3 boxes was a bit more complicated but basically two nested for loops did the trick. One problem we had to solve was how to set the initial values of both loop counters. We discovered that the indexes of the upper left cell of each box—that's where the search should start—are multiples of three. So we only had to subtract from the row index the remainder after division of the row index by three, and subtract from the column index the remainder after division of the column index by three to get the start position. This is the complete solution:

```
//Checks a 3x3 box:
iStart = row - row % 3;
jStart = col - col % 3;
for (i = iStart; i < iStart + 3; i++) {
  for (j = jStart; j < jStart + 3; j++) {
  if (initial[i][j] == number) {
      exists = true;
    }
  }
}
console.log(exists);
```

Professor: You did a great job indeed.

Your solution, however, is just a small part of what we have to write in order to get a complete working Sudoku. It's high time we started thinking about how to pack individual portions of code neatly into functions to keep things under control.

10.2   Writing Function Definitions

Professor: You already know that a function has a name to which you append a pair of parentheses in order to *call* or *invoke* it. Sometimes you pass one or more comma-separated *arguments* to the function, which you do by putting the arguments inside the parentheses. You also know that a function invocation is an expression, which returns a value like all expressions do. What you do not know is how to define your own function.

One way of defining a JavaScript function is by using a *function expression*:

```
var fun_name = function(par1, par2, ..., parN) {
  statements
};
```

A function expression begins with the `function` keyword followed by a pair of parentheses with an optional comma-separated *parameter* list (*par1* to *parN*). At the end of a function expression there is a function *body*, which comprises any number of JavaScript statements within curly brackets. Note that, unlike statement blocks of `while` and other statements, the function body requires curly brackets. They cannot be omitted even when the function body is composed of only a single statement.

Because a function is an object in JavaScript, a function expression in fact returns a function object, a reference to which you usually assign to a variable. In the above definition, the returned reference is assigned to a variable *fun_name*. Later, you can use *fun_name* to call, or *invoke* the function, which will effectively execute the code placed in its body. Notice a semicolon at the end of the function expression, which is needed to make a statement out of the expression.

Maria: Can you give us a simple example?

Professor: All right, here's a really simple one, without any parameters for starters:

```
var cheer = function() {
  var i;
  for (i = 0; i < 2; i++) {
    document.write("Hip ");
  }
  document.write("Hooray!");
};
```

You invoke this function by referring to the variable `cheer` using an additional pair of parentheses. Like this:

```
cheer(); //Writes Hip Hip Hooray!
```

When you call a function, the main program execution sequence is temporarily suspended and the code within the function body starts to execute. After the function has terminated, the execution returns to the main program. The following diagram shows what happens when `cheer()` is invoked.

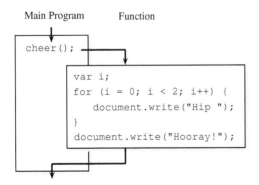

Alternatively, you can define a function using a *function definition statement*:

```
function fun_name (par1, par2, ..., parN) {
  statements
}
```

Maria: Are there any differences between both forms of function definition?

Professor: The most important difference is that you cannot call a function before you define it if you use a function expression. This limitation is, however, rarely an issue and we will consistently use function expressions to define our functions.

Mike: Are you saying that it is possible to call a function before it is defined when you use a function definition statement? How can you call a function before it is defined?

Professor: In JavaScript terminology we say that the function definition statement is *hoisted* to the top of the containing script so that it is available to all the code right from the beginning of the script. With a function expression, however, only the variable declaration is hoisted. The variable initialization code, which represents the actual function definition, is not hoisted.

Mike: I don't think I quite follow you. Can you give us an example?

Professor: OK, here's a short one. In the following code, the test() function doesn't execute because the function is not yet defined at the time of its invocation:

```
test(); //Error: undefined is not a function
var test = function() {
  document.write("Hello!");
};
```

If you open the JavaScript Console and examine more closely what is happening, you get the message "undefined is not a function." Try deleting the second line of the example and you get the message "test is not defined." When the second line of the code is present, the test variable is already declared in the first line because its declaration is hoisted to the top of the containing <script> element. Still, it has the undefined value because its initialization code—which is the function definition proper—is not hoisted.

The situation is identical to the following code:

```
var test;
test(); //Error: undefined is not a function
test = function() {
  document.write("Hello!");
};
```

If you instead use a function definition statement, then the invocation of test() in the following code will work because the function definition is hoisted:

```
test(); //Writes Hello!
function test() {
  document.write("Hello!");
}
```

Maria: How do you use function parameters?

Professor: You can think of each parameter as a variable used inside the function body. When a function is invoked, the provided arguments serve as initial values for these variables.

Maria: If parameters are variables, then I suppose the rules for naming them are the same as they are for variables.

Professor: That's true. Exactly the same.

As an example, let's write a function that takes two arguments and prints the larger of both to the browser window. Here is the definition:

```
var writeLarger = function(x, y) {
  document.write(x > y ? x : y);
};
```

You can invoke this function like this:

```
writeLarger(5, 13);  //Writes 13
```

When you pass arguments to a function, you must be careful that their order matches that of the parameters used in the function definition. In the above example, the argument value 5 is assigned to the x parameter, while the argument value 13 is assigned to the y parameter. Both parameters are used within the function body to calculate the value of the conditional expression x > y ? x : y, which is then written to the browser window.

The following diagram shows how the argument values are transferred to the function parameters x and y just before the function body starts to execute.

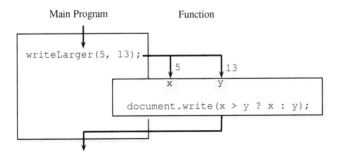

Do you see anything wrong with this example?

Maria: Well, no. It works.

Professor: It works all right, but there's something fundamentally wrong with it. Imagine a more elaborate example when you write a function that computes and prints the arithmetic mean of an array of numbers. Would such a function be useful?

Mike: Maybe not so much. For example, if you only want to compute the arithmetic mean and then do something else with it, not print it, then that function would be completely useless.

Professor: Precisely. One of the important programming skills is to be able to write functions that don't do more than one otherwise independent task in a single package. Giving your functions descriptive names is of great help in managing that. As soon as you discover that a name hints at two different and generally independent tasks, make two functions. For example, the name `writeLarger` indicates that the function does two things—it determines which value is larger and writes that value to the browser window.

Another advantage of using descriptive names for functions is that your code will be more readable and thus easier to maintain.

Mike: I guess that the rules for naming functions are generally the same as they are for variables.

Professor: That's right. You may use any legal JavaScript identifier as a function name so long as it is not a reserved keyword.

Maria: But if a function doesn't write out what it computes, then I think it should at least be able to return the computed value, shouldn't it?

Professor: Exactly. You can use the `return` keyword to accomplish that. The value of the expression that you associate with `return` is the return value of the function. If you don't use `return` inside the function body—or you use `return` without any expression—then the function returns the `undefined` value. Note that you can put `return` anywhere inside the function body and not just at the end—the function will stop its execution and return as soon as the `return` keyword is reached.

Let's now split the above `writeLarger()` function into two independent functions. The first one determines and returns the larger of two arguments, and the other simply writes out the value of the passed argument:

```
var larger = function(x, y) {
  return x > y ? x : y;
};

var write = function(x) {
  document.write(x);
};
```

The first function returns the value of the expression that appears after the `return` keyword. In order to write the larger of two values to the browser window, we may invoke both functions in the following manner:

```
write(larger(5, 13)); //Writes 13
```

The following figure shows how both functions are executed.

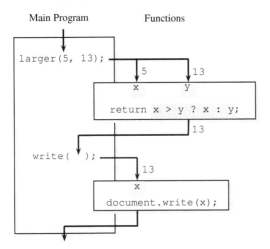

The function `larger()` is called first and it returns 13, which is the value of the expression x > y ? x : y. This value is immediately used as an argument of the function `write()`, which is called next and writes the value to the browser window.

Maria: I don't see why `write()`, which only calls `document.write()`, is useful.

Professor: Later you may decide to change how you write things out, in which case it is very convenient that you already have a function defined for writing. That way, you don't have to change the code all over your program, but you only have to rewrite a single function definition. Sometimes a function like that is called a *placeholder* because you define it primarily to replace it later with different code.

Mike: What happens if you provide a different number of arguments when you call a function? We learned, for example, that a number of arguments passed to a `Date()` constructor decides about what a constructor actually does. Can we also implement behavior like that in our functions?

Professor: JavaScript won't mind if you invoke a function with either fewer or more arguments than declared parameters. It's the programmer's responsibility to handle those situations.

If you provide fewer arguments than declared parameters, then the missing arguments are assumed to have the `undefined` value. You can use that fact to write a function that accepts optional parameters. For example, we can write a function that returns a Euclidean norm of either a real number or a two-dimensional vector, depending upon the number of passed arguments. In not-so-mathematical language that means the function returns either the absolute value of a real number or the length of a vector. The latter is calculated as the square root of the sum of squares of the vector components according to the Pythagorean theorem. This is the function:

```
var norm = function(x, y) {
  if (y === undefined) {
  //Provides a default value if the second argument is omitted.
    y = 0;
  }
  return Math.sqrt(x * x + y * y);
};
```

The second argument of the above function is optional. Whether you pass it or not, the function knows how to handle either situation. This is how you can invoke the function:

```
norm(-3)    //Returns 3: the absolute value of a real number
norm(3, 4) //Returns 5: the length of a two-dimensional vector
```

Note that you must always put optional parameters at the end of the parameter list.

It is also possible to provide more arguments than there are declared parameters. In that case you can use the arguments object, which is automatically created and can be accessed within the function body. You can think of the arguments object as an array whose elements are equal to the arguments passed to the function. You can use a numeric index in square brackets to access the passed arguments, and there is of course also the length property. Technically speaking, arguments is not an array, though.

We can rewrite our norm() function to accept any number of arguments and thus be able to calculate the length of a vector of arbitrary dimension. We are going to omit parameters from the function definition altogether because we can access all the arguments exclusively through the arguments object. The first argument is accessible via arguments[0], the second via arguments[1], and so on. This is the function:

```
var norm = function() {
  var i, sum = 0;
  //Calculates the sum of squares of the passed arguments:
  for (i = 0; i < arguments.length; i++) {
    sum += arguments[i] * arguments[i];
  }
  //Calculates and returns the square root of the sum:
  return Math.sqrt(sum);
};
```

You can use this function to compute a space diagonal of a cube with a side length of two, for example:

```
norm(2, 2, 2) //Returns 3.4641016151377544
```

Functions like this one that you can pass any number of arguments go under the names *variadic functions*, *variable arity functions*, or *varargs functions*.

Note that the `arguments` object is always there for you even if you define named parameters. If a function has named parameters, the arguments are accessible within the function by parameter names or as elements of `arguments`. You can even have some parameters that are named and handle the optional ones using the `arguments` object.

10.3   References to Function Objects

You are already familiar with the fact that functions are objects and that you can call a function by appending a pair of parentheses (with necessary arguments, if any) after its name. The name of a function without parentheses is nothing more than a reference to the function object. Of course, it is possible to assign that reference to any object that can hold a value like, for example, a variable, an object property, an array element, or even a function parameter. You can invoke a function through any of these objects simply by appending parentheses with any necessary arguments. That allows you to do feats like these, for example:

```
var square = function(x) { return x * x; };
var cube = function(x) { return x * x * x; };
var otherSquare = square;
var moreOfThem = [square, cube];

console.log(otherSquare(3));       //Writes 9: same as square(3)
console.log(moreOfThem[1](2));     //Writes 8: same as cube(2)
```

The first two lines are simple function definitions. In the third line I declared a variable `otherSquare`, to which I assigned a reference to the first of the above functions. This makes `otherSquare` and `square` references to one and the same function object, which means that the expression `otherSquare === square` will return `true`. Because of that, writing either `square(3)` or `otherSquare(3)` invokes one and the same function object. A similar situation occurs in the last line of the code, where the expression `moreOfThem[1]` is a reference to the second of the above function objects.

So long as you accept the fact that a function value is a reference, your possibilities don't end so quickly. For example, you can assign a function definition directly to an array element not needing any special name for the function:

```
var funcArray = [];
funcArray[0] = function() { /* Do something useful */ };
```

By the way, a function defined with no specific name is sometimes called an *anonymous* function.

Maria: So I can call the above function that does something useful like this:

```
funcArray[0]();
```

Professor: Precisely.

Armed with that knowledge, we are now able to finish our lucky lottery number generator, which uses the `sort()` method of the `Array` object to sort the numbers. By default, `sort()` orders array elements alphabetically, which isn't exactly what we want in our example. It is actually possible to sort array elements using an arbitrary comparison criterion, which is supplied as an argument to the `sort()` method. The argument is simply a reference to a function implementing the comparison criterion. The function must take two arguments, *arg1* and *arg2*, and return one of the following:

- A value less than zero, if *arg1* should appear before *arg2* in the sorted array.
- Zero, if both arguments are equal in view of this sort and their relative positions don't matter.
- A value greater than zero, if *arg1* should appear after *arg2* in the sorted array.

This is a solution that will work for our lottery generator:

```
var order = function(x, y) { return x - y; };
//...
drawnNumbersList.sort(order);
```

If you're puzzled how this might work, imagine that `sort()` uses the passed reference to call our criterion function many times, each time comparing two elements of the array and switching their positions when necessary. The scheme defining the fashion in which elements are selected for a comparison is described by a sorting algorithm, which you don't need to know in order to be able to use `sort()`.

Mike: It's beginning to worry me that with several function definitions in my program I might accidentally choose a parameter name that already exists as a variable elsewhere in my code. That could easily happen as the size of my code grows, couldn't it?

Professor: Luckily, you are not the only one who has thought of that, so there exist so-called *scoping rules* to help you manage the name conflicts like the one you hypothesized.

10.4   Variable Scope

Professor: You know, not every variable is accessible anywhere in your program source code. The *scope* of a variable specifies precisely in which parts of the code the variable is defined. If you declare a variable outside of any function, then that is a *global* variable, which has global scope. Global variables are accessible anywhere in your program. On the other hand, if you declare a variable within a function body, then you declare a *local* variable, which has local scope. This is also called *function scope* in JavaScript because such a variable is defined only within the function body in which it is declared. Note that function parameters act as local variables and are thus only accessible within the function body.

Now, a question arises: what if there are two variables, one local and the other global, using the same name? According to the just-mentioned rules, both should be accessible within the function body in which the local variable is declared. To solve the conflict, there's an additional rule which says that a local variable takes precedence over a global variable with the same name. If you hence declare a local variable, or a function parameter, for that matter, choosing the same name as that of some global variable, then you hide that global variable from the code within the function body. The next short example illuminates the whole idea:

```
var scope = "global";
var f1 = function() { var scope = "local"; return scope; };
var f2 = function() { return scope; };

console.log(scope); //Writes global
console.log(f1()); //Writes local
console.log(f2()); //Writes global
```

There are two variables named `scope` in the example. The first one is declared outside of any of the functions and is hence a global variable. The second one is declared within the body of the `f1()` function, which makes it a local variable of that function. The first call to `console.log()` takes place outside of any function and can therefore only access the global `scope` variable. The second call to `console.log()` writes out whatever `f1()` returns, which is the value of its local `scope` variable. Namely, `f1()` has no access to the global `scope` variable, which is hidden by the local variable with the same name. Finally, the third call to `console.log()` writes out what `f2()` returns, which is the value of the global `scope` variable since this is the only variable accessible within the body of `f2()`.

To help you better visualize the situation, here's a graphical representation of the scopes of the two `scope` variables.

The white area indicates the scope of the local `scope` variable, whereas the scope of the global `scope` variable is shaded gray.

Mike: I should probably have asked this before, but are there any limitations on where you can declare variables? You always declare them at the beginning of your code or a function body, although I think I saw them declared elsewhere as well.

Professor: You can in fact declare a variable anywhere you want. In any case the declaration will be hoisted to the top of the containing script or the function body for local variables. The variable initialization, however, is not hoisted.

Maria: That's the same as it is with a function expression, which is also not hoisted.

Professor: Absolutely.

Let me show you some examples. Consider the following code:

```
var scope = "global";
var f1 = function() {
  var tmp = scope;
  var scope = "local";
  return tmp;
};
```

It looks like the local `scope` variable has not yet been declared at the time when we assign it to the `tmp` variable, and that the global `scope` variable is visible instead. That is, however, not the case. The local `scope` variable declaration is hoisted to the top of the `f1()` function definition so that the variable is already declared at the time when the `var tmp = scope;` statement is executed. On the other hand, the *initialization* of the local `scope` variable takes place exactly where it is written—in the second line of the function body. That means that the local `scope` variable has the `undefined` value at the time we assign it to the `tmp` variable.

Maria: So the function will return `undefined`?

Professor: Precisely. In fact, the above code is equivalent to this one:

```
var scope = "global";
var f1 = function() {
  var scope;
  var tmp = scope;
  scope = "local";
  return tmp;
};
```

That's why I always put variable declarations at the beginning of their scope. That way, what is actually going on is more evident.

Let's examine this second example in the DevTools. You should add an `f1()` function invocation at the end of the code so the function will execute, and set a breakpoint at the first line of the function body. After that, you reload the page and expand the Scope Variables menu item at the right side of the DevTools window. This is what you should see.

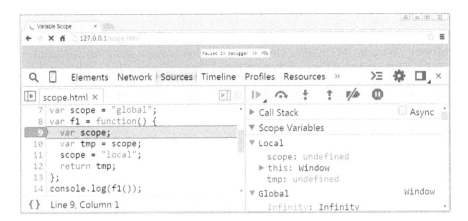

Under the Local item you'll find both local variables, scope and tmp of the current local scope. The current scope is that of f1() because that's where the debugger has paused. You can also see the this keyword, which you'll learn about next week. Below, you can see the Global section, under which there's quite a bit of stuff listed. There is, for example, the familiar Infinity global variable, and if you scroll further down, you'll find other recognizable items like Array, Date, Document, NaN, and so forth. These are all globally accessible variables and functions. They are, technically speaking, all members of a global object, which is called Window in client side JavaScript. That's why there's "Window" written at the right of the Global item. Don't worry about all that, though. The important thing is that you can also find our global scope variable on the list of global identifiers. As you can observe on the right side of the DevTools on the next screenshot, its value is "global", just like it should be.

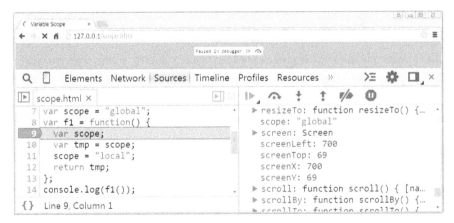

If you expand the Watch Expressions menu item and add the scope variable to the list of watch expressions, you'll see that the variable has the undefined value. That's because we're currently inside the local scope of f1() and Watch Expressions uses only variables that are accessible within the current scope.

10.5 Passing Function Arguments by Reference

Professor: We now take a little closer look at what happens when an argument is passed to a function by reference. Consider the following example:

```
var x = [3, 2, 5];
var myFunc = function(y) {
  y[0] = 10;
};
myFunc(x);
console.log(x);  //Writes [10, 2, 5]
```

Because x is an object, a reference to the object is copied to the function parameter y when we call the `myFunc()` function. Inside the function, we assign the value 10 to the first element of an array. Since y refers to the same global array as x does, the global array is changed and the last line of the code prints the changed array to the JavaScript Console.

Let me give you a slightly modified example:

```
var x = [3, 2, 5];
var myFunc = function(y) {
  y = 10;
};
myFunc(x);
console.log(x);
```

What do you think happens now?

Mike: I guess that the array is now overridden by the value 10, so the program writes out 10.

Professor: Does it really? The statement `y = 10;` indeed overrides the array. However, it does not override the array proper, it only changes one of the two references to the array. The result is that y simply does not refer to the array object any more. By changing the value of y we therefore do not change either x or the array object it refers to, and the program writes out the unchanged array `[3, 2, 5]`.

Maria: That is a little confusing. How can I tell the difference?

Professor: Let's use our analogy of boxes once more. The variable x is a box holding a piece of paper, on which the location of the array `[3, 2, 5]` is written. When we call the `myFunc()` function, we effectively copy that piece of paper and put it into another box named y. Now, in our second example, we simply replace that copy with the value 10, which obviously cannot change the original object. In the first example, however, we do not replace the contents of the box with another value. In the first example, the program is ordered to do something with the object to which the paper in the box refers. The situation is the same every time when you use an access operator on an object. In the following program, for example, the original array is truncated and then sorted:

```
var x = [3, 2, 5];
var myFunc = function(y) {
  y.length = 2;
  y.sort()
};
myFunc(x);
console.log(x);  //Writes [2, 3]
```

10.6 The Scope Chain and Closures

Professor: Closures are not the most trivial concept, but they are important in Java-Script, so it's good to be at least familiar with the basic idea.

Consider the following function definition:

```
var count = function() {
  var i = 0;
  return function() {
    return i++;
  };
};
```

The first unusual thing you will notice is that one function is defined within the other. If you haven't been exposed to programming languages like C before, then this may not even seem so strange to you, but anyway: If you want to understand what the above code is all about, you should first understand the *scope chain*.

You are already familiar with the basic scoping rules that JavaScript uses to search for the closest variable accessible inside the current scope. If the current scope is global, then it looks for a global variable with that name. If, however, the scope is local, then it first looks in the local scope and if there's no local variable by that name, it also looks in the global scope. The chain of the two objects—a function object and the global object—that define variables within the two mentioned scopes is called the *scope chain* of that variable.

In the above example, I created yet another scope when I defined a function within a function, and added that function at the end of the existing scope chain. When the variable i is referenced within the inner (anonymous) function body, the local scope of that function is first checked for the definition of i. Since the definition is not found, the search continues down the scope chain. The next scope searched is the scope of the count() function, where the definition of i is eventually found. The real fun, however, begins when we actually invoke the count() function. Like this, for example:

```
var countThis = count();
```

Mike: Doesn't this only assign a reference to the count() function to a variable countThis?

Professor: It does, but not a reference to the count() function. Note that the expression count() is not a reference but a function invocation because it has parentheses at the end. So, in fact, the *return value* of the count() function is assigned to the variable countThis, which is a reference to the inner anonymous function.

In JavaScript, not only a function definition but also a function invocation creates a new function object. The new function object holds the local variables for that specific invocation and is added to the existing scope chain, which represents the scope for that particular function invocation. The trick of the count() function is that it defines another function within its body and returns a reference to it. As long as you keep a reference to the returned anonymous function, the anonymous function object persists in memory and so must the count() function object because it is a part of the anonymous function's scope chain.

We can invoke the count() function once more, and again store a reference to the returned anonymous function, this time to a variable countThat:

```
var countThat = count();
```

By doing that, we create yet another scope chain and append it to the existing one. Note that the existing scope chain only contains the global object because count() is invoked within global scope. Exactly the same was true with the first invocation of the count() function, and we end up with two scope chains, both starting with the global object. Here's an illustration of the two scope chains of the two objects created by the two invocations of count().

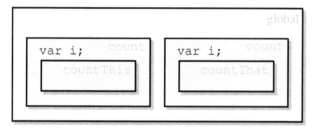

The two gray areas represent scopes of both i variables, and countThis and count That are references to the two anonymous function objects that we created when we called count() two times. What we can do now is call any of the two anonymous functions, which sit on the top of the two scope chains. A call of the left one will increment and return the left i and a call of the right one will increment and return the right i. For example:

```
countThis(); //The left i is now 1
countThat(); //The right i is now 1
countThis(); //The left i is now 2
countThis(); //The left i is now 3
console.log(countThis());//Writes 3 and increments the left i to 4:
console.log(countThat());//Writes 1 and increments the right i to 2:
```

Maria: Let me check something. It looks like `countThis` and `countThat` refer to one and the same function. In fact they are references to two distinct objects, both representing the same function definition. If they are really distinct, then the next expression should return `false`:

```
countThis === countThat   //Returns false
```

Professor: It does indeed return `false`.

Maria: Then you said that a function invocation also creates a function object. The next code should therefore perform the same:

```
var count = function() {
  var i = 0;
  return i++;
};
var countThis = count;
var countThat = count;
countThis();
countThat();
countThis();
countThis();
console.log(countThis());
console.log(countThat());
```

Professor: Not really. To begin with, the variables `countThis` and `countThat` now both refer to the same object, which is the `count()` function definition. It's true that when you invoke `count()` through either of the two references, you create a new object. But because you do not keep—and there's no way you can keep—a reference to that object, the object together with its local variable `count()` is deleted from the memory as soon as the function execution finishes. Each one of the above invocations creates a new object, which has no memory of the past invocations of `count()`, and therefore all invocations return zero.

In languages like C such behavior can be implemented using so-called static variables, but JavaScript does not know static variables, so you must use closures instead.

Mike: I guess that's useful whenever you need to store the state of the program after the last function call. But you can also use global variables for that, can't you?

Professor: Technically, that would work. But cluttering your program with too many global variables is a sure sign of bad style. The main reason for that is that global variables are accessible from everywhere and some buggy chunk of your code might accidentally change the value of an important variable. If that important variable is kept local and you provide a special function to control it, then unwanted changes to it are much less likely. The rule of thumb is that you try to avoid global variables unless there's a really good reason for them.

Maria: You mentioned closures. What exactly is a closure?

Professor: Oh, sorry, I forgot to give the formal definition. A *closure* is a function together with scope from which it can use variables. If you want, you can also check which variables are accessible inside the current closure in the DevTools:

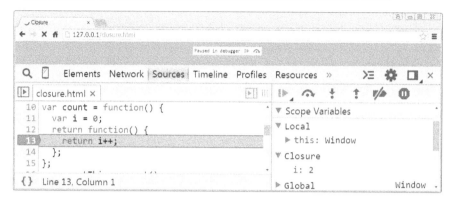

You should place a breakpoint inside the anonymous function and reload the page to see the situation like on this screenshot. Notice how the variable i is not a local variable of the anonymous function but is accessible inside its closure as you can see at the right side of the DevTools window. Incidentally, if you want i to get incremented to two, then you must step over several function calls by pressing F10 several times. However, if there was no breakpoint inside the function, then you wouldn't be able to enter the function using the Step Over command. To enter a function, you can use the Step Into debugger command or press F11.

10.7   Homework

Professor: For homework I want you to write and use a function that will help you to generate a math worksheet for kids to practice addition, subtraction, multiplication, and division. The function should accept two arguments: the desired operation and the maximum allowed value for the two operands. Let the function generate and return a random equation in the form of a string together with the corresponding solution. You can pack both in a two-element array. Additionally, take care that the result of a subtraction is never negative and that the result of a division is always an integer with no remainder.

The last thing for today is, of course, a list of keywords:

> **In this meeting**: function, call/invoke, argument, parameter, function definition statement, function expression, `return`, function body, function name, hoisting, optional parameters, `arguments` object, reference to function, passing arguments by reference, passing function as argument, `sort()`, variable scope, local variable, function scope, global variable, scope chain, closure

Building Your Own Objects

11.1   Homework Discussion

Professor: Are we getting a little arithmetic practice today?

Mike: I don't know. Our generator still needs some cosmetic touch-ups but I think it's working.

Professor: Don't worry about that. Please, go on, show me what you've got.

Mike: This is our function that produces a single equation:

```
/*
 * Returns a randomly generated equation together with the solution.
 * Parameters:
 *   oper: the desired operator in form of a string ("+", "-",
 *         "*", or "/")
 *   maxVal: the maximum allowed value for the two operands
 * Returns:
 *   An array with two elements:
 *       -The generated equation as a string
 *       -The solution as a number
 * Example:
 *   generateEquation("+", 10, 10) may return the array
 *   ["8+2=_____", 10]
 */
var generateEquation = function(oper, maxVal) {
  var output = [];
  var firstOper = Math.round(Math.random() * maxVal);
  var secondOper = Math.round(Math.random() * maxVal);
  var solution, tmp;

  switch (oper) {
    case "+":
      solution = firstOper + secondOper;
      break;
```

```
case "-":
  if (firstOper < secondOper) {
  //The difference shouldn't be negative, so swap the operands.
    tmp = firstOper;
    firstOper = secondOper;
    secondOper = tmp;
  }
  solution = firstOper - secondOper;
  break;
case "*":
  solution = firstOper * secondOper;
  break;
case "/":
  if (secondOper == 0) {
  //Shouldn't divide by zero.
    secondOper++;
  }
  //Uses the product in place of the dividend, so the result is
  //always an integer with no remainder.
  tmp = firstOper * secondOper;
  solution = firstOper;
  firstOper = tmp;
}
output[0] = firstOper + oper;
output[0] += secondOper + "=_____";
output[1] = solution;
return output;
};
```

Professor: Very nice indeed! I like how you commented the function.

Maria: We found some recommendations about how to write comments for functions on the Internet. We discovered that it is extremely useful to have our functions commented in this way, so we don't have to wonder about how they work every time we want to use them.

Professor: I don't think the code needs any further explanation.

Mike: In the end we tested the function in the following manner:

```
var i;
for (i = 0; i < 20; i++) {
  document.write("<div>" + generateEquation("-", 10)[0] + "</div>");
}
```

Because the function returns an array, we used an array access operator directly on the function call in order to extract equations and leave out the solutions. Here is how the resulting worksheet looks.

```
8-6=_____
3-0=_____
7-2=_____
8-7=_____
```

Professor: Magnificent! I'm now even more excited about next week, when we're going to make your math worksheet generator interactive.[1]

11.2 JavaScript Objects Revisited

Professor: You already know plenty about objects in JavaScript. You know that an object is a composite value in the sense that it combines multiple values. It can incorporate either primitives or other objects, which act like its properties or members. If a property is a function object, which you can invoke to do something, then such a property is often called a method. You access an object's properties by their names using the property access operator in the form of either a dot or a pair of square brackets. It is important that objects with the exception of strings are mutable. You also know that JavaScript objects are manipulated by reference rather than by value, and that you create an object using the `new` operator in combination with an appropriate constructor.

That said, you can also create an object by means of an *object literal*, which is simply a comma-separated list of name and value pairs enclosed within a pair of curly brackets. A property name may either be a JavaScript identifier or a string literal, and must be separated from its value by a colon (:). A value can be any JavaScript expression, whose value—either primitive or object—becomes the value of the property. Here's an example:

```
//An object with two properties representing a complex number
//1 + 3i:
var complex = {re: 1, im: 3};
//The conjugate of the above number:
var conjugate = {re: complex.re, im: -complex.im};
```

I created two objects, an object `complex` with the real part set to one and the imaginary part set to three, and its conjugate. I used more complicated expressions for property values to construct the conjugate. You recall complex numbers from high school algebra, don't you?

Maria: Vaguely, but yes. Is this the square root of minus one?

Professor: The square root of minus one is the imaginary unit i. A complex number is composed of the real and imaginary parts and can be written as $a + bi$, where a is the real part and b is the imaginary part. You get the conjugate of a complex number simply by multiplying its imaginary part by minus one.

---

[1]The math worksheet example continues on page 249.

You can now access the properties `re` and `im` of the two above objects as follows:

```
complex.re    //Returns 1
complex.im    //Returns 3
conjugate.re  //Returns 1
conjugate.im  //Returns -3
```

An object can also be seen as a mapping of strings, or *keys* to values, which is sometimes called a "hash table," "associative array," or "dictionary." You can in fact construct a dictionary quite easily. Here, for example, is a tiny fraction of an English–German dictionary:

```
var englGerm = {
  "programming language": "Programmiersprache, die",
  "summer": "Sommer, der"
};
```

Because the first of the dictionary entries is not a valid JavaScript identifier, the only way it can be accessed is using square brackets:

```
//Returns a string "Programmiersprache, die"
englGerm["programming language"]
```

If, however, a string satisfies the rules that apply to JavaScript identifiers, then either a dot or a pair of square brackets can be used. That is true regardless of the fact that the name was originally specified in string form rather than as an identifier:

```
englGerm["summer"] //Returns "Sommer, der"
englGerm.summer    //Returns "Sommer, der"
```

Remember, you have already seen something like this in association with accessing the properties of the `Math` object on page 153.

Note that JavaScript objects are *dynamic*, which means that you can add properties to objects on the fly later in the program. For example, you can easily add a new entry to the `englGerm` dictionary like this:

```
var english = prompt("Enter an English word.");
var german = prompt("Enter the German translation.");
englGerm[english] = german; //Adds and sets a new property
```

Since we're going to get more and more involved with objects, let me point out that JavaScript basically knows three types of objects:

- First, there are *native* objects, which are those defined by the ECMAScript specification. For example, arrays, functions, and dates are all native objects.

- Second, *user-defined* objects are all objects defined by the programmer and created during the execution of JavaScript code.

- The last group are *host* objects, which are defined by the host environment in which the JavaScript interpreter is running. In our case, host objects will be the browser objects because we run our programs within a browser.

11.3   Classes

Professor: The objects `complex`, `conjugate`, and `englGerm`, which we just created, were all unique, each having its own set of properties. It is, however, often useful to define not just properties but also a *behavior* of an object. For complex numbers, for instance, you could define arithmetic operations. Note that it wouldn't make much sense to define arithmetic operations for each complex object separately as we did for the `im` and `re` property values, which must be unique for each object. In such cases it is more useful to define a *class* of objects. You may then create *members* or *instances* of the defined class, which have their own unique property values that specify their individual state, and methods that define their behavior and are shared by all the members of the class.

The following picture shows an imaginary class named Dog and an object instance of this class named myDog.

Class

Properties:	Methods:	Constructor:
name	sit()	Dog()
length	come()	
sitting	bark()	
barking	quiet()	

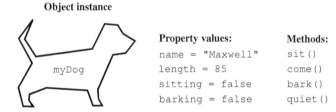

Object instance

Property values:	Methods:
name = "Maxwell"	sit()
length = 85	come()
sitting = false	bark()
barking = false	quiet()

You can see how the object instance has the same methods as the class object, while it specifies its own values for the properties. The class also defines a constructor, which is used for instantiating member objects.

To a certain extent, you are already familiar with these concepts. `Date`, for example, is a class, and you create a `Date` instance using the `Date()` constructor, like this:

```
var today = new Date(); //today is an instance of the Date class
```

Maria: I remember that you once said a class was a blueprint for creating objects.

Professor: That's exactly what it is. I will now show you how to define your own class.

Mike: But no complex numbers, please.

Professor: OK, let me think of something really simple. Are you familiar with elementary algebraic operations on two-dimensional vectors?

Maria: Can you please just briefly review the basics?

Professor: OK, let's take vector addition and dot product. Assume you have two vectors, $\mathbf{v}_1 = (x_1, y_1)$ and $\mathbf{v}_2 = (x_2, y_2)$. You compute the sum of two vectors by simply adding the corresponding components, so that the x component of the resulting vector is the sum of the x components of both vectors, and the y component of the resulting vector is the sum of the y components of both vectors:

$$\mathbf{v}_1 + \mathbf{v}_2 = (x_1 + x_2, y_1 + y_2)$$

The dot product is the sum of the products of the corresponding components of each vector:

$$\mathbf{v}_1 \cdot \mathbf{v}_2 = (x_1 x_2 + y_1 y_2)$$

Notice that the sum of two vectors is again a vector while the dot product of two vectors is a single number.

We are going to define a class `Vector`, for which we first need a constructor. Whenever you invoke a constructor using `new`, an object instance is automatically created. You don't need to worry about creating one by yourself. A constructor only needs to set the initial state of the created object instance by setting values of its properties. You can set those values using the `this` keyword, which acts as a *reference* to the object instance just being created. Here's how a definition of a `Vector` constructor looks:

```
var Vector = function(x, y) {
//Create and initialize the properties x and y of the created
//instance of class Vector.
  this.x = x;
  this.y = y;
};
```

Notice that I capitalized the name of the constructor. Because constructors define classes, it is a common convention that they have names that begin with capital letters. In contrast, regular functions and variable names begin with lower-case letters.

When the body of a constructor starts to execute, a new object instance has already been created and `this` is a reference to that object instance. With the two assignment statements inside the `Vector()` constructor, we created the properties x and y and

---

gave them the values of the arguments passed to the constructor. Note that x on the left-hand side of the assignment operator is not the same as the one on its right-hand side. The first is a property of the newly created object instance while the second is a declared parameter of the constructor. The same holds for y, of course.

With a constructor like that, it is possible to create some vectors:

```
var v1 = new Vector(0, 2);
var v2 = new Vector(1, 0);
console.log(v1.x); //Writes 0
console.log(v1.y); //Writes 2
console.log(v2.x); //Writes 1
console.log(v2.y); //Writes 0
```

Maria: If you called the constructor without arguments, would the properties x and y be created just the same?

Professor: They would. Recall that the function parameters for which you don't provide arguments get the undefined values. So both properties would be created and initialized to the undefined values.

Note that there's also the arguments object working behind the scenes inside a constructor just as it is inside any function body. That allows you to write a constructor that initializes an object in more than a single way depending on a number of passed arguments.

11.4 Constructor Overloading

Professor: In many object-oriented languages, there's a special mechanism involved when you pass a different number of arguments to functions, called *function overloading*, or *constructor overloading* if you are overloading constructors. In a nutshell, this means that you can define two or more different constructors by the same name and they will be discerned from each other depending exclusively on the number of arguments passed to them. JavaScript does not support overloading but you can still implement it with the help of either the arguments object or the fact that parameters that do not get argument values are set to the undefined value. For example, you can write a constructor for our vector class that will behave differently when you provide no arguments:

```
var Vector = function(x, y) {
  if (arguments.length == 0) { //If no arguments are passed, then
    this.x = 0;                //set both components to zero.
    this.y = 0;
  }
  else { //Otherwise, copy the values of the passed arguments.
    this.x = x;
    this.y = y;
  }
};
```

You can now construct a `Vector` object either by explicitly providing the initial values for both components or omitting initial values altogether:

```
var vec1 = new Vector(4, -15); //Sets x to four and y to -15
var vec2 = new Vector();       //Sets x and y to zero
```

11.5   Factory Methods

Professor: Constructor overloading may be an elegant solution for many needs but there is a problem when you want to write two different constructors that both happen to have the same number of necessary arguments. Consider, for example, that you want to construct a vector using polar instead of Cartesian coordinates as arguments. Namely, instead of giving x- and y-coordinates denoting an end point of a vector in a Cartesian coordinate system, you can define the same end point in a polar coordinate system by specifying its distance (d) from the origin and the angle (α) made between the vector and the positive x-axis. The polar coordinates d and α can be converted to the Cartesian coordinates x and y by means of the trigonometric functions sine and cosine. The next drawing shows the relation between both coordinate systems.

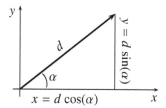

Note that a vector in the polar coordinates is also specified by two values just as it is in the Cartesian coordinates and it is impossible to write a single constructor that will know which of the two coordinate systems you've had in mind when you pass two arguments. JavaScript allows you to have two constructors with different names but that is not a recommended technique. An entirely better approach is to use a *factory method*.

A factory method is an ordinary class method, only that it initiates and returns an instance object of the class. For example, the next definition makes `polar()` a factory method of the `Vector` class that creates a `Vector` object instance initialized with polar coordinates:

```
Vector.polar = function(d, alpha) {
//Converts degrees to radians.
   var rads = Math.PI * alpha / 180;
//Converts polar coordinates to Cartesian and calls the original
//constructor.
   return new Vector(d * Math.cos(rads), d * Math.sin(rads));
};
```

Recall that trigonometric functions of the `Math` object use radians rather than degrees. If we want our factory function to accept degrees, then the conversion performed in the first line of the function body is necessary. The last line of the method converts polar coordinates to Cartesian and passes the latter to the original `Vector()` constructor. The object created by the constructor is then returned from the `polar()` factory method. Note that because we're just adding a new property to the existing `Vector` class, we don't use the `var` keyword before the `polar()` method definition.

Maria: How on Earth can we invoke this factory method? We're only just creating a vector object instance with it and there exists as yet no object on which to invoke the method. Or must we construct one in order to be able to use the above factory method?

Professor: I see what you mean. Recall that everything you see in JavaScript is an object. When we defined the `Vector()` constructor, an object was created for that definition and a reference to the object was assigned to `Vector`. Then we appended the `polar()` method to the existing `Vector` class object. As a consequence, we can invoke `polar()` on the `Vector` class just as if the class was an object instance:

```
var v3 = Vector.polar(2, 90);
console.log(v3.x); //Writes 1.2246467991473532e-16
console.log(v3.y); //Writes 2
```

Mike: Is the above `polar()` method like the static methods `now()` and `UTC()` of the `Date` class?

Professor: Exactly. You can call a static method on a class object but there are limitations on what such a method can do. Note that there's no object connected with a static method other than the class definition object. What's more, `this` inside a static method refers to the static method itself and is therefore rarely of any use. That means that a static method cannot directly access any properties of object instances of the class that it belongs to.

Mike: Wait a minute! How then can a constructor store property values?

Professor: A constructor is different because it actually creates an object instance, something that a factory method cannot do. Notice that a factory method still needs to call a constructor in order to create an object instance. Furthermore, you shall not confuse a constructor definition object with an object instance that is created as a result of invoking the constructor. The `this` keyword used inside a constructor refers to the object instance that is just being created by the constructor. That means that property values are stored in the newly created object instance and not in the class object itself.

Maria: By the way, I think that the x component in the above example should be exactly zero and not just a very small number. Is that due to the lack of precision?

Professor: That's true. `Math.PI` is, as are all numbers in JavaScript, limited to the precision of 15 significant digits. Because of that, an angle of 90 degrees is not converted exactly to $\pi/2$, which means that the cosine of the angle isn't precisely zero as well. In cases like ours such errors are not difficult to control and usually a simple rounding[2] does the trick. Sometimes, however, limited precision can become a real nuisance and it takes quite a bit of knowledge and experience to be able to handle it.

11.6 The prototype Object

Professor: As you may have guessed, the polar() method that we added directly to the Vector class object is only accessible through the Vector class object itself. Instances of Vector do not have this method. They also do not need it because it doesn't do anything to individual object instances to begin with. The story is, however, quite different if we want to add methods that perform some operations on vector instances like computing the dot product of two vectors, for example. Functions like that should be able to access the properties of individual object instances and should therefore be defined on object instances rather that on the class object itself. That's where the prototype object comes into play.

The prototype object is the central element of a class because all the methods that you add to this object will be *inherited* by all the object instances created with the constructor of the class. Inheritance is crucial to object programming languages because it allows you to reuse your code very simply many times.

Let me shed a little light on this by defining the addition and dot product operations for our class of vectors. All we have to do is assign the function definitions for both operations to corresponding properties, which we make the properties of the prototype property of the Vector class:

```
Vector.prototype.add = function(vec) {
  return new Vector(this.x + vec.x, this.y + vec.y);
};

Vector.prototype.dot = function(vec) {
  return this.x * vec.x + this.y * vec.y;
};
```

Both methods are now inherited by instances of the Vector class, which means that you can invoke them on every object instance created by the Vector() constructor. You already know that such methods are called instance methods. Unlike static methods, however, instance methods cannot be invoked on the Vector class itself. Of course, it wouldn't make much sense to invoke them on the Vector class object.

Each of the methods add() and dot() operates on two vectors. The first vector is the one on which a method is invoked, while the second one is passed as a method's argument. Note that this inside a method refers to the particular object instance on which a method is invoked. Because we do not want to change the original vectors involved in operations, the add() method creates and returns a new vector, which holds the result of vector addition.

You can use the above methods add() and dot() like this, for example:

---

[2]See the example of rounding to two decimal places on page 182.

```
var v4 = new Vector(2, 7);
var v5 = new Vector(-3, 6);
var dot = v4.dot(v5); //Computes the dot product of v4 and v5
v4 = v4.add(v5);       //Computes the sum of v4 and v5
console.log(v4.x);     //Writes -1
console.log(v4.y);     //Writes 13
console.log(dot);      //Writes 36
```

Mike: It occurred to me that if x and y are properties of Vector instances, they should be in scope inside the methods of the Vector class, shouldn't they? What do we need this for?

Professor: The truth is that only *function* objects create scope for their local variables. Properties of an object are of course not the same as local variables. For one thing, you first need access to an object in order to be able to access its properties. This is true even when you want to access the object's own properties from within its methods.

So, we do not speak of scope in connection with object properties, but rather of an *invocation context*. Although it is not the same as scope, an invocation context nevertheless serves a similar purpose—it defines an environment from which you can access properties.

An invocation context is simply the object instance that invoked the method in question and to which the this keyword inside the method refers. It can be seen as an implicitly passed additional argument of a method. Suppose, for instance, that we add another method to our Vector class, which sets the x and y components of a vector, and name it set(). We can invoke the method on the above defined vector v4 like this:

```
v4.set(3, 7);
```

Alternatively, we can write an ordinary function that does exactly the same thing. In order for the function to be able to access the vector v4, we must pass the vector as an additional argument of the function:

```
set(v4, 3, 7);
```

Can you perhaps write definitions for both, the set() method and function?

Maria: I think so. Here they are:

```
//The set() method
Vector.prototype.set = function(x, y) {
  this.x = x;
  this.y = y;
};
```

```
//The set() function
var set = function(vec, x, y) {
  vec.x = x;
  vec.y = y;
};
```

Professor: Very well. You can see clearly from your code that the `this` keyword behaves precisely as if it were a regular parameter of the `set()` method. In the case of the `set()` function, this role is taken by the actual `vec` parameter. In both cases we're dealing with a reference to the v4 object instance, only that in the first case v4 represents the invocation context accessible through `this`, while in the second case it is passed as a function argument accessible through the parameter `vec`.

Mike: Is it also possible to add methods to built-in JavaScript classes?

Professor: Of course it is possible. As an example, we can add a method to the `Array` class that returns the maximum value stored in an array or `-Infinity` if the array is empty. If any of the array elements cannot be converted to a number, then let the method return NaN. I think you can do it on your own if you deploy some of your little gray cells. Just imagine that the complete `Array` class is your own work and you only need to add one more method.

Mike: Are you saying that it doesn't matter whether you add methods to a user-defined or built-in class? Yes, I think I can do it. Something like this, perhaps:

```
Array.prototype.max = function() {
  var i, max, n;
  max = -Infinity; //No one element can be smaller than -Infinity.
  for (i = 0; i < this.length; i++) {
    n = Number(this[i]);
    if (n != n) {
    //If the element cannot be converted to a number, return NaN.
      return NaN;
    }
    if (n > max) {
    //If the element is larger than the largest so far, then store
    //it as a new maximum value.
      max = n;
    }
  }
  return max;
};
```

Professor: That's it. Of course, you were able to use an array access operator on `this` because `this` refers to an array object.

Maria: Isn't there an error in Mike's code? A variable `max` has the same name as the method.

Professor: This isn't a problem in this particular case. Because the max() method is being defined as an instance method of the Array class, the method is never accessible without an object reference. Hence, if you want to access the max() method from within the same method, you should use the expression this.max().

Maria: I see. Can we test our max() method to see how it works?

Professor: No problem. Here are some examples:

```
var a1 = [7, 3, 8, -1, 99, 5];
var a2 = [];
var a3 = [5, "w", 66, 2];
var a4 = [-3, -12, ""];
console.log(a1.max()); //Writes 99
console.log(a2.max()); //Writes -Infinity
console.log(a3.max()); //Writes NaN
console.log(a4.max()); //Writes 0
```

Before we continue, I should mention that it is not always possible to add methods in this way to classes implemented by different host environments, like browsers, for example. These are classes that are not defined by the core JavaScript language and are characterized as part of client-side JavaScript in the JavaScript reference at the end of the book.

11.7 More on Setting and Querying Object Properties

Professor: One of the key concepts in the object-oriented programming paradigm is *encapsulation*, which, formally, bears more than a single meaning. One definition of encapsulation is that it is a mechanism for *information hiding*. In practice this means that the internal representation of an object is separated and hidden from a programmer who uses the object. Take, for example, our Vector class. The class is internally represented by the Cartesian coordinates x and y, but that need not always be the case. You could just as well decide to store vectors in polar form. In order to hide that detail from the programmer, you can provide special methods that allow the programmer to access the object data indirectly. If the object's internal data representation ever changes, then you simply reprogram the access methods to perform necessary data transformations so that they can be used in exactly the same way as they had been before the change.

For example, if you want to write the value of the x component of the v4 vector from the example on page 214 to the JavaScript Console, you currently have no other option but to access the component directly:

```
console.log(v4.x);
```

That is not such a good idea, though, especially when direct access like this one is sprinkled all over the code. Consider, for example, that you write, or even get from another author the new version of the Vector class, whose internal representation of vectors has been changed from Cartesian to polar. There's a lot of work awaiting you in bringing the whole program up to date with the change.

Instead of directly accessing properties, you can write methods to *set* and *query* the object properties. By the way, JavaScript has a special mechanism that allows you to write so-called *getters* and *setters* to set and query the object properties. However, I personally don't like to use it because it is not immediately evident from the property access expression that in fact a method is called behind the scenes.

Anyway, you can still define your own methods to set and query the x property of Vector. This is an example:

```
Vector.prototype.setX = function(x) {
  this.x = x;
};

Vector.prototype.getX = function() {
  return this.x;
};
```

The x property of the v4 vector can now be manipulated like this:

```
v4.setX(42);
console.log(v4.getX()); //Writes 42
```

Now, if the internal representation of the Vector class changes, you only need to rewrite the getX() and setX() methods, while all the code using these methods stays perfectly OK. For example, this is how the two methods might look if the internal Vector representation should change to polar:

```
Vector.prototype.getX = function() {
  return this.d * Math.cos(this.alpha);
};

Vector.prototype.setX = function(x) {
  var y = this.getY();
  this.d = Math.sqrt(x * x + y * y);
  this.alpha = Math.atan2(y, x);
};
```

Notice that in order to be able to compute the values of both polar coordinates (d and alpha) from the given *x*-coordinate, you also need the getY() method. So here are the getY() and setY() methods as well:

```
Vector.prototype.getY = function() {
  return this.d * Math.sin(this.alpha);
};
```

```
Vector.prototype.setY = function(y) {
  var x = this.getX();
  this.d = Math.sqrt(x * x + y * y);
  this.alpha = Math.atan2(y, x);
};
```

You can use these modified methods in exactly the same way as the original ones, and the two lines of code from the above example still work just the same:

```
v4.setX(42);
console.log(v4.getX()); //Writes 42
```

Note that what I have just showed you is not the only advantage of having special methods for setting and querying the object's properties. Using methods instead of directly setting the values of properties also enables you to check the integrity of the values before setting them. Imagine that you want to have a class representing the natural numbers and prevent the programmer from erroneously using a negative or non-integer numbers with that class. You can easily add this kind of checking to a method that the programmer will use for setting values of members of the natural number class.

I think we can now return to our Sudoku puzzle helper.

11.8 Sudoku Puzzle Helper

Professor: You already did some work on the Sudoku helper when you implemented a checking of whether a number can be placed on the playing grid. Our final goal, however, is more ambitious than that. We are going to design a web page that will assist the visitor in solving a Sudoku puzzle. Our product will consist of a grid of edit boxes, into which the visitor will enter the missing numbers. Each time the entered number conflicts with the numbers already on the puzzle board, the number will be colored red. Apart from that, our Sudoku helper will allow the visitor to enter more than a single number into a cell, thus creating a list of solution candidates for that cell. We will display solution candidates smaller and will not check whether or not they are placed wrongly.

Note that our program won't attempt to actually *solve* a puzzle. It will only check whether the most recently placed number is in conflict with the numbers already on the board, regardless of whether those numbers have been placed correctly or not.

Let's now begin programming our puzzle. Because you already know enough about objects, we can opt for *object-oriented design*. Basically, we need to identify different objects that will help us to construct the program and then define the interactions of these objects. Can you perhaps think of any part of a Sudoku puzzle that we can implement as an independent object?

Maria: A cell could be an object.

Professor: Definitely. So let's start out with a cell.

However, before we begin to program a cell object, we should be able to answer two very important questions about it:

- What information should the object store?
- How should the object interact with its environment? By environment I mean the rest of the program.

Mike: Those are properties and methods, aren't they?

Professor: Precisely. Information about the object is stored in its properties, while methods enable the object to respond to actions from its environment. You can perform a simple test to see whether you've picked the right candidate to serve as an object in your application. If you cannot think of any obvious piece of information that the object could store, or you cannot think of any obvious action that an object could respond to, then you probably haven't picked the right candidate.

What kind of information do you think our cell should store?

Maria: Obviously a number. And perhaps a Boolean property specifying whether or not the number is permitted to change. Namely, the original numbers that define a particular Sudoku puzzle—I think they are called *givens* or `clues`—should not be changed.

Professor: This is no doubt essential information. Let's write it down in a list:

- `number`: a number that is written in the cell or `null` if the cell is empty.
- `writable`: a Boolean property controlling the permission to change the number.

Apart from these, a cell will of course have methods to set and query its properties, except the property `writable` won't need a special set method as there's no need for this property ever to change. Its value will be set in the cell's constructor. Note that names of methods used for querying Boolean properties sometimes start with "is" rather than "get" in order to suggest that the property is of the Boolean type. Namely, if the method is named `isWritable()`, then the name can be interpreted as a question whether or not the object is writable.

I guess that everything is quite straightforward and we can produce code for the `Cell` class right away:

```
var Cell = function(num) {                    //The Cell() constructor
  this.number = num;
  if (num == null) {
    this.writable = true;
  }
  else {              //If the initial value is not null, then this
                      //is a given and is not allowed to change.
    this.writable = false;
  }
};
```

```
Cell.prototype.setNumber = function(num) {
  if (this.writable) {
  //Sets the value only if this cell is writable.
    this.number = num;
  }
};

Cell.prototype.getNumber = function() {
  return this.number;
};

Cell.prototype.isWritable = function() {
  return this.writable;
};
```

The Cell() constructor sets both properties of Cell to their initial values. Recall that you once used the null value to denote an empty cell in your homework and we will stick to this convention throughout the project. All the values that are initially not null are locked for writing in the constructor by setting their writable property to false. These cells represent clues for a puzzle. Notice how I used the property writable inside the setNumber() method to prevent writing a value to a cell that is not writable.

The Cell class is thus completed and we're ready to move on. The nature of Sudoku is such that cells do not interact with each other directly, so it comes naturally to build an object named Sudoku and make cells properties of that object. The Sudoku object will be responsible for ordering the cells and controlling their behavior.

In Sudoku, there are 91 cells altogether, arranged in nine rows and nine columns, which can be represented in a matrix form as follows:

$$
\begin{pmatrix}
C_{0,0} & C_{0,1} & C_{0,2} & C_{0,3} & C_{0,4} & C_{0,5} & C_{0,6} & C_{0,7} & C_{0,8} \\
C_{1,0} & C_{1,1} & C_{1,2} & C_{1,3} & C_{1,4} & C_{1,5} & C_{1,6} & C_{1,7} & C_{1,8} \\
C_{2,0} & C_{2,1} & C_{2,2} & C_{2,3} & C_{2,4} & C_{2,5} & C_{2,6} & C_{2,7} & C_{2,8} \\
C_{3,0} & C_{3,1} & C_{3,2} & C_{3,3} & C_{3,4} & C_{3,5} & C_{3,6} & C_{3,7} & C_{3,8} \\
C_{4,0} & C_{4,1} & C_{4,2} & C_{4,3} & C_{4,4} & C_{4,5} & C_{4,6} & C_{4,7} & C_{4,8} \\
C_{5,0} & C_{5,1} & C_{5,2} & C_{5,3} & C_{5,4} & C_{5,5} & C_{5,6} & C_{5,7} & C_{5,8} \\
C_{6,0} & C_{6,1} & C_{6,2} & C_{6,3} & C_{6,4} & C_{6,5} & C_{6,6} & C_{6,7} & C_{6,8} \\
C_{7,0} & C_{7,1} & C_{7,2} & C_{7,3} & C_{7,4} & C_{7,5} & C_{7,6} & C_{7,7} & C_{7,8} \\
C_{8,0} & C_{8,1} & C_{8,2} & C_{8,3} & C_{8,4} & C_{8,5} & C_{8,6} & C_{8,7} & C_{8,8}
\end{pmatrix}
$$

Apart from rows and columns, there are also nine 3×3 boxes, delimited by the two horizontal and two vertical lines in the matrix. Notice that each cell $C_{i,j}$ has two indexes that stand for the row (i) and column (j) in which the cell is placed. It is very useful that we have a scheme like that so we know exactly which cells we are referring to.

There's not much work for the Sudoku() constructor. It only initializes the board property of the Sudoku class:

```
var Sudoku = function() {
  this.board = [];
};
```

The board property represents the two-dimensional grid of Cell objects and is for now set to an empty array.

Next, we need to write a method to initialize the playing board. The following method calls the Cell() constructor for each of the 91 cells and fills the board property of the puzzle with them:

```
Sudoku.prototype.setClues = function(initial) {
  for (i = 0; i < 9; i++) {
    this.board[i] = [];
    for (j = 0; j < 9; j++) {
      this.board[i][j] = new Cell(initial[i][j]);
    }
  }
};
```

The parameter initial is expected to be a two-dimensional array of primitive values like the one on page 185, which you used for your homework. Because the problem is two-dimensional, there are two nested for loops inside the method definition. The outer loop iterates rows, and for each row it first creates an empty array representing that row. With the help of the inner for loop it then creates and inserts nine cells into that row. Notice how a corresponding value is passed as an argument to the Cell() constructor.

Maria: I'm not sure if I quite follow you. Can you please explain how you created the two-dimensional array board?

Professor: No problem. In the constructor I already created an empty array object:

```
this.board = [];
```

At the beginning of the outer for loop in the setClues() method there's this statement:

```
this.board[i] = [];
```

The statement creates another empty array object, a reference to which is assigned to the ith element of the board array. Elements of board are therefore also arrays. By the way, because the ith element does not yet exist at the time of the statement execution, it is automatically appended to the board array.

It is clear now that the expression board[i] is an array representing the *i*th row of the board. Inside the inner for loop the *j*th element of this array is assigned a reference to a cell object created at the right side of the assignment operator. That leads us to the fact that the expression board[i][j] is from now on a reference to the cell object representing the cell $C_{i,j}$.

For clarity, here is a drawing of the board with its nine rows and all the cells they refer to.

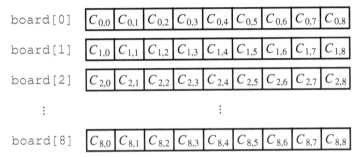

The last of the Sudoku methods that we're going to write today will take care of entering numbers into the cells, and we will name it setNumber(). As it happens, you already did most of the job of checking whether a number can be placed into the grid. We will only need to modify your code a bit to suit our current needs.

The setNumber() method will take as arguments the number that we want to enter and a reference to the cell that should accept the number.

Mike: We already implemented the same method on the Cell class. Why do we need another method that does exactly the same thing?

Professor: We need it because the setNumber() method of the Sudoku object will do a lot more than just enter a number to a cell. Since the Sudoku object has an overview of the complete grid of cells, it can also check whether the entered number already exists in a particular row, column, or 3 × 3 box.

Here is the implementation of the method:

```
Sudoku.prototype.setNumber = function(number, row, col) {
  var i, j;
  var iStart, jStart;

  //Removes the current number
  this.board[row][col].setNumber(null);

  //Checks a row
  for (i = 0; i < 9; i++) {
    if (this.board[row][i].getNumber() == number) {
      return false;
    }
  }
```

```
//Checks a column
for (i = 0; i < 9; i++) {
  if (this.board[i][col].getNumber() == number) {
    return false;
  }
}

//Checks a 3x3 box
iStart = row - row % 3;
jStart = col - col % 3;
for (i = iStart; i < iStart + 3; i++) {
  for (j = jStart; j < jStart + 3; j++) {
    if (this.board[i][j].getNumber() == number) {
      return false;
    }
  }
}

this.board[row][col].setNumber(number);
return true;
};
```

You will recognize pieces of code from your homework on page 187. I had to modify the part that checks the rows because we now don't have primitive values but rather object references stored in rows. That means we cannot use the indexOf() array method to search for a value. You'll also notice that I left out the auxiliary variable exists because it is not necessary. The method can simply return false if and as soon as it discovers that a number cannot be placed. If, however, the number can be placed, then all the for loops terminate normally, the number is stored into the cell, and the method returns true. It is important that we return true upon successfully entering a number because the code that calls this method sometimes needs to know whether or not the operation was successful.

Mike: The setNumber() method called from the second to the last line belongs to the Cell class, doesn't it?

Professor: Indeed it does.

Maria: What's the use of the setNumber(null) call placed at the beginning of the method?

Professor: It can easily happen that the player enters the same number into the same cell twice after changing his/her mind several times. Notice that our method checks all the nine cells of a certain section, including the one whose value we're just setting. Thus, we first have to remove the old number by setting the cell to null or else the entered number is also compared to the one it is going to replace anyhow. Had we not done this, entering the same number into the same cell twice would be recognized as an invalid action. Apart from that, this line deletes the old number from the cell even if the new number cannot be placed.

11.9 Homework

Professor: For homework, I want you to test all the Sudoku code that we wrote today. Try to identify as many different representative situations as you can and work out test examples for each one of them. Use your imagination and write additional methods for testing objects if you find it necessary to do so.

And, oh, yes, review HTML and CSS because we're going to use them next week.

Otherwise, that's it for this week.

> **In this meeting**: objects, properties, methods, constructor, object literal, hash table, associative array, dictionary, property access operator, dot, square brackets, native objects, user-defined objects, host objects, class, member, object instance, this, invocation context, constructor overloading, factory function, prototype object, inheritance, encapsulation, getters, setters

Using JavaScript to Control the Browser

12.1   Homework Discussion

Professor: What did you find out? Is our code operational?

Maria: We did some testing and it seems all right. At least we didn't find any errors.

Professor: How did you test the code?

Mike: First we wanted to have a method that would print the whole Sudoku puzzle so we could see what numbers are in it. Here's the code:

```
Sudoku.prototype.print = function() {
  var i, j;

  document.write("<table border='1'>");

  for (i = 0; i < 9; i++) {
    document.write("<tr>");
    for (j = 0; j < 9; j++) {
      document.write("<td>")
      if (this.board[i][j].getNumber() != null) {
        document.write(this.board[i][j].getNumber());
      }
      document.write("</td>")
    }
    document.write("</tr>");
  }

  document.write("</table>");
};
```

The code is nothing special. It simply writes out all the numbers that it finds inside the puzzle in form of an HTML table.

Professor: That's nice. What else did you do?

Maria: We couldn't think of any systematic approach that would perform an exhaustive test. We simply tried to insert some wrong numbers that violated only one of the three conditions. If a number had violated more than one condition, then we would only have been testing a condition that is violated first inside the setNumber() method of Sudoku. We did the same test against the initial clues as well as against a manually entered number. At the end we also tried to enter the same number in the same cell twice.

This is our code:

```
var initial = [
  [null, 5,    7,    null, 4,    null, null, 2,    null],
  [null, 3,    8,    null, null, 2,    7,    null, null],
  [null, 1,    null, 7,    3,    null, null, null, null],
  [null, 7,    null, 2,    8,    null, null, null, 3   ],
  [3,    null, 4,    null, null, null, 1,    null, 6   ],
  [1,    null, null, null, 6,    4,    null, 7,    null],
  [null, null, null, null, 2,    7,    null, 1,    null],
  [null, null, 1,    9,    null, null, 2,    3,    null],
  [null, 4,    null, null, 1,    null, 5,    9,    null]
];

var test = [];

var S = new Sudoku();
S.setClues(initial);

test.push(S.setNumber(2, 0, 0));
test.push(S.setNumber(8, 0, 0));
test.push(S.setNumber(9, 0, 3));
test.push(S.setNumber(6, 0, 0));
test.push(S.setNumber(6, 0, 5));
test.push(S.setNumber(6, 3, 0));
test.push(S.setNumber(6, 2, 2));
test.push(S.setNumber(6, 0, 0));

console.log(test);

S.print();
```

Mike: All except two of the calls to the setNumber() method return false, which was expected. At the end, the whole puzzle is written to the browser window and you can see that number six was indeed successfully inserted to the upper left cell of the puzzle.

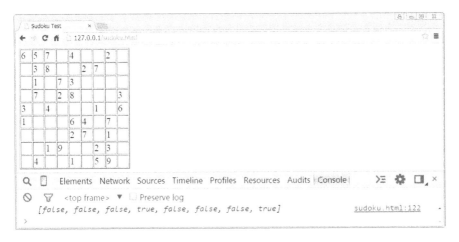

Professor: Good job. I guess you must be eager to make your Sudoku interactive, so there's no time to lose.

12.2 Deeper into the Browser

Professor: Our work so far focused primarily on the core JavaScript language, and even though we ran all our examples within a web browser, the code itself had little to do with the specific context of a web browser. Today, you are finally going to see how JavaScript comes alive within the browser and how you integrate it with HTML and CSS. Because that part of JavaScript is designed to run exclusively on the side of a client—a computer that runs the web browser—you'll often hear the term *client-side JavaScript* connected with it.

The very basic object of client-side JavaScript is the Window object, which represents a web browser's window and takes the role of the client-side JavaScript global object. You can refer to the Window object using the `window` identifier, although you rarely need to do that explicitly. Because the Window object is in fact the global object, all its properties and methods are eventually treated as global variables and functions, and that is actually all you need to know for now. For example, if you want your web browser to load a page from another URL, you simply assign the URL to the global `location` variable:

```
location = "http://www.checkthisout.com"
```

Or, if you want to display a message to the visitor inside a dialog box, you use the global `alert()` function:

```
alert("Let's rock and roll!");
```

You can do that because both `location` and `alert()` are members of the Window object. You can find a short selection of other properties and methods that the Window object defines in the reference on page 427. All in all, you can do lots of things with `window`, but I'll let you discover them on your own.

As soon as we become aware of the environment in which our programs run, it becomes of paramount importance to know where to put the code and how the code is executed. So far we put our examples into `<script>` elements embedded within HTML code, which is convenient for learning and quick testing, but discouraged in modern web design. Recall that HTML, CSS, and JavaScript each play their own individual role inside a web page by separately defining content, presentation, and behavior. Also recall how you separate CSS code from HTML by putting it to a separate file and linking both by means of the `<link>` HTML element. You can separate JavaScript from HTML in the same way and put it in an external file, which you connect with HTML using the `src` attribute of the `<script>` element. JavaScript detached this way is also called *unobtrusive* JavaScript, and for an obvious reason—it makes HTML code much more readable because there are no blocks of JavaScript code sitting in between the sections of HTML code. This is not the sole advantage of such a separation, though. For example, if multiple HTML files share the same JavaScript code, you can maintain only a single copy of the code, rather than having to edit each of the HTML files whenever you want to change it.

Mike: If you put JavaScript code into an external file and include it by means of a `<script>` element, then I suppose that everything else stays the same. I mean the position of the `<script>` element doesn't change, does it?

Professor: If you want to implement the concept properly, then `<script>` elements shouldn't be scattered all over HTML code either. This is a problem so long as you use `document.write()` to write to the browser window. Namely, even if JavaScript code is placed to an external file, it is still executed *synchronously*. This means that it is executed exactly when the `<script>` element referring to it is encountered during the document parsing. The parser will wait until the code has executed before proceeding with the parsing of the HTML that follows. If you use any `document.write()` methods, the text produced by those methods will be inserted into the current location in the input stream of HTML code that is being parsed. So you cannot move JavaScript code that contains any calls to `document.write()` without also changing the location of the output they produce.

Another problem with `document.write()` is that you cannot use it after the document parsing process has completed. Calling it after the document has closed will erase the document completely. As soon as you want to design a page that is responsive to user actions, this is not the kind of behavior that you would want. That's why `document.write()` is rarely, if at all, used in modern web pages.

Maria: I suppose that you will show us alternatives.

Professor: Indeed I will, we're coming to that shortly. But first things first: I need to tell you about events so we can start executing JavaScript code *asynchronously*.

12.3   Events

Professor: When the document loads completely and the web page is displayed, the story is usually not finished. On the contrary, it's only now that interesting things begin to happen. After the document has been parsed and loaded, the JavaScript enters its asynchronous mode, which means that it is executed as needed in response to user and other *events*. An event is any action that a computer program detects and *handles*. For example, an event may be triggered when a visitor clicks a mouse button or presses a key, or when a timer expires.

It is fairly easy for the programmer to handle an event because all objects that could detect and react to events have predefined *event-handler properties* attached to them for every possible event. All you have to do as a programmer is to define an *event-handler function* and assign it to the corresponding property. The event-handler function is then invoked automatically by the program whenever a corresponding event occurs. Event-handler properties can be easily recognized because they have names that begin with "on." For example, if you want to do something whenever the object named `clickMe` is clicked, then you should write code like this:

```
clickMe.onclick = function() { /* Do something awesome */ };
```

Maria: What's that `clickMe` object?

Professor: It is an imaginary object that could represent any HTML element on the page that your visitor can click. You'll soon learn how to capture actual HTML elements from inside JavaScript in order to manipulate them. What I wanted to show you right now is how to connect a function definition with the `onclick` event-handler property of an arbitrary object that can actually be clicked. Once you have written the above code, something awesome will be done each time someone clicks the element represented by the `clickMe` object. Because you don't know in advance when and how many times, if at all, `clickMe` will be clicked, you also don't know when, if at all, something awesome will be done. That's why we say that JavaScript events are handled asynchronously, in contrast to the timely, synchronous execution of code during the web-page parsing.

Because JavaScript actually invokes a function as a response to an event, you can also think of the JavaScript event-handling mechanism as simply calling the appropriate function every time an event occurs. For example, if you click three times on the HTML element represented by the `clickMe` object, the method `onclick()` is invoked three times as shown on this picture.

Event	Response
click	`clickMe.onclick()`
click	`clickMe.onclick()`
click	`clickMe.onclick()`

time

It's about time to look at a concrete example of handling an event. One of the most important event-handler properties is the `onload` handler of the Window object. It is triggered immediately after the document parsing and loading is complete and the document is visible within the browser window. You will frequently write JavaScript code right inside the `onload` handler.

Imagine that you have the following simple HTML:

```
<!DOCTYPE html>
<html lang="en">
  <head>
    <meta charset="utf-8">
    <title>External JavaScript</title>
    <script src="/scripts/tools.js"></script>
  </head>
  <body>
    <p>I'm plain HTML.</p>
  </body>
</html>
```

Notice that there's no JavaScript in this code but the document includes an external JavaScript file named *tools.js*. External JavaScript code is referenced using the `src` attribute of the `<script>` HTML element, whose value represents the URL of the desired file. Notice also that because `<script>` is not a void element it needs the closing tag even though it has no content.

Inside the *tools.js* file there is the following code:

```
window.onload = function() {
  alert("I handle the load event.");
};
```

Note that an external JavaScript file should contain only pure JavaScript code without any `<script>` tags, just like an external CSS file only contains CSS code. By convention, the name extension of a JavaScript file is *.js*.

As you can see, there's really not much to the above JavaScript code. The function definition, which only calls the global `alert()` function, is assigned to the `onload` event handler. Right after the HTML document has loaded completely, a load event fires and the function attached to the `onload` handler is invoked. This opens the dialog displaying the message "I handle the onload event." If you reload the document, you'll get the message again, of course. Be careful to put the *tools.js* file into the right folder so that the `src` attribute of the `<script>` element finds it.

Maria: And that's inside the *scripts* sub-folder of our root folder, right?

Professor: Precisely.

Maria: Why did you use the `window` identifier? You said that it is not needed, and that its properties can be used as though they were global variables, didn't you?

Professor: That's true. The above example would by all means work just as well without explicitly referring to `window`. It is, however, the usual convention to use `window` with the `onload` handler to make it more obvious that the handler is for the Window and not some other object.

Mike: Are there any rules such as where to put a `<script>` element referring to JavaScript code that will execute asynchronously?

Professor: There is a general agreement that you place it inside the head of the document. You'll also come across the advice that `<script>` elements should be placed after any `<link>` elements that refer to external CSS. That is, however, irrelevant for JavaScript that is executed asynchronously because the style information will already have been available at the time of JavaScript code execution no matter where the actual code is included. That's because styles are needed if the page is to be rendered correctly.

Mike: Why is that important? I mean what has JavaScript to do with CSS?

Professor: You will soon learn that you can manipulate styles dynamically using JavaScript, which is called *style scripting*. If you want to manipulate styles, it is important that they are available, of course. That's why CSS should be loaded before JavaScript executes. But more on this later.

12.4 Scripting Documents

Professor: The most fundamental role of client-side JavaScript is to make static HTML documents responsive to user and other actions and turn them into either dynamic documents or web applications.

If you use JavaScript in web documents only to augment HTML with behavior, then you get what is usually called *Dynamic HTML* or DHTML. In such cases, JavaScript should only be used to enhance a visitor's browsing experience. The actual content that a visitor can read, however, should by no means be dependent on JavaScript.

On the other hand, if you go beyond that limitation of merely enhancing a visitor's experience, you get what is called a *web application*. The HTML5 specification defines a whole lot of objects that allow you to perform application-like feats such as manipulating graphics or storing and retrieving data. Such sets of objects are called *application programming interfaces* or APIs because programmers use them for interacting with different (software and hardware) components. There's simply too much of everything, so we're not going to plunge into details. I'll just leave you some clues at the end of this course as to which directions you can take. They are gathered in Appendix B of this book.

It doesn't matter whether you're up to developing dynamic documents or web applications—there's one API you will constantly work with. It is called the *Document Object Model* API, or DOM API. Think of DOM as being an object representation of the collection of all the HTML elements and strings of text positioned on a web page. By using DOM API, the programmer can manipulate or query any element or string of text on a web page indirectly by manipulating or querying a corresponding DOM object.

The fundamental object of the DOM API is the Document object, which represents whatever is displayed in the browser's window. The Document object is a property of the Window object and can hence be accessed as a global property under the name document.

Maria: Has this got anything to do with the document.write() that we've been using?

Professor: Indeed it has. We've been using the Document's write() method in order to write text to the document, which we are no longer going to do. Instead, we're going to use Document's other methods and properties in order to work with its content.

One of the more important methods of the Document object is getElementById(), which returns a reference to an Element object that represents a specific HTML element. The method makes use of the id HTML attribute to identify the HTML element you want to fetch. Because the value of the element's ID must be unique within the document, you'll always know exactly which element you have got a reference to.

Don't panic, here comes an example. Imagine you have the following line of HTML code somewhere in your document:

```
<div id="click-me">Click me, please.</div>
```

If you pass the value of the id attribute as an argument to getElementById(), you get a reference to the Element object representing the above <div> element:

```
var myDiv = document.getElementById("click-me");
```

The following picture shows how a DOM Element object gives JavaScript access to different parts of an HTML element through its corresponding properties.

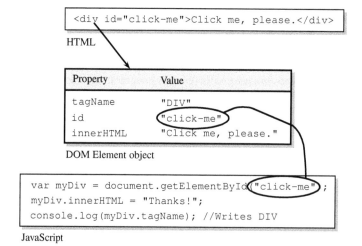

The most important property is of course id, which allows the getElementById() method to return a reference to the correct Element object. In the above example, a reference to the Element object representing <div> is stored in the myDiv variable. Once you have obtained this reference, you can either set or query the Element object's properties in the usual way, thus indirectly manipulating the underlying HTML element. For example, in the last line of JavaScript code in the above picture, the value of the `tagName` property is written to the JavaScript Console.

If you browse through the properties of the the Element class in the JavaScript reference, you'll find the `innerHTML` property on page 399. This property represents whatever is placed between the opening and closing tags of the element, and you can read as well as set this property. If you set the `innerHTML` property to a new string, the parsed representation of this new string will be shown in the browser window instead of whatever has been inside the element before that. In the above picture, the second line of the JavaScript code changes the text that is displayed on a page from "Click me, please." to "Thanks!"

Of course, it doesn't make much sense to immediately change the value of `innerHTML`. Instead, we can allow our visitor to change the text from "Click me, please." to "Thanks!" by clicking the `<div>` element. For this purpose we simply define the `onclick` event handler on the `myDiv` Element object.

One question that needs to be answered before we do this is how you can refer to the Element object from within the event-handler function definition. Any idea?

Maria: It seems very similar to an ordinary method definition to me. If an event-handler function is invoked as a method of a specific Element object, then this object acts as the invocation context of the handler and we can probably use `this` to access it.

Here's the complete `onclick` event-handler definition:

```
myDiv.onclick = function() {
  this.innerHTML = "Thanks!";
};
```

Professor: Splendid! And don't forget to put all the JavaScript code for this last example inside the `onload` event handler.

Incidentally, in this particular case you could just as well use `myDiv` instead of `this` inside the function definition, because the function is used on a single object instance. This is, however, not possible as soon as you want to use the same function definition on more than one object instance. Consider, for example, that you have two Element objects, `obj1` and `obj2`, and you want to thank each one of them individually. This is what you could do:

```
var thankYou = function() {
  this.innerHTML = "Thanks!";
};
obj1.onclick = thankYou;
obj2.onclick = thankYou;
```

In this case `this` is the only possible solution.

Apart from their specific properties and methods, Element objects also have properties that mirror the HTML attributes of corresponding HTML elements. For example, there exists a `title` HTML attribute, which is mirrored in every Element object as a property named `title`. The `title` attribute represents textual information about the element, which is typically displayed in the form of a tooltip appearing whenever a mouse pointer hovers over the element. You can dynamically set the `title` attribute of the above `<div>` element simply by assigning a text string to the `title` property of the `myDiv` object:

```
myDiv.title = "Click the text and see what happens.";
```

The element has thus obtained a tooltip.

Mike: Is it also possible to add new elements to a document?

Professor: There's virtually no limit to what you can do with the document. There exist specialized methods to add and remove elements from the document dynamically, some of which I will show you later. Nevertheless, you can also use `innerHTML` for that purpose. For example, you can insert two paragraphs inside our `<div>` element like this:

```
myDiv.innerHTML = "<p>Number nine</p><p>Number nine</p>";
```

Maria: Doesn't that delete the previous content of the `<div>` element?

Professor: As a matter of fact, it does. You can use a concatenation operator to actually add to the existing content. For example, this adds the third Number nine paragraph to the element:

```
myDiv.innerHTML += "<p>Number nine</p>";
```

12.5   Timer Events

Professor: One interesting type of events are events fired by timers. Sometimes you may want certain portions of your code to execute later or even to execute many times at regular time intervals. For example, you may want to remind your visitor that the time has expired, or to update a clock on your web page every second. There are two important global methods, `setTimeout()` and `setInterval()`, which let you register a function that you want to invoke at a later time once or many times. Although functions registered this way are invoked asynchronously as a response to the timer-elapse event, they are registered differently than other event handlers and we often call them *callbacks*.

Both methods are called with two arguments, the first of which is a reference to a function you want to schedule for later invocations, and the second is a number of milliseconds after which the function should be invoked. The difference between both functions is that `setInterval()` repeats the function invocation in intervals of the stated number of milliseconds, while `setTimeout()` schedules the function for a single invocation. Both functions return a value that you can pass to the `clearInterval()` function in order to cancel all the future invocations of the scheduled function.

To get an idea of how this works, let's take our 2016 Summer Olympics countdown page from page 162 and make it alive. This is the HTML portion of the example:

```
<!DOCTYPE html>
<html lang="en">
<head>
  <meta charset="utf-8">
  <title>2016 Summer Olympics Countdown</title>
  <script src="/scripts/countdown.js"></script>
</head>
<body>
  <p>
    Time to 2016 Summer Olympics: <span id="counter"></span>
  </p>
</body>
</html>
```

The JavaScript part looks like this:

```
window.onload = function() {
  var updateCounter = function() {
    var now = new Date();
    var opening = new Date(Date.UTC(2016, 7, 5, 23));
    var seconds;
    var output = "";
    var counter = document.getElementById("counter");
    if (opening.getTime() > now.getTime()) {
      seconds = opening.getTime() - now.getTime();
      seconds = Math.floor(seconds / 1000);
      output += Math.floor(seconds / 86400) + " days, ";
      seconds = seconds % 86400;
      output += Math.floor(seconds / 3600) + " hours, ";
      seconds = seconds % 3600;
      output += Math.floor(seconds / 60) + " minutes, ";
      output += Math.floor(seconds % 60) + " seconds.";
    }
    else {
      output = "0 seconds."
    }
    counter.innerHTML = output;
  };
  setInterval(updateCounter, 1000);
};
```

The first part of the `onload` event handler simply defines the `updateCounter()` function, which is registered as a callback in the last line of the handler. Note that `setInterval()` must be called after the `updateCounter()` definition or else you will be registering an `undefined` reference. Remember that function expressions are not hoisted.

If you now load this page, you get the running countdown to the Summer Olympics, which updates every 1000 milliseconds as specified by the second argument of the `setInterval()` function.

12.6 Scripting Styles

Professor: In the same way as HTML, it is, of course, possible to dynamically change the document styling as well. You might have guessed that Element objects also have the `style` property. We're not going to use that property, though, because it doesn't allow for separating the necessary CSS code from JavaScript. A much better approach is to define all the CSS rules that you need in a separate CSS file and then only switch between relevant class names in order to obtain the desired effect. Incidentally, because `class` is a reserved word in JavaScript, the JavaScript property name reflecting the `class` HTML attribute is `className` instead.

I would now like to take you to Kubla Khan's summer garden at Xanadu, which vividly rises before us in Samuel Taylor Coleridge's poem, "Kubla Khan." The poem is so powerful that I didn't have the nerve to shorten it for the purposes of the next example. Nevertheless, we are going to use JavaScript and CSS to hide part of the poem and make it available on demand by clicking a text, "Read the whole poem." This is the HTML part of the example:

```
<!DOCTYPE html>
<html lang="en">
<head>
  <meta charset="utf-8">
  <title>Kubla Khan</title>
  <link href="/styles/kublakhan.css" rel="stylesheet">
  <script src="/scripts/kublakhan.js"></script>
</head>
<body>
  <hgroup>
    <h1>Kubla Khan</h1>
    <h2>By Samuel Taylor Coleridge</h2>
  <hgroup>
  <p>
    In Xanadu did Kubla Khan                          <br>
    A stately pleasure-dome decree:                   <br>
    Where Alph, the sacred river, ran                 <br>
    Through caverns measureless to man                <br>
<span class="stray">Down to a sunless sea.</span>    <br>
    So twice five miles of fertile ground             <br>
    With walls and towers were girdled round;         <br>
    And there were gardens bright with sinuous rills, <br>
```

```
      Where blossomed many an incense-bearing tree;      <br>
      And here were forests ancient as the hills,        <br>
      Enfolding sunny spots of greenery.
   </p>

   <div id="show-hide">
      <p>
         But oh! that deep romantic chasm which slanted      <br>
         Down the green hill athwart a cedarn cover!         <br>
         A savage place! as holy and enchanted               <br>
         As e'er beneath a waning moon was haunted           <br>
         By woman wailing for her demon-lover!               <br>
         And from this chasm, with ceaseless turmoil seething, <br>
         As if this earth in fast thick pants were breathing, <br>
         A mighty fountain momently was forced:              <br>
         Amid whose swift half-intermitted burst             <br>
         Huge fragments vaulted like rebounding hail,        <br>
         Or chaffy grain beneath the thresher's flail:       <br>
         And mid these dancing rocks at once and ever        <br>
         It flung up momently the sacred river.              <br>
         Five miles meandering with a mazy motion            <br>
         Through wood and dale the sacred river ran,         <br>
         Then reached the caverns measureless to man,        <br>
         And sank in tumult to a lifeless ocean;             <br>
         And 'mid this tumult Kubla heard from far           <br>
         Ancestral voices prophesying war!                   <br>
<span class="stray">The shadow of the dome of pleasure</span> <br>
<span class="stray">Floated midway on the waves;</span>        <br>
<span class="stray">Where was heard the mingled measure</span><br>
<span class="stray">From the fountain and the caves.</span>    <br>
         It was a miracle of rare device,                   <br>
         A sunny pleasure-dome with caves of ice!
      </p>
      <p>
         A damsel with a dulcimer                            <br>
<span class="stray">In a vision once I saw:</span>            <br>
<span class="stray">It was an Abyssinian maid</span>           <br>
<span class="stray">And on her dulcimer she played,</span>     <br>
<span class="stray">Singing of Mount Abora.</span>             <br>
<span class="stray">Could I revive within me</span>            <br>
<span class="stray">Her symphony and song,</span>              <br>
<span class="stray">To such a deep delight 'twould win me,</span>
                                                             <br>
         That with music loud and long,                     <br>
         I would build that dome in air,                    <br>
         That sunny dome! those caves of ice!               <br>
         And all who heard should see them there,           <br>
         And all should cry, Beware! Beware!                <br>
         His flashing eyes, his floating hair!              <br>
         Weave a circle round him thrice,                   <br>
         And close your eyes with holy dread                <br>
         For he on honey-dew hath fed,                      <br>
```

```
        And drunk the milk of Paradise.
      </p>
    </div><!-- show-hide -->
    <div id="more-less"></div>
  </body>
</html>
```

Notice how I used <br> for breaking lines. In poems you are allowed to use breaks for this purpose because line breaking in poems doesn't serve as a purely presentational tool. Line breaks convey some auxiliary meaning as, for example, a change of movement or intensification or softening of rhyme. The <span> elements that I used with their class attribute set to stray also act as semantic devices. You know, Kubla Khan is a dream-generated poem and Coleridge wrote it as a transcript of his dream. Unfortunately, he was interrupted at doing it and a significant part of the poem was lost. Coleridge allegedly indented the lines that would be later assembled from stray lines and verses that he somehow managed to recall after the interruption.

Mike: I suppose we will use CSS to specify how much those stray lines should be indented.

Professor: Yes, we're coming to that in a moment. There are two more things I'd like you to pay attention to in the above HTML: the <div> element marked by the ID show-hide, and the <div> element marked by the ID more-less. These elements have no immediate effect but their role is to set the stage for JavaScript to be able to work with the poem later. This is in line with the notion of a dynamic document where user experience shouldn't be dependent on but merely enhanced by JavaScript code. In our example, the visitor simply sees the whole poem if JavaScript is not enabled.

CSS and JavaScript, which follow, add presentation and behavior to the above HTML.

In the *styles* sub-folder of our website's root, there's a file *kublakhan.css* comprising the following styles:

```
hgroup h1 {
  margin-bottom: 0em;
}
hgroup h2 {
  margin-top: 0em;
  color: gray;
}
.stray {
  margin-left: 1em;
}
.hidden {
  display: none;
}
.visible {
  display: block;
}
```

```
#more-less {
    color: blue;
    cursor: pointer;
}
```

I think the first three rules are obvious and need no explanation.

The key purpose of this example is to initially show the visitor just the first stanza, offering him/her an option to show or hide the rest of the poem. The CSS classes hidden and visible will be used by JavaScript to hide and show pieces of the text, simply by assigning them to the className property of a corresponding Element object. This way, presentation stays separated from behavior.

The last of the above rules is here simply to make it more obvious to a visitor that the element is clickable by changing its color to blue and the mouse cursor to a hand pointer, which are the normal attributes of a link.

The most intriguing part of our example is coded in JavaScript, which pulls everything together. The following is the content of the *kublakhan.js* file, which should be placed into the *scripts* sub-folder:

```
window.onload = function() {
    var textMore = "Read the whole poem";
    var textLess = "(Hide)";
    var whatToHide = document.getElementById("show-hide");
    var moreLess = document.getElementById("more-less");
    whatToHide.className = "hidden";
    moreLess.innerHTML = textMore;
    moreLess.onclick = function() {
        if (this.innerHTML == textMore) {
            this.innerHTML = textLess;
            whatToHide.className = "visible";
        }
        else {
            this.innerHTML = textMore;
            whatToHide.className = "hidden";
        }
    };
};
```

After specifying the text that should appear under the collapsed or expanded poem, the code grabs the element whose visibility we want to control (represented by the object whatToHide) and the element used to toggle the visibility (represented by the object moreLess). In the fifth line, a part of the poem is initially hidden by setting the whatToHide object's className property to hidden and, in the sixth line, the text "Read the whole poem" is written to the toggle element so that the visitor knows what will happen when he/she clicks the element.

The last part of the code defines the onclick handler on the moreLess Element object. The whole handler consists of a single if/else statement, whose job is to check what text is currently written inside the element and to change that text and the corresponding visibility state to their opposites.

Let's try how this works. Initially you can only read the first stanza as seen on the next screenshot.

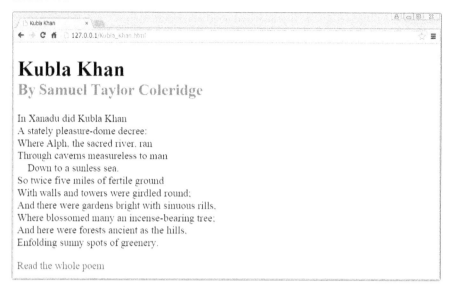

After you click the "Read the whole poem" link, you see the complete poem with the "(Hide)" link at the end.

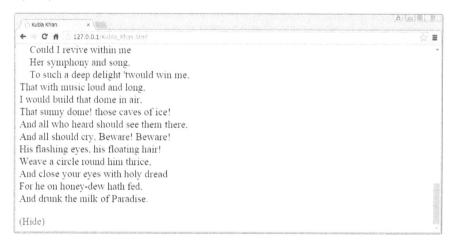

12.7 Introducing Forms

Professor: What we just did was already an interactive web page, although the amount of interaction available to the visitor was minimal. In order to be able to make our Sudoku helper and math worksheet generator interactive, we need at least some controls to capture textual input from the user. There exist certain HTML elements that create controls for user input, and we'll take a glimpse at some of them.

Basically, to implement a user interface that collects data from a visitor, the `<form>` element is used. This element helps to send the collected data to some kind of server, which then processes them. Although we're not going to send data to a server, it's still a good idea to use `<form>`. We'll also need the `<input>` element to create text boxes and buttons, and the `<label>` element to create captions for text boxes.

The following is a summary of the HTML elements and attributes that we are going to use:

- The `<form>` element represents a part of a document containing interactive controls that a visitor can use to enter information.

- The `<input>` element creates some of the interactive controls for web pages. The `type` attribute of the `<input>` element decides what kind of control this element will represent. If `type` is set to `text`, then the element will appear as a text box, and if `type` is set to `button`, then the element will show as a button. Apart from that, you should mark each `<input>` element with the `id` attribute if you want to access it from JavaScript or associate a label with it.

  Note that `<input>` is a void element and therefore has no content. That means that the corresponding Element object has no `innerHTML` property and the text that is written on a button or inside a text box is accessible through the `value` property. It is very important to remember that the `value` property always returns a string even if you enter a numeric value inside the text box.

- The `<label>` element is used to equip a control with a caption. The `for` attribute of the `<label>` element is used to associate a label with an `<input>` element. One practical consequence of associating a label with a control is that when the visitor clicks a label, a click event is triggered on the associated control as well. That's especially convenient for tiny controls like checkboxes as the user need not click on a control itself but can just as well click on the associated label, which is usually much larger.

Maria: I assume that a JavaScript property reflecting `for` should have a different name because `for` is a reserved word in JavaScript.

Professor: That's right. The corresponding JavaScript property is named `htmlFor`.

12.8   Homework

Professor: For homework, you will rewrite your math worksheet generator to make it an object class. Let the class have methods for setting the operator and maximum operand values, and methods for fetching the generated equation and the correct solution. The user should be able to interact with your worksheet via the interface specified by the following HTML:

```
<div id="ws1">
  <form class="worksheet">
    <div class="equations">
      <div>
        <label for="eq0" id="eq0-label"></label>
```

```
            <input type="text" id="eq0">
          </div>
        </div>
        <input type="button" value="Submit" id="submit">
      </form>
      <div id="score"></div>
    </div>
```

Basically, the worksheet is an HTML form containing one or more pairs of `<label>` and `<input>` elements, each pair representing an equation. For the homework you will only work with a single equation but in general there will be more of them. The question part of the equation is represented by a label, while the part accepting the answer is implemented as a text box. At first glance, some of the elements in the above document look superfluous but they are in fact there to make your life easier. The innermost `<div>` element, for example, allows you to treat the label-input pair as a separate entity, which is especially useful for positioning the equations.

Let the finished application initially display an equation, which should be contained in the `<label>` element. Each time the user clicks the submit button, a solution entered into the text box should be checked and the total score updated accordingly. You use the `<div>` element with the ID `score` to display the score. Finally, a new equation should be generated and displayed.

Before we go home, here's a list of today's keywords:

> **In this meeting**: core JavaScript, client-side JavaScript, Window object, global object, global variables and functions, `<script>`, `src`, synchronous execution, asynchronous execution, event, event-handler property, `onclick`, `onload`, DHTML, web application, API, DOM API, Document object, `getElementById()`, Element object, `innerHTML`, timers, `setTimeout()`, `setInterval()`, callback, class, `className`, `<form>`, input controls, `<input>`, `type`, `text`, `button`, `value`, `for`, `htmlFor`, `<label>`

User Interface

13.1 Homework Discussion

Professor: Did you succeed in connecting your generator to the user interface?

Maria: Yes, we did. We started by reshaping our generator from page 205 into an object class. This is the code:

```
var MathWorksheet = function() {
  this.equation = null;
  this.solution = null;
  this.operator = null;
  this.maxVal = null;
};

MathWorksheet.prototype.setOperator = function(op) {
  this.operator = op;
};

MathWorksheet.prototype.setMaxVal = function(val) {
  this.maxVal = val;
};

MathWorksheet.prototype.getEquation = function() {
  return this.equation;
};

MathWorksheet.prototype.getSolution = function() {
  return this.solution;
};

MathWorksheet.prototype.generate = function() {
  var firstOper = Math.round(Math.random() * this.maxVal);
  var secondOper = Math.round(Math.random() * this.maxVal);
  var tmp;
```

```
    switch (this.operator) {
      case "+":
        this.solution = firstOper + secondOper;
        break;
      case "-":
        if (firstOper < secondOper) {
          tmp = firstOper;
          firstOper = secondOper;
          secondOper = tmp;
        }
        this.solution = firstOper - secondOper;
        break;
      case "*":
        this.solution = firstOper * secondOper;
        break;
      case "/":
        if (secondOper == 0) {
          secondOper++;
        }
        tmp = firstOper * secondOper;
        this.solution = firstOper;
        firstOper = tmp;
    }
    this.equation = firstOper + this.operator + secondOper + "="
  };
```

Professor: There's nothing to discuss here, I guess. What about the second part? How did you connect the generator to the HTML that I gave you?

Mike: We needed to call the `getElementByID()` method many times so we decided to make the code a bit shorter by defining the next wrapper function:

```
var getElement = function(id) {
  return document.getElementById(id);
};
```

Using this function, we produced the following code:

```
window.onload = function() {
  var ws = new MathWorksheet();
  var score = 0;

  ws.setOperator("*");
  ws.setMaxVal(10);
  ws.generate();
  getElement("eq0-label").innerHTML = ws.getEquation();
  getElement("submit").onclick = function() {
    if (ws.getSolution() == parseInt(getElement("eq0").value)) {
      score++;
    }
```

```
      ws.generate();
      getElement("eq0-label").innerHTML = ws.getEquation();
      getElement("score").innerHTML = score;
   };
 };
```

Checking whether the answer was correct gave us quite a headache. Initially, we used the Number() function to convert the answer from string to number. The score was sometimes increased even though we entered no answer. Then we realized that an empty string is, of course, converted to zero. So we used parseInt() instead.

Professor: Excellent! You did a great job. I believe that everything else is quite obvious and we're ready to move on.

13.2   Using Family Relations to Manipulate Elements

Professor: Unlike the Kubla Khan poem, which was an example of a dynamic document, the math worksheet generator is a web application. It is therefore not necessary that we limit ourselves to a fixed number of equations. However, to be able to programmatically set the number of equations that are displayed within the worksheet, you should know some more about the DOM manipulation methods that JavaScript has in store for you.

Recall when we discussed family relations of HTML elements, and how we represented those relations in the form of a family tree. If you want, you can refresh your memory by looking at the tree on page 61. It will come as no surprise to you that DOM objects—because they represent HTML elements—reflect exactly the same relations within a tree, and you can refer to these objects in terms of relations like child or sibling as well. The DOM API is in fact quite complex but you don't need to know much to be able to start using it to your advantage. The important thing to know is that every HTML element is represented by an *Element* object, which is a subtype of the *Node* class. What that means in practice is that you can use methods and properties of the Node as well as those of the Element class when working with HTML elements.

You have already learned how to get hold of an HTML element using the getElement ById() method of the Document object and manipulate it using the innerHTML property, or properties that mirror HTML attributes. You also learned that you can define event handlers for elements, and you can even insert other elements inside existing ones dynamically by simply assigning appropriate HTML code to innerHTML. You can do some pretty amazing stuff with all that already but sooner or later you'll want more flexibility.

While you can grab an element by its ID, this may not be very convenient if you want to get more elements that logically belong together and to manipulate them all in the same way. This is exactly what we need in our math worksheet generator as soon as we decide to include more than one equation into the worksheet. One way of getting several elements at the same time is fetching elements by their names. You can do this with the getElementsByTagName() method of an Element or Document object. Since the Document object is the root object of DOM, invoking this method on the Document object will get you all the corresponding elements within the entire document. Invoking `getElementsByTagName()` on an Element object, however, will get you only the corresponding children of that specific element.

A similar method is `getElementsByClassName()`, which searches elements by their class name instead.

You can also take advantage of the family relations between elements and use the `firstChild` or `nextSibling` properties of the Node class if you want to get the first child or the next sibling of the current Node object, respectively. Or, you can get the parent Node object using the `parentNode` property.

When you work with a Node object, you must know that nodes represent not only elements but also text between and inside them. Therefore, a Node object obtained by `firstChild` or `nextSibling` isn't necessarily an element but can be plain text as well. If you are primarily interested in elements rather than text within them, then you'll use the corresponding properties of the Element class instead. They have similar names with "Element" inside them like, for example, `firstElementChild` or `nextElementSibling`. Note, however, that text cannot have children and that's why `parentNode` will always return an Element object.

Mike: Pardon me, I'm not following you any longer. Can you show us some examples?

Professor: Yes, I'm sorry. I really overdid it. Consider, for example, that we add more equations to our math worksheet. Like this:

```
<div id="ws1">
  <form class="worksheet">
    <div class="equations">
      <div>
        <label for="eq0"></label>
        <input type="text" id="eq0">
      </div>
      <div>
        <label for="eq1"></label>
        <input type="text" id="eq1">
      </div>
      <div>
        <label for="eq2"></label>
        <input type="text" id="eq2">
      </div>
    </div>
    <input type="button" value="Submit" id="submit">
  </form>
  <div id="score"></div>
</div>
```

If you want to get an Element object that represents the `<div>` element used for displaying the score, you can refer to it by its ID, of course, but you can also refer to it as the last child of the outermost `<div>` element:

```
var ws = document.getElementById("ws1");
ws.lastElementChild.innerHTML = "You scored 10 out of 10.";
```

You can get an array of the three text boxes from the above HTML like this:

```
var ws = document.getElementsByClassName("equations")[0];
var eqs = ws.getElementsByTagName("input");
```

There is only a single element with the class name `equations` but the `getElements ByClassName()` method nevertheless returns an array. That's why I had to use an array access operator after the method invocation.

As soon as the above two lines have finished execution, `eqs` holds an array of all text boxes within the `<div>` element with the class name `equations`.

Let's see if you can get, say, the solution that was entered in the last of the text boxes.

Maria: Because `eqs` is an array, I must use an index to select the wanted text box and then access its `value` property using the dot operator:

```
//Prints the solution entered in the third text box:
console.log(eqs[2].value);
```

Professor: Precisely. What if you didn't know how many paragraphs there were?

Maria: Oh yes, the `length` property probably works as well:

```
//Prints the solution entered in the last text box:
console.log(eqs[eqs.length - 1].value);
```

Professor: Perfect. However, we are not only going to access HTML elements in our final math worksheet generator but we will also generate them dynamically. We need a different set of methods to do so, of course, and they are coming in a moment.

13.3   Completing Math Worksheet Generator

Professor: Because we'll generate the whole worksheet dynamically, the HTML document that we'll actually use is very plain:

```
<!DOCTYPE html>
<html lang="en">
  <head>
    <meta charset="utf-8">
    <title>Math Worksheet Generator</title>
    <link href="/styles/mathWorksheet.css" rel="stylesheet">
    <script src="/scripts/mathWorksheet.js"></script>
  </head>
```

```
<body>
  <div id="ws1"></div>
</body>
</html>
```

Notice that all that is left of the static HTML defining the worksheet is the outermost `<div>` element. This element will serve as a placeholder to mark a position at which the worksheet will appear.

The styling part of our project is stored within a file *mathWorksheet.css*, whose content also isn't much of a wonder:

```
.equations {
  margin-bottom: .2em;
  text-align: right;
}

.worksheet {
  border-top: solid 1px black;
  border-bottom: solid 1px black;
  margin-bottom: .3em;
  padding: .2em;
  width: 10em;
  text-align: center;
}

.worksheet input {
  width: 4em;
  position: relative;
  bottom: .1em;
  font-size: .8em;
}

#submit {
  width: 100%;
}
```

As you can see, there are four rules. The first two define the classes `equations` and `worksheet`. The third rule, which uses a descendant selector, styles the `<input>` elements that are descendants of any element whose `class` property is set to `worksheet`. The last rule extends the submit button over the whole width of the form, which I think is more sexy than having a smaller button either centered or aligned at either side of the form.

For the JavaScript part we will use the `MathWorksheet` class object that you wrote for your homework, and we only need to rewrite the `onload` handler. Among other things, the `onload` handler will also dynamically produce the HTML for our worksheet. As a starting point we will use the static HTML code on page 248 so to have a clearer picture of what we need to produce.

The `<div>` element with the ID `ws1` and all of its content are represented by the following DOM tree.

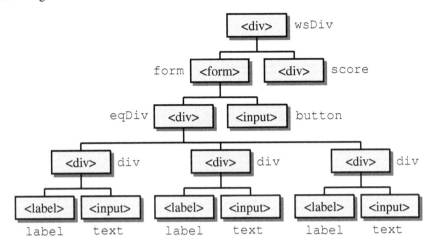

Note that only Element nodes are shown in this tree, while other nodes such as text nodes are left out. The identifiers written beside some of the boxes are the names of corresponding Element objects that we will use in our JavaScript code. For example, we'll use `wsDiv` as a reference to an Element object representing the `<div>` element at the root of the above DOM tree. Incidentally, this is the element with the ID `ws1` in HTML on page 248.

Mike: I am comparing this tree to the source HTML and I see that the last two levels of nodes represent the equations. I mean, the three innermost `<div>` elements containing the label-input pairs. I suppose that they are distinct elements but you assigned them to one and the same reference named `div` in the above drawing of the DOM. Why is that?

Professor: Some of the references indicated in the tree will refer to respective Element objects just as long as we don't append them to the document tree, and we will reuse the same references many times. To access these objects later, we will get them using the `getElementsByTagName()` method.

The following code is the beginning of our `onload` handler:

```
window.onload = function() {
  var wsDiv, scoreDiv, eqDiv;
  var form, div, text, label, button;
  var i;
  var ws;

  ws = new MathWorksheet();
  ws.setOperator('/');
  ws.setMaxOper(10);

  wsDiv = document.getElementById("ws1");
  //...
```

Apart from the declaration of variables that will refer to the above mentioned DOM objects, the code also creates and initializes a `MathWorksheet` object instance `ws`. The last line of the code fragment fetches the `<div>` element that will hold the whole worksheet.

It's now time to create the elements of the worksheet. You can do that using the Document's `createElement()` method, passing the name of the element you wish to create as its argument. However, `createElement()` only creates an element, it doesn't insert it into the document tree. You can insert the created element to the right position within the document tree using one of the appropriate methods like, for example, the Node's `appendChild()` method. The element will be appended as a child of the Node or Element object on which you invoke this method.

The next lines of code first create a form and append it as a child of the outermost `<div>` element, and then create a `<div>` element that will hold all the equations and append it as a child of the form:

```
//...
form = document.createElement("form"); //Creates a <form> element
form.className = "worksheet";             //Sets its class attribute
wsDiv.appendChild(form);                  //Makes form a child of wsDiv

eqDiv = document.createElement("div");   //Creates a <div> element
eqDiv.className = "equations";             //Sets its class attribute
form.appendChild(eqDiv);                  //Makes eqDiv a child of form
//...
```

By this time, we're ready to generate and insert the equations:

```
//...
for (i = 0; i < 5; i++) {       //Creates five equations
//Generates an equation and a <div> element to hold it:
  ws.generate();
  div = document.createElement("div");
  eqDiv.appendChild(div);
//Creates a label and inserts the equation into it:
  label = document.createElement("label");
  label.innerHTML = ws.getEquation();
  label.htmlFor = "eq" + i;
  div.appendChild(label);
//Creates a text box and attaches a solution to it:
  text = document.createElement("input");
  text.type = "text";
  text.id = "eq" + i;
  div.appendChild(text);
  text.dataset.solution = ws.getSolution();
}
//...
```

The code is more or less self-explanatory except maybe for the last line. HTML5 allows you to attach your own data to elements using so-called *custom data attributes*, which are mapped to properties of the JavaScript property named `dataset`. I used this possibility to associate the correct solution of the equation with the text box that will later contain an answer provided by the user. This will come quite handy when we eventually compare the answers with the correct solutions.

Maria: Why do we have to use `dataset`? You told us once that JavaScript objects were dynamic. If I remember correctly, that means you can add any property to any object at any time, can't you?

Professor: That's true. And if you try to add a property directly to the above `text` object, you'll discover that this in fact works. However, it is basically not recommended that you add any custom properties or methods to the DOM objects. One of the main reasons is that DOM objects are host objects, which are subject to quite different implementation rules. So, user-defined properties and methods are not guaranteed to work properly with DOM objects and, worse still, sometimes you won't even notice that something went wrong until much later.

Let's continue with work. Once we have created all the text boxes, it is convenient that we place them into a single array, and we can use the `getElementsByTagName()` method to get the job done. We are going to check this array later to see how many correct answers the user has submitted.

The following line fetches all the `<input>` elements contained within `eqDiv` and stores the returned array to `answers`:

```
//...
answers = eqDiv.getElementsByTagName("input");
//...
```

Mike: If I understood everything correctly, then at this point the worksheet is not completed yet and there's no way that the user could have entered any answers. I suppose we should include this line into the `onclick` handler of our submit button instead.

Professor: We could do that as well. It is, however, just as good if we do it right away. You know, the array returned by `getElementsByTagName()` is *live*, which means that it is automatically updated whenever the DOM tree is updated. The values stored in the array `answers` will therefore always be the most recent ones without any need to call `getElementsByTagName()` again.

There are only just two more elements left to create and the worksheet is completed. The following is the code that creates and inserts into the document tree the submit button and the `<div>` element for displaying a score:

```
//...
button = document.createElement("input");
button.type = "button";
```

```
button.value = "Submit";
button.id = "submit";
form.appendChild(button);

scoreDiv = document.createElement("div");
wsDiv.appendChild(scoreDiv);
//...
```

Note that the code we've written so far is already operational in the sense that it will produce a worksheet of five equations and even allow the user to enter solutions. You can try it if you like, just don't forget to type the closing curly bracket of the `onload` handler. I haven't typed it because we're not finished with the handler yet.

Let's now breathe some life into our generator by making it responsive to user clicks. We're going to do this by means of the `onclick` handler of the submit button. The handler will compare the submitted answers with the correct solutions and display the score each time the user clicks the button. The code is amazingly simple:

```
//...
button.onclick = function() {
  var i, score = 0;
  for (i = 0; i < answers.length; i++) {
    if (answers[i].dataset.solution ==
                          parseInt(answers[i].value)) {
      score++; //Counts correct answers
    }
  }
  scoreDiv.innerHTML = "You scored " +
                        score + " out of " + i + ".";
}; //Closing brace of the onclick handler
}; //Closing brace of the onload handler
```

Note that the array `answers` is accessible inside the `onclick` event handler because it is defined within its closure and stays available so long as the page is loaded in the browser. To check the answers it is only necessary to compare the properties `dataset.solution`, and `value` of each of the Element objects stored in `answers`.

Finally, this is how our completed generator works:

13.4   Completing Sudoku Puzzle Helper

Professor: I hope you still have energy to finish Sudoku, which is our last project in this course. Before doing it, let's briefly refresh our memory of what we produced so far. We wrote two objects, Cell and Sudoku, and this is the list of things we need to know to be able to continue the work:

- The Sudoku() constructor creates a Sudoku object instance.
- The setClues() method of the Sudoku object takes as an argument a 9×9 array of initial clues for the cells. Each clue can be either a numeric value between one and nine, or the null value if the cell is empty.
- The setNumber() method of the Sudoku object accepts three arguments: a number that is to be written into the puzzle, and the row and column index of the cell into which the number is to be written. The method returns false if the same number already exists in the same row, column, or 3×3 box. Otherwise, it inserts the number into the puzzle grid and returns true.
- The board property of the Sudoku object is a 9×9 array of references to Cell object instances.
- The getNumber() method of the Cell object returns the number that is currently written in the cell. If the cell is empty, then the method returns null.
- The isWritable() method of the Cell object returns true if you can write numbers into the cell, and false otherwise.

Now, let's get the job done. We are going to generate the HTML part of the puzzle dynamically, of course. Nevertheless, here is static HTML code for the puzzle grid to get an impression of how the final document should look:

```
<form>
<table class="sudoku-table">
  <tr>
    <td><input type="text"></td>
    <td><input type="text"></td>
    <!--  ...  -->
  </tr>
  <tr>
    <td><input type="text"></td>
    <td><input type="text"></td>
    <!--  ...  -->
  </tr>
  <!--  ...  -->
</table>
</form>
```

The comments containing ellipses hint that parts of the code are omitted. Everything in fact repeats verbatim nine times and the whole puzzle is simply a 9×9 table containing text boxes inside its cells.

This is the actual HTML that we're going to use for our Sudoku:

```
<!DOCTYPE html>
<html lang="en">
  <head>
    <meta charset="utf-8">
    <title>Sudoku</title>
    <link href="/styles/sudoku.css" rel="stylesheet">
    <script src="/scripts/sudoku.js"></script>
  </head>
  <body>
    <div id="sudoku-puzzle"></div>
  </body>
</html>
```

Notice that the code is practically the same as it was for the math worksheet generator, and you will probably have guessed that the whole Sudoku puzzle goes into the only <div> element inside the document body. That said, the style part is a little more intriguing. The *sudoku.css* file, to which the above HTML refers, starts with the following rule:

```
.sudoku-table {
  border: 2px solid;
  border-collapse: collapse;
}
```

This rule simply sets a border around the puzzle and eliminates space between cells. Now comes the interesting part: we need to draw two additional vertical and horizontal lines to visually separate the 3 × 3 boxes. We can do this using the :nth-child() selector:

```
.sudoku-table td:nth-child(3n) {
  border-right: 2px solid;
}
.sudoku-table tr:nth-child(3n) {
  border-bottom: 2px solid;
}
```

The first of the above rules selects every third <td> child element that is a descendant of an element whose class attribute is set to sudoku-table, and makes its right border solid and two pixels thick. Similarly, the second rule selects every third <tr> child element and makes its bottom border thicker.

Mike: How exactly does the :nth-child() selector work?

Professor: The :nth-child() selector accepts as an argument an expression of the form $an+b$, which reads: "Select every ath child starting with the bth child." For example, suppose you want to select every second child, starting with the first one. You can do that with the following selector:

```
:nth-child(2n+1)
```

Or, if you just want to select every second child, you can drop the +1 part:

```
:nth-child(2n)
```

Incidentally, you can achieve the same effect using the keywords odd and even instead of 2n+1 and 2n.

When you place another selector directly before this selector—without any space, that is—then you limit the :nth-child() selector only to children that match that other selector as well. Therefore, the expression td:nth-child(3n) only selects every third child that is at the same time a <td> element.

Maria: But <td> is a child of a row, not a child of a table.

Professor: What do you mean?

Maria: I don't understand this rule:

```
.sudoku-table td:nth-child(3n) {
  border-right: 2px solid;
}
```

A <td> element cannot be a child of a table.

Professor: Oh, I see your point. The above rule uses a descendant selector, all right. However, :nth-child() itself does not imply children of any specific element. When used alone, it simply refers to *any* children. You should read the above rule as: "Select every third child that is a <td> element, and is at the same time a *descendant* of an element with the class attribute set to sudoku-table." You should not understand it as: "Select every third *child* of the element with the class attribute set to sudoku-table." This is a big difference.

The *sudoku.css* file continues with the following two rules, which format the <td> and <input> elements within a Sudoku table:

```
.sudoku-table td {
  width: 2em;
  height: 2em;
}

.sudoku-table input {
  width: inherit;
  height: inherit;
  text-align: center;
  font-size: 2em;
}
```

Note that, by default, the properties `width` and `height` are not inherited. That's why I used the `inherit` keyword to force the `<input>` element to inherit these two properties from its parent, which is `<td>` in our case.

Mike: Why didn't you simply set the size of the `<input>` element?

Professor: If I did that, then it wouldn't be possible to set font size of the `<input>` element without also affecting the size of the `<input>` element itself. Remember that the em unit is relative to the size of the font of the element itself, rather than its containing block. I could have, of course, used other units but then the cell wouldn't resize accordingly if the base font would change.

Mike: I still don't see the problem.

Professor: If font size were the same for all `<input>` elements then this indeed wouldn't be a problem. However, we will later set smaller font size for the elements displaying solution candidates.

Mike: Oh, I see.

The last four rules take care of the appearance of numbers to reflect their meaning. The numbers that represent the initial clue will be shown as black and in bold, the numbers entered by the user as answers will be shown gray, the numbers entered by the user that are not placed according to the rules will be colored red, and the solution candidates will be rendered smaller than normal numbers:

```
.sudoku-table .clue {
  color: black;
  font-weight: bold;
}
.sudoku-table .answer {
  color: gray;
}
.sudoku-table .wrong {
  color: red;
}
.sudoku-table .candidate {
  font-size: .7em;
}
```

This completes the styling portion of our project and we're ready for the JavaScript part. We are going to add another method to the Sudoku class called `start()`, whose job will be twofold: to build the appropriate document structure and to connect the `oninput` handlers of all the `<input>` elements to the event-handler function that we're going to write later. The input event is triggered every time when the value of the `<input>` element changes, which is just perfect for us because we'll be able to react upon every change inside the puzzle grid. These are the first lines of the `start()` method:

```
Sudoku.prototype.start = function(id) {
  var row, cell, text, number;
  var puzzleContainer = document.getElementById(id);
  var form = document.createElement("form");
  var puzzle = document.createElement("table");
  puzzle.className = "sudoku-table";
  puzzleContainer.appendChild(form);
  form.appendChild(puzzle)
//...
```

The method accepts a single argument, which is the ID of the HTML element that will contain the whole puzzle. The code begins with some local variable declarations, followed by getting the puzzle container. The last five lines create Element objects representing a form and table, and insert the created elements into the document tree.

We can now create rows and table cells and append them to the table object `puzzle`. The following two nested `for` loops do this:

```
//...
  for (i = 0; i < 9; i++) {
    row = document.createElement("tr");
    puzzle.appendChild(row);
    for (j = 0; j < 9; j++) {
      cell = document.createElement("td");
      row.appendChild(cell);
      text = document.createElement("input");
      text.type = "text";
      text.className = "answer";
      cell.appendChild(text);
//...
```

The code is nothing special, really. It creates nine table rows and appends them to `puzzle`. Before going back to create the next row, the current row is filled with nine table cells, each containing a text box.

After each of the text boxes has been created and put in place, it's time to write a number into it. The next code fragment puts a clue into a corresponding text box if the cell is read-only and hence includes an initial clue. Of course, it also sets the appropriate value for the `class` attribute and prevents the user from changing the contents of the text box by specifying the `readonly` attribute:

```
//...
    if (!this.board[i][j].isWritable()) {
      text.value = this.board[i][j].getNumber();
      text.className = "clue";
      text.readOnly = true;
    }
//...
```

Maria: Why did you camelCase the `readOnly` property, while the name of the original HTML attribute is `readonly`?

Professor: Nothing escapes your eye, does it? As it happens, the `readonly` attribute is changed to `readOnly` inside the DOM. That's because JavaScript just loves to camelCase its identifiers and multi-word HTML attributes are mirrored to DOM property names that are camelCased.

The last line of the `start()` method connects the `oninput` event-handler property of the corresponding `<input>` element to the event-handler function that we're going to write in a moment. Finally, three closing braces finish our job with this method:

```
      text.oninput = handleInput(this, i, j);
    } //Closes the inner for loop
  }    //Closes the outer for loop
};    //Closes the start() method
```

If you look carefully at the above code, you'll notice that I didn't actually define any function but I rather invoked one. What do you think the `handleInput()` function should return?

Mike: Since an event-handler property should be a reference to a function, I guess that `handleInput()` must return a reference to some kind of function nested within it.

Professor: Exactly! The `handleInput()` function is simply used to create a closure for the event-handler function, a reference to which it returns. This enables us to actually pass arguments to an event-handler function. You know, because event handlers are invoked automatically by the system, it is not possible to pass arguments to them in the usual way.

This is the `handleInput()` function definition:

```
var handleInput = function(sdk, row, col) {
  return function() {
    var v = parseInt(this.value);
    this.className = "answer";
    if (v >= 1 && v <= 9) {
      if (!sdk.setNumber(this.value, row, col) {
        this.className = "wrong";
      }
    }
    else {
      sdk.setNumber(null, row, col);
      this.className = "candidate";
    }
  };
};
```

As you can see, the function indeed defines and returns another, anonymous function. It also creates closure for this anonymous function, inside which it stores the three passed arguments: a reference to the Sudoku object, and the row and column indexes of the cell. Note that a separate anonymous function-definition object is created for each of the text boxes, whose closure stores the row and column indexes of the cell associated with the current text box.

The anonymous function nested within `handleInput()` is quite simple. First, it parses the value currently written in the text box that actually triggered the input event. Recall that `this` is a reference to a method's invocation context, which in our case is the Element object representing the `<input>` element that triggered the event. After parsing the value, the function resets the element's `class` attribute to `answer`.

If the entered number is between one and nine inclusive, then the function tries to insert that number into the puzzle grid using the `setNumber()` method of the Sudoku object. If the operation is not successful, then the element's class is changed to `wrong` so that the entered number will be colored red. Otherwise, if there is anything other than a number between one and nine entered in the text box, the old number is removed from the corresponding Sudoku cell and the element's class is set to `candidate`. However, the entered value is still shown in the text box but is rendered smaller.

The last missing piece of code is the Window's `onload` event-handler definition, which incorporates a two-dimensional array containing the initial clue, a Sudoku object creation and initialization, and an invocation of the `start()` method:

```
window.onload = function() {
  var initial = [
    [null, 5,    7,    null, 4,    null, null, 2,    null],
    [null, 3,    8,    null, null, 2,    7,    null, null],
    [null, 1,    null, 7,    3,    null, null, null, null],
    [null, 7,    null, 2,    8,    null, null, null, 3   ],
    [3,    null, 4,    null, null, null, 1,    null, 6   ],
    [1,    null, null, null, 6,    4,    null, 7,    null],
    [null, null, null, null, 2,    7,    null, 1,    null],
    [null, null, 1,    9,    null, null, 2,    3,    null],
    [null, 4,    null, null, 1,    null, 5,    9,    null]
  ];

  var s = new Sudoku();
  s.setClues(initial);
  s.start("sudoku-puzzle");
}
```

That's it. We're ready to play Sudoku.

Maria: If I want to take a break from the puzzle and continue solving it at some later time, is there a way to save my partial solution?

Professor: You can save data locally using the `localStorage` object, whose description you will find in Appendix B on page 280. It's not very difficult to use it and I think you shall be able to figure it out by yourself.

Mike: Have we learned enough to be able to write a program that will generate a Sudoku puzzle?

Professor: I think so. You know what? I'll let you do it for homework. Because we are not going to see each other any more, I'll leave a solution for you in Appendix A. However, you should really try to do it on your own and use the solution only if you really, really have to. Promise?

Mike: Promise.

Maria: Can you at least give us some directions, please?

Professor: I'll do that, of course.

13.5   Homework

Professor: Basically, there are three steps in generating a Sudoku puzzle. The easiest one is to produce a puzzle. Somewhat more difficult is to make sure that the puzzle has a unique solution. This is essential, or else it is not possible to solve it without guessing. The most difficult part is to determine the puzzle's difficulty level. I suppose that it's OK if you implement the first two steps for your homework.

To produce a puzzle, you simply start out with any completely solved puzzle and shuffle the numbers to get a new one. There are some simple swapping operations you can perform on any valid Sudoku to get a different Sudoku that is still valid. You can swap:

- Any two columns within any of the three columns of 3 × 3 boxes.

- Any two rows within any of the three rows of 3 × 3 boxes.

- Any two columns of 3 × 3 boxes.

- Any two rows of 3 × 3 boxes.

For example, you can swap the first and second row of cells, or the fourth and sixth column of cells, but you cannot swap the first and fourth column of cells. You can also swap all three 3 × 3 boxes in the first row with the three boxes in the last row.

If you swap randomly selected rows and columns according to the above rules many times, you can get almost any conceivable puzzle. Finally, you remove some numbers from the grid, and the numbers that are left represent clues for the new puzzle.

Mike: How many clues should we leave?

Professor: Recently it has been shown that it is not possible to construct a Sudoku with only 16 clues, so 17 is an absolute minimum. However, an average number of clues is usually around 25, and I suggest that you use 30 clues if you are not implementing an estimate of difficulty level. This way your puzzle will most probably be very easy or easy to solve, and it won't take too much time to generate it.

A puzzle that you get by randomly removing numbers from a completely solved Sudoku will very likely have more than a single solution. Hence your next step is to check how many solutions your puzzle has. If it has more than one, then you should try another one, and another one, until you find a puzzle having a single solution.

To check whether a Sudoku has more than one solution, you can simply try to systematically put every possible number into each of the empty cells and see if a combination solves the puzzle. This is a possible algorithm:

Construct a one-dimensional array of all the empty cells.
Set n to zero (i.e., the position of the first cell in the array of empty cells).
Check the nth empty cell:

    Repeat for every number from one to nine:

        If you can place the number into the current empty cell **Then**

            If this is the last of the empty cells **Then**
                A solution has been found.
            Else Check the $(n + 1)$th empty cell.
            End If

        End If
        If more than one solution has been found so far **Then**
            Stop checking.
        End If

    End Repeat
    Clear the nth cell back to `null`.

End Check

The first line of this algorithm should find all the empty cells in the grid and place them into a one-dimensional array. Note that you will need the row and column indexes of cells to be able to place them into the puzzle grid, which `Cell` objects don't store. To get around this, you can either add row and column indexes to the `Cell` class object or simply create an array of pairs of indexes storing positions of empty cells.

Note that Check in the above algorithm is a function, and you will notice a curious thing: that this function in fact calls itself from within the innermost if/else statement. A function that calls itself is called a *recursive* function, which offers an elegant solution for types of problems like this one. What you have to do is first try to place number one in the first empty cell. If that is not possible, then try it with number two, and number three, and so on, until either you find a proper number or you run out of numbers. If you successfully place a number into the grid, then repeat the whole story with the next empty cell. If, however, no one number can be placed into the current cell, then you should go back to the previous empty cell and try to replace its number with a bigger one.

If you use a recursive function to solve this problem, then lots of bookkeeping is done automatically for you. Each time the function is called, a new function object is created and appended to the current scope chain. If the function also creates local variables remembering the index of the cell and the number that has been placed most recently into that cell, then there is not much work left for you. Whenever no further number can be placed in the current cell, the current function stops executing, which automatically removes it from the scope chain and the checking continues with the previous cell from the exact point where it left it.

I admit that this second part of the assignment is quite a challenge, but you shall try to write a program nonetheless. In case you don't succeed, there's a solution in Appendix A.

Which brings us to the end of our course. You guys have been terrific and I hope you enjoyed it as much as I did. Take care and keep up the good work!

In this meeting: `getElementsByTagName()`, `getElementsByClassName()`, `firstElementChild`, `firstChild`, `lastElementChild`, Element object, Node object, Document object, `parentNode`, `appendChild()`, `oninput`, passing arguments to event handlers, recursion, `inherit`, `createElement()`

Solution to the Last Homework:
Sudoku Generator

The following code implements the SudokuGenerator class used for generating a Sudoku puzzle, which is not guaranteed to have a unique solution. At the end of the code there is a definition of the rand() function used for generating integer random numbers.

```
/*
 * Constructs a SudokuGenerator object instance. After constructing
 * an object instance, you should call the methods reset(),
 * shuffle(), removeNumbers(), and getClues().
 *
 * Example:
 *    var clues, g = new SudokuGenerator();
 *    g.reset();
 *    g.shuffle();
 *    g.removeNumbers(30);
 *    clues = g.getClues();
 */
var SudokuGenerator = function() {
  this.clues = null;
};

/*
 * Resets a SudokuGenerator object. Whenever you want to re-create
 * a puzzle, this method should be called first.
 */
SudokuGenerator.prototype.reset = function() {
  this.clues = [
    [8, 2, 7, 1, 5, 4, 3, 9, 6],
    [9, 6, 5, 3, 2, 7, 1, 4, 8],
    [3, 4, 1, 6, 8, 9, 7, 5, 2],
    [5, 9, 3, 4, 6, 8, 2, 7, 1],
    [4, 7, 2, 5, 1, 3, 6, 8, 9],
    [6, 1, 8, 9, 7, 2, 4, 3, 5],
```

```
      [7, 8, 6, 2, 3, 5, 9, 1, 4],
      [1, 5, 4, 7, 9, 6, 8, 2, 3],
      [2, 3, 9, 8, 4, 1, 5, 6, 7],
    ];
};

/*
 * Shuffles the numbers in the puzzle grid.
 */
SudokuGenerator.prototype.shuffle = function() {
  var i;
  for (i = 0; i < 500; i++) {
    this.swapBlockCols(rand(3), rand(3));
    this.swapBlockRows(rand(3), rand(3));
    this.swapColsInBlock(rand(3), rand(3), rand(3));
    this.swapRowsInBlock(rand(3), rand(3), rand(3));
  }
};

/*
 * Swaps two rows.
 * Parameters:
 *    r1, r2: the indexes of rows to be swapped (indexes between
 *            zero and eight)
 */
SudokuGenerator.prototype.swapRows = function(r1, r2) {
  var i, tmp;
  for (i = 0; i < 9; i++) {
    tmp = this.clues[r1][i];
    this.clues[r1][i] = this.clues[r2][i];
    this.clues[r2][i] = tmp;
  }
};

/*
 * Swaps two columns.
 * Parameters:
 *    c1, c2: the indexes of columns to be swapped (indexes between
 *            zero and eight)
 */
SudokuGenerator.prototype.swapCols = function(c1, c2) {
  var i, tmp;
  for (i = 0; i < 9; i++) {
    tmp = this.clues[i][c1];
    this.clues[i][c1] = this.clues[i][c2];
    this.clues[i][c2] = tmp;
  }
};
```

```
/*
 * Swaps two rows within a row of 3x3 blocks.
 * Parameters:
 *   br: the index of a row of 3x3 blocks (an integer between zero
 *       and two)
 *   r1, r2: the indexes of rows to be swapped. Indexes are
 *           relative to the given row of 3x3 blocks and can be
 *           integers between zero and two.
 */
SudokuGenerator.prototype.swapRowsInBlock = function(br, r1, r2) {
  this.swapRows(br * 3 + r1, br * 3 + r2);
};


/*
 * Swaps two columns within a column of 3x3 blocks.
 * Parameters:
 *   bc: the index of a column of 3x3 blocks (an integer between
 *       zero and two)
 *   c1, c2: the indexes of columns to be swapped. Indexes are
 *           relative to the given column of 3x3 blocks and can be
 *           integers between zero and two.
 */
SudokuGenerator.prototype.swapColsInBlock = function(bc, c1, c2) {
  this.swapCols(bc * 3 + c1, bc * 3 + c2);
};


/*
 * Swaps two rows of 3x3 blocks.
 * Parameters:
 *   r1, r2: the indexes of rows to be swapped (integers between
 *           zero and two)
 */
SudokuGenerator.prototype.swapBlockRows = function(r1, r2) {
  for (i = 0; i < 3; i++) {
    this.swapRows(r1 * 3 + i, r2 * 3 + i);
  }
};


/*
 * Swaps two columns of 3x3 blocks.
 * Parameters:
 *   c1, c2: the indexes of columns to be swapped (integers between
 *           zero and two)
 */
SudokuGenerator.prototype.swapBlockCols = function(c1, c2) {
  for (i = 0; i < 3; i++) {
    this.swapCols(c1 * 3 + i, c2 * 3 + i);
  }
};
```

```
/*
 * Randomly removes numbers from the puzzle grid by setting them to
 * null.
 * Parameters:
 *   numClues: the number of clues to be left in the grid
 */
SudokuGenerator.prototype.removeNumbers = function(numClues) {
  var i, j;
  var remain = 81;

  while (remain > numClues) {
    i = rand(9);
    j = rand(9);
    if (this.clues[i][j] != null) {
      this.clues[i][j] = null;
      remain--;
    }
  }
};


/*
 * Returns a two-dimensional array of clues. Empty cells are set to
 * null.
 */
SudokuGenerator.prototype.getClues = function() {
  return this.clues;
};


/*
 * Returns a non-negative random integer.
 * Parameters:
 *   n: the upper limit of a returned number
 * Returns:
 *   A random integer between zero and n - 1
 */
var rand = function(n) {
  return Math.floor(Math.random() * n);
}
```

The following two methods are additions to the Sudoku class. They are used to check whether a puzzle has a unique solution:

```
/*
 * Checks whether a puzzle has a unique solution.
 * Returns:
 *   If the puzzle has a single solution, then the method returns
 *   true.
 */
```

```
Sudoku.prototype.isSolutionUnique = function() {
  var i, j;
  this.empty = [];
  this.solutions = 0;
  //Search for all the empty cells:
  for (i = 0; i < 9; i++) {
    for (j = 0; j < 9; j++) {
      if (this.board[i][j].getNumber() == null) {
      //Store the row and column indexes of the cell into the array
      //empty:
        this.empty.push({row: i, col: j});
      }
    }
  }
  this.checkCell(0);
  //The checkCell() recursive function stores the number of
  //solutions in the property solutions. If there is more than one
  //solution, then solutions is set to two.
  if (this.solutions == 1) {
    return true;
  }
  else {
    return false;
  }
};


/*
 * This is a recursive function used by the isSolutionUnique()
 * method.
 */
Sudoku.prototype.checkCell = function(n) {
  var number;
  for (number = 1; number < 10; number++) {
  //Try to insert numbers from one to nine, one at a time.
    if (this.setNumber(number, this.empty[n].row,
                                      this.empty[n].col)) {
    //If the insertion was successful.
      if (n == this.empty.length - 1) {
      //If the last of the empty cells was filled, we have a
      //solution.
        this.solutions++;
      }
      else {
      //Otherwise, go and check the next of the empty cells
      //recursively.
        this.checkCell(n + 1);
      }
    }
```

```
    if (this.solutions > 1) {
    //More than one solution already found: no need to continue.
      return;
    }
  }
  //All the numbers have been tried for this cell, so clear the
  //cell before returning:
  this.setNumber(null, this.empty[n].row, this.empty[n].col);
};
```

Finally, this is an example of generating a Sudoku puzzle using the above-defined SudokuGenerator class and the isSolutionUnique() method of the Sudoku class:

```
window.onload = function() {
  var s = new Sudoku();
  var g = new SudokuGenerator();
  do {            //Repeats until we have a single-solution Sudoku.
    g.reset();    //Loads a completely solved puzzle.
    g.shuffle();  //Swaps rows and columns randomly.
    g.removeNumbers(30);        //Leaves 30 clues in the puzzle.
    s.setClues(g.getClues());   //Initializes the Sudoku using
                                //the obtained clues.
  } while (!s.isSolutionUnique()); //Checks whether this is
                                   //a single-solution Sudoku.
  s.start("sudoku-puzzle");        //Shows the puzzle to the user.
};
```

Ways to Continue

In this appendix, we glance through some of the more exciting technologies that are not part of the core JavaScript language and therefore change and develop more rapidly. The intention is not to write an exhaustive reference but rather a gentle introduction to those technologies, to broaden your horizons and make it easier for you to find your way through the formidable and ever-growing load of available sources on the topic.

B.1 Graphics with Canvas

JavaScript can be used to dynamically draw client-side graphics, which is an important feature in modern web application design, where clients are responsible for presenting data provided by servers. Because composing and rendering graphics from data in real-time uses a lot of computational time, this takes a substantial computational burden from servers, possibly saving on hardware equipment. Apart from that, raw images are normally much bigger than the code used to produce graphics on the client side, which significantly reduces needed bandwidth.

There are two quite powerful technologies for generating client-side graphics. The first one is based on the <canvas> HTML5 element using a JavaScript-based API for drawing. Another graphics technology is called Scalable Vector Graphics (SVG), which is conceptually quite different from <canvas> in that you create drawings by composing a tree of XML elements instead of invoking JavaScript methods. Each of the two technologies has its strengths and weaknesses but essentially there's nothing that one could do which the other couldn't. Because the <canvas> element is probably easier to grasp for a beginner than SVG, we only put some focus on the <canvas> element.

The <canvas> HTML5 element has no visual display of its own. Rather, it sets up a surface within a document on which you can draw using a powerful API that the element exposes to client-side JavaScript in the form of a Canvas object. The surface created by Canvas is a bitmap, which operates in a so-called *immediate mode*. In practice, that means that your program renders graphics directly to the Canvas's bitmap and has no "memory" of the graphical objects drawn on the canvas—the only thing that stays behind is the resulting bitmap. A direct consequence of this fact is that if you want, for example, to remove a graphical object from the bitmap, you must redraw the whole bitmap by yourself or at least redraw the part of the bitmap affected by the removed object.

To set up a canvas in HTML, you most commonly provide the id attribute for easier reference from JavaScript code, and the width and height attributes to set physical dimensions of the element. The <canvas> element can also have content, which will only be displayed if the visitor's browser doesn't support it. This is an example of HTML code defining a 400×200 px canvas:

```
<canvas id="someCanvas" width="400" height="200">
  Your browser doesn't seem to support HTML5 Canvas.
</canvas>
```

As with any other HTML element, you can of course also set Canvas dimensions through CSS or even dynamically using JavaScript.

From within JavaScript, you fetch a reference to a canvas element in the same fashion as you fetch a reference to any other HTML element. For example, you can use the getElementById() method:

```
var canvas = document.getElementById("someCanvas");
```

A peculiar thing about Canvas is that most of the time you don't actually use Canvas for drawing. Instead, you use its *drawing context* object. You get this object by calling the getContext() method on a Canvas object using the argument "2d":

```
var context = canvas.getContext("2d");
```

This way, you obtain a CanvasRenderingContext2D object, which can be used for two-dimensional drawing on the canvas. If you want to use three-dimensional graphics, you can use WebGL (Web Graphics Library) JavaScript API, which is the newest addition to the <canvas> element and is already supported by most major browsers. If you want to use WebGL, you can refer to the *developer.mozilla.org/en-US/docs/Web/WebGL* where you'll find a quite exhaustive collection of useful WebGL articles and other resources.

Drawing Lines

Drawing a line on a canvas takes some programming effort because to draw even a single line you need to define a *path*. A path is a series of one or more *subpaths* and a subpath is a series of one or more lines. When you're done defining subpaths, you can *stroke* them and you can also fill the individual areas enclosed by the subpaths. A new path and subpath begin with the beginPath() and moveTo() methods, respectively. You don't have to end a path or subpath because it closes automatically when a new path or subpath starts. Note that the moveTo() method starts a new subpath and at the same time defines the starting point for that subpath. For example, the following code fragment draws a single line from point (10, 190) to point (390, 10) on our previously defined canvas:

```
context.beginPath();      //Begins a new path
context.moveTo(10, 190);  //Begins a new subpath starting at
                          //(10, 190)
context.lineTo(390, 10);  //Adds a line from (10, 190) to (390, 10)
context.stroke();         //Strokes the path
```

This is what the code will produce in the browser window.

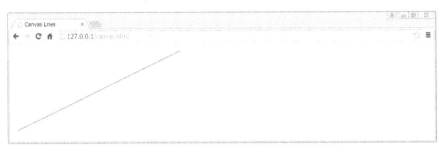

The line may strike you as odd if you're accustomed to the standard, positively oriented Cartesian coordinate system. Namely, the default canvas coordinate system is negatively oriented (i.e., with the positive y-axis pointing down) with the origin (0, 0) in the upper left corner of the canvas. That's why the point with the y-component value 190 is placed lower than the point with the y-component value 10.

The next example is a bit more complex. There is still a single path, but this time composed of two subpaths. The first of the subpaths makes up a rectangle and the second one a triangle:

```
context.beginPath();       //Begins a new path
context.moveTo(180, 20);   //Begins the first subpath: a rectangle
context.lineTo(30, 20);
context.lineTo(30, 180);
context.lineTo(180, 180);

context.moveTo(370, 20);   //Begins the second subpath: a triangle
context.lineTo(220, 100);
context.lineTo(370, 180 );
context.closePath();       //Closes the second subpath
```

Observe a call to the closePath() method at the end of the second subpath. The closePath() method closes the most recently added subpath so that it connects the last point in the subpath to the first point of the same subpath. That said, nothing is drawn in the browser window until we call either of the methods stroke() or fill(). As already seen in the previous example, stroke() simply draws all the lines of the path, while fill() fills the areas defined by individual subpaths with the current fill color. These are the lines that finalize the drawing:

```
context.fillStyle = "lightgray"; //Sets the fill color
context.fill();   //Fills the areas enclosed by the subpaths with
                  //the current fill color
context.stroke(); //Strokes the path
```

This is the drawing you will see in the browser window.

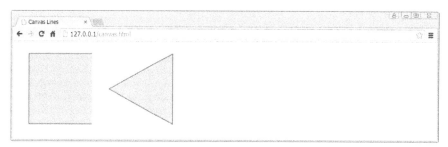

Notice that the rectangle is in fact enclosed by only three lines, each produced by one of the calls to lineTo(), while the triangle has all of its sides, the last one being the result of the call to closePath(). The gray areas in the drawing are produced by fill() and you can observe that the rectangle is filled as though the subpath were closed. Note that calling closePath() explicitly is only required for stroke() because fill() automatically closes a subpath to get a closed area to fill.

Canvas State

In our last example you probably noticed how the fill color was set as a property of a CanvasRenderingContext2D object. Graphics parameters like color and line width are not passed as arguments to drawing methods but are rather part of the generic *graphics state* of the canvas. You can think of this partition of the graphics state and drawing methods as being like the partition of presentation and content shared between CSS and HTML, respectively.

Technically, you set the state of the canvas by assigning appropriate values to certain properties of its context object. For dimensions, CSS pixels are assumed, while colors can be specified as CSS color strings, such as "#0FA" or "green". For instance, we could use the following values to change the look of our triangle and rectangle:

```
context.lineWidth = 2;        //Sets the width of lines
context.lineStyle = "black";  //Sets the color of lines
context.fillStyle = "#0FA";   //Sets the fill color
```

Note that the canvas state is used only by the stroke() and fill() methods, and that the state which is in effect at the time of calling stroke() or fill() is used. That means that if line width changes from, say, two to four between two distinct calls to lineTo(), both lines will be drawn four pixels wide because at the time of calling stroke() this is the current width.

Drawing Rectangles and Circles

You can draw a rectangle also by simply calling the rect() method of the context object. The method accepts four arguments, the first two specifying the position of the rectangle's upper left corner and the next two the rectangle's width and height. Alternatively, you can specify any other corner of the rectangle and pass negative width and/or height accordingly. The created rectangle represents a subpath of its own and cannot be connected to any other subpath.

If you search for a method to draw a circle, you won't find it because there's no special method for drawing circles. Instead, you can use the arc() method, which draws a part of a circle. The method's five arguments are: *x*- and *y*-coordinates of the center point, a radius, a start angle, and an end angle. The angles define the start and end points along the circle between which an arc should be drawn and are given in radians. Unlike rect(), arc() adds an arc to the current subpath. arc() also accepts an optional Boolean parameter, which causes an arc to be drawn in anticlockwise direction if set to true.

For example, the following code draws a circle inscribed into a square:

```
context.beginPath();

//Adds a full arc (0 to 2 * PI rads) to the subpath. The last
//argument (true) specifies that the arc direction is
//anticlockwise, so the fill will apply to the circle's outside:
context.arc(70, 70, 50, 0, 2 * Math.PI, true);

//rect() begins a new subpath:
context.rect(20, 20, 100, 100);

context.lineWidth = 2;
context.lineStyle = "black";
context.fillStyle = "lightgray";
context.fill();
context.stroke();
```

This is the drawing produced by the above code.

Apart from rect(), there are three more methods for drawing a rectangle defined by CanvasRenderingContext2D. The fillRect() and strokeRect() methods fill and stroke the specified rectangle, respectively. While doing so, fillRect() uses the current `fillStyle`, and `strokeRect()` uses the current `strokeStyle`. The third method, `clearRect()`, is like `fillRect()`, except that it uses a transparent black color for fill, which is the default color for an empty canvas. One important fact about these three methods is that they do not form a subpath nor do they need any calls to `fill()` or `stroke()`. Even when any of these methods is invoked between any of the subpath-forming functions, it does not affect the current path in any way.

Running Clock with Hands Example

Drawing on HTML5 Canvas is quite fast, which makes it very suitable for animations. In this section you'll see how a simple clock with hands can be animated using Canvas.

In terms of its basics, an animation is a constant repeating of two operations: erasing a drawing area and showing a new, slightly modified image, which delivers an illusion of movement. Our job is therefore to produce graphics for a complete clock and update it every second.

The HTML part of an example is nothing special, of course:

```
<!DOCTYPE html>
<html lang="en">
<head>
  <meta charset="utf-8">
  <title>Clock With Hands</title>
  <script src="/scripts/clock.js"></script>
</head>
<body>
  <canvas id="clock" width="300" height="300"></canvas>
</body>
</html>
```

The first of the JavaScript functions is the `eraseCanvas()` function:

```
/*
 * Paints a canvas with a light gray color.
 * Parameters:
 *   id: the ID of the canvas which to paint.
 */
var eraseCanvas = function(id) {
  var canvas = document.getElementById(id);
  var context = canvas.getContext("2d");
  context.fillStyle = "lightgray";
  context.fillRect(0, 0, canvas.width, canvas.height);
};
```

This function is used to "erase" the canvas as the name suggests. It uses `fillRect()` to paint the whole canvas with a light gray color, which becomes a background color for the clock. Since the call to the `fillRect()` method fills the entire canvas area, this also erases all of the clock's hands. The `eraseCanvas()` function accepts the ID of the `<canvas>` element to be erased. This makes the function more generic as it can erase any canvas. Notice that the function automatically detects the canvas dimensions by using the `width` and `height` properties of the Canvas object.

The drawing part of our program is implemented in two levels. First, we write a `drawHand()` function to draw a single hand, and then we assemble a clock by calling this function three times, each time specifying a necessary width, length, and angle of the hand. This is the definition of the `drawHand()` function:

```
/*
 * Draws a single hand from the center of a canvas.
 * Parameters:
 *    id: the ID of the canvas on which to draw.
 *    width: the width of the hand in pixels.
 *    length: the length of the hand as a fraction of the radius of
 *            the clock's face.
 *    angle: the angle of the hand specified in the minutes of an
 *            hour rather than in degrees (i.e., values from zero to
 *            60 define a full circle).
 * Example:
 *    Draws a hand with a length of 60 percent of the radius of the
 *    clock's face, pointing at three o'clock:
 *    drawHand("clock", 2, .6, 15);
 */
var drawHand = function(id, width, length, angle) {
  var canvas = document.getElementById(id);
  var context = canvas.getContext("2d");

  //Calculates the center point of the canvas:
  var centerX = canvas.width / 2;
  var centerY = canvas.height / 2;

  //Converts "length" to CSS pixels and "angle" to radians:
  length = Math.min(centerX, centerY) * length;
  angle = Math.PI * angle / 30;

  context.beginPath();
  context.moveTo(centerX, centerY);
  context.lineTo(centerX + length * Math.sin(angle),
                 centerY - length * Math.cos(angle));

  context.lineWidth = width;
  context.strokeStyle = "black";
  context.stroke();
};
```

Because we are dealing with a clock, it is more convenient to think in terms of the minutes of an hour, rather than degrees when specifying the angle at which the hand should be drawn. That's why we measure the angle in minutes of time (i.e., from 0 to 60) rather than in degrees, so that an angle of 15 actually means 90 degrees or $\pi/2$ radians. To calculate the end point of the hand in Cartesian coordinates, the trigonometric functions sine and cosine are used. Notice the subtraction in the second argument of the lineTo() method, which is due to the fact that the positive *y*-axis actually points down.

We put it all together by defining the following function:

```
/*
 * Draws a clock showing the current time.
 * Parameters:
 *   id: the ID of the canvas on which to draw.
 */
var showClock = function(id) {
  var canvas = document.getElementById(id);
  var context = canvas.getContext("2d");
  var now = new Date();
  eraseCanvas(id);
  drawHand(id, 1, .9, now.getSeconds()); //The second hand
  drawHand(id, 4, .8, now.getMinutes()); //The minute hand
  drawHand(id, 4, .6, now.getHours() * 5 +
                      now.getMinutes() / 12); //The hour hand
};
```

Notice how easy it is to use the drawHand() function for the second and minute hands—for an angle we simply use whichever value the methods getSeconds() and getMinutes() return. We have a little more work with the hour hand, though. Because hours run from 0 to 12, we must multiply these values by 5 to extend them around the whole circle of the clock's face. In addition, we should be able to draw the hour hand to also point between the hour marks. We accomplish this by adding the number of minutes divided by 12, as we need to shrink the whole circle of 60 minutes to an arc between two neighboring hour marks.

Finally, the Window's onload handler takes care of the clock being drawn every second:

```
window.onload = function() {
  //Show the clock immediately:
  showClock("clock");
  //Schedule clock update for each second in the future:
  setInterval(showClock, 1000, "clock");
};
```

This is what our clock looks like when it is just about time for the morning coffee.

Further Exploration

The `<canvas>` HTML5 is an exciting technology, one that offers considerably more than I have had a chance to put on these few pages. You can also use a canvas to work with other graphic elements such as patterns, curves, pixels, text, and images. You can make elements transparent, add shadows, and you can apply coordinate system transformations. And yes, you can detect mouse clicks over certain areas of the canvas to create interactive graphics. I therefore encourage anyone who feels he/she could make a great use of this tempting technology to search among countless sources of information to find out more about it. For example, the book *HTML5 Canvas* by Steve and Jeff Fulton offers a decent start.

B.2   Local Data Storage through Web Storage

As soon as you're connected to the web, there are two places where you can store your data: on the web server (server-side) or on your own computer (client-side). Both have their pros and cons, and you'll learn which type of storage works best for which purpose as you progress in your web programming proficiency. However, as you can read in the paragraph about security concerns later in this section, sensitive data definitely belongs to a web server and not to a client computer.

Web applications and even web documents can improve tremendously if you furnish them with some client-side memory. You can program web pages to store data such as user preferences or even save the complete state of the page, so your visitor always finds a page exactly as he/she left it at the end of his/her last visit. In the early days of the Internet, storing user data was only possible on a server. This was soon recognized as a drawback because of excessive storage and retrieval load on servers for tasks that could be managed locally by a client. As early as in 1994, Netscape Communications used cookies (a type of client-side data storage) to implement a reliable virtual shopping cart to store items a user selected from different pages of a site, possibly across multiple visits. Online shopping sites today almost exclusively store selected items in a database on the server side. For many web applications, however, local data storage capability has become indispensable, especially when it comes to running web applications offline. This makes web applications look and feel more like real applications.

For a long time, cookies were the only client-side data storage mechanism around. However, they were designed for use by server-side scripts and it was quite awkward to utilize them from client-side programs. As the HTML5 *Web Storage* API became universally available, cookies slowly turned back to their initial role as a client-side storage instrument for server-side scripts.

Apart from the Web Storage, there exist some other local storage technologies, but none of them succeeded in getting the level of standardization that Web Storage did. The potentially exciting *Filesystem API*, conceptualized to work with a private local filesystem utilizing file-based storage mechanisms familiar to most programmers, was pronounced dead in April 2014 because of little interest shown by the majority of browsers. Also of interest was a client-side database support, whose standardization efforts also failed and that is not implemented equally in all browsers.

`localStorage` and `sessionStorage` Objects

The Web Storage API is composed of two objects, `localStorage` and `session` `Storage`, which are defined as properties of the Window object. Both objects are in fact associative arrays that store key-value pairs and implement exactly the same set of methods and properties. The only difference between them is their *scope* and *persistence*, which define who can access the data and for how long.

Data stored in a `sessionStorage` object—as the name suggests—are roughly limited to a single browser tab, both in their scope and persistence. If you open the same document in two different tabs, each of them will have its own `sessionStorage` object, which means that scripts from one tab cannot access (neither read nor write) data stored by scripts from the other tab. Data stored in a `sessionStorage` object will be deleted as soon as you close the tab associated with them.

On the other hand, data stored in the `localStorage` object stay on the user's computer forever, or at least until a web application explicitly deletes them or a user clears them through the browser's user interface. Besides, `localStorage` has a broader scope than `sessionStorage`—it is scoped by the origin of a document. The document origin is determined by a protocol, a hostname, and a port, and a change to any of these also changes the document origin. For example, the following four URLs represent four different document origins:

> *http://www.nowheresite.com*
>
> *https://www.nowheresite.com* (A different procol)
>
> *http://www.nowhere-site.com* (A different hostname)
>
> *http://www.nowheresite.com:42* (A different port)

All documents with the same origin share data from a single `localStorage` object, which means that they can read as well as override each other's data.

Note that physical memory used to store local data is set aside by the browser and cannot be accessed by browsers of other vendors. This effectively limits `localStorage` scope to a specific browser vendor as well. So, if you visit a site using one browser, you cannot access data stored during that visit through another browser, even if you are viewing the same site within the same origin.

Storage API

As already mentioned, the `localStorage` and `sessionStorage` objects store associative arrays of key-value pairs. One way of using these objects is that you simply set and query their properties as you can do with any other JavaScript object. For example, you can store a counter with a value of one in either of the following two ways:

```
localStorage.counter = 1;
localStorage["counter"] = 1;
```

Note that all values are eventually stored as strings even if you provide other data types. So, if you want to increment the stored counter, be careful to convert it to a number before incrementing its value:

```
localStorage.counter += Number(localStorage.counter) + 1;
```

While storing and retrieving data by setting and querying properties of storage objects is convenient, it is often better to use special methods to manage data storage. For one thing, they are safer and provide more functionality. To store and retrieve a value, you use the methods setItem() and getItem(), respectively:

```
localStorage.setItem("counter", 1);  //Stores a counter with a
                                      //value of one
c = localStorage.getItem("counter"); //Retrieves the counter value
```

Don't forget that values are always stored as strings, so getItem() returns a string too.

You can also delete a particular value or even delete all stored values using the methods removeItem() and clear(), respectively. For example:

```
localStorage.removeItem("counter"); //Removes the counter from
                                     //storage
localStorage.clear();                //Removes all stored values
```

Like regular JavaScript arrays, localStorage and sessionStorage also have the length property. But this property works differently in that it always holds the number of stored values. Recall that the length property of Array is one bigger than the largest array index, and is not equal to the number of elements when an array is sparse. The key() method of the localStorage and sessionStorage objects allows you to enumerate the names of all stored values by passing numbers from zero to length - 1 to it. As an example, the next code fragment writes to the JavaScript Console all the values stored in localStorage:

```
for (i = 0; i < localStorage.length; i++) {
  name = localStorage.key(i);
  console.log(localStorage.getItem(name));
}
```

An Example

As a more realistic example, imagine that you want to add a memo utility to your website, so that your visitor could write down his/her thoughts and be able to read them during his/her next visit.

To do that, you first need to include an <textarea> HTML element:

```
<textarea id="memo" rows="5" cols="100">
```

The JavaScript part of the example needs to do three things: First, it has to get a reference to the `<textarea>` element. Second, it needs to read the previously stored text from `localStorage` (if it exists) and write it to the `<textarea>` element. Finally, it has to react to any change to the text inside the `<textarea>` element and save it to `localStorage`. Here's complete code implementing all three tasks:

```
window.onload = function() {
  m = document.getElementById("memo");

  //If the item exists, get its value from localStorage and show it
  //inside <textarea>:
  if (localStorage.getItem("text")) {
    m.value = localStorage.getItem("text");
  }

  //Every time the content of <textarea> is changed, store the new
  //content to localStorage:
  m.oninput = function() {
    localStorage.setItem("text", this.value);
  };
};
```

Synchronizing Pages with Storage Events

Each time the data stored in `localStorage` or `sessionStorage` changes, a storage event is fired on any tabs or windows that are currently displaying pages with the same origin. However, it is not fired on the window that caused the change. A storage event can conveniently be used to synchronize all the pages a user is viewing, so that a change that a user makes in one window is immediately reflected in other windows or tabs as well. A programmer can utilize this event by implementing the `onstorage` event handler, which is a property of the Window object. While Web Storage is amongst the better supported HTML5 features, the storage event itself is not fully and equally supported in all browsers. For example, if you want your code to also work in Firefox, you should attach the storage event using the Window's `addEventListener()` method, instead of directly defining the `onstorage` event handler. However, the `addEventListener()` method is not covered in this book.

Security Concerns

When a web browser offers to remember a password for you, it stores it securely in encrypted form on your hard drive. That said, the Web Storage mechanism does not encrypt data before saving it, and stored data is freely accessible to uninvited snoopy users sharing access to the computer as well as to malicious software that might lurk in the system. For this reason, never use any form of client-side storage for storing sensitive user data like passwords or credit card numbers.

Further Reading

If you want to read more about Web Storage, there's quite an elaborate and beginner-friendly coverage of the topic inside the book titled *HTML5* by Matthew MacDonald. You may also want to look inside *JavaScript: The Definitive Guide* by David Flanagan for more elaborate coverage of the subject. Both books also cover offline applications quite thoroughly.

Online, there's a decent article available at *www.sitepoint.com/an-overview-of-the-web-storage-api*. For more comprehensive reading, though, you might want to dig into *developer.mozilla.org/en-US/docs/Web/API/Web_Storage_API*.

B.3   Ajax

In the not-so-distant past, to retrieve new data from a web server, browsers had to reload the whole web page. Not only was this slow due to the substantial amount of data that had to be sent back and forth, but it was also uncomfortable for a user as the whole page disappeared and reappeared again even for updating the smallest chunk of information from the server. Then, in 2005, the term *Ajax* was fashioned, which originally stood for Asynchronous JavaScript and XML. Ajax was first used by Google in its magical Google Suggest, the other name for Google's auto-complete functionality. It works so that when a visitor starts to enter letters in a search field, a dropdown menu appears, suggesting for the user possibly interesting terms. Although the information is coming from the web server, the whole page stays put during the operation. This is in fact the only—albeit extremely important—advantage of Ajax: it permits a web page to be updated with information from the web server without ever reloading the page. That way, web pages feel more responsive, greatly enhancing the user experience. How is that accomplished?

Ajax isn't a technology in and of itself, but rather an interaction of more technologies working together. You need JavaScript, an XMLHttpRequest object, and some server-side scripting. The latter is, unfortunately, not covered in this book. Nevertheless, you will get a basic idea of how Ajax works.

All the traffic between web browsers and servers is specified by the Hypertext Transfer Protocol (HTTP), which is usually not controlled by scripts but happens as a consequence of user actions such as following a link or typing a URL. That said, the XMLHttpRequest object defines an API that allows a programmer to control HTTP requests through JavaScript code. Basically, JavaScript uses an XMLHttpRequest object to send a request to a web server, which responds by sending some data back. JavaScript then collects the received data, again through the XMLHttpRequest object, and does with them whatever it likes.

Before you can use the just-mentioned HTTP API, you must create an XMLHttpRequest object:

```
var myRequest = new XMLHttpRequest();
```

Now you're ready to make an HTTP request, which is made by calling the open()

method of the XMLHttpRequest object. The two required parts of the request are the method and URL:

```
myRequest.open("GET", "greetings.txt");
```

The first argument of the open() method selects the desired HTTP method and the second one specifies the URL to which the request is to be addressed. The HTTP method is most often GET or POST. In a nutshell, you can use GET to get data from the server, and POST to modify the server data (e.g., update a database). There's also a technical difference between both methods in that GET completely specifies the requested resource within the URL, while POST expects additional data in the request body. You have probably seen URLs like *.../index.php?search=holidays*, where there's additional information added at the end of the regular URL (after the question mark). This is the GET method's way of specifying additional information needed by a server-side script to process the request. For instance, if the web page requests the server to run the *index.php* script, which returns a list of pages containing some specific keywords, the script also needs the keywords to search for ("holidays" in our example).

Note that the URL given as a second argument in our example of the open() method is a plain text file and the myRequest object will simply get the whole content of the file from the server. From a practical point of view, this is rarely useful, but it nevertheless demonstrates the basic procedure that is to be performed by a JavaScript program to process an HTTP request. In reality, you would instead provide a URL of a file containing some script that would intelligently process the HTTP request. A very common example would be a PHP script communicating further with a MySQL database.

After composing the request, all you have to do is send it to the server:

```
myRequest.send();
```

No matter what kind of technology sits on the other side of an HTTP request, you more or less only need to provide the correct URL and wait for the response after sending the request. While it is possible to wait for the response synchronously, this is not recommended since the send() method will block the browser until the request is completed. A much better approach is to handle the HTTP response asynchronously. To do that, you simply define the onreadystatechange handler of the XMLHttpRequest object, which listens to readystatechange events:

```
myRequest.onreadystatechange = function() {
  if (this.readyState == 4 &&   //Is the response complete?
      this.status == 200) {     //Was the request successful?
    //Do something with this.responseText
  }
};
```

Because the readystatechange event can be fired (depending on the browser) at any change of the state of the request, you must check the `readyState` property of the XMLHttpRequest object before processing the returned data. Only if the value returned by `readyState` is four, the response is complete and ready to process. It is also important that the request be successful on the whole, which results in the `status` property having the value 200 ("OK"). If the provided URL doesn't match any of the resources on the server, then `status` returns the infamous error value 404 ("Not Found").

If everything went OK, then the text returned from the server is to be found in the `responseText` property of the XMLHttpRequest object, and from now on you're on familiar ground. The following is a complete example which, when a button is clicked, demands the *greetings.txt* file through an HTTP request and places the received text inside a `<div>` element:

```
<html lang="en">
<head>
  <meta charset="utf-8">
  <title>Ajax</title>
  <script>

window.onload = function() {
  var myRequest = new XMLHttpRequest();
  myRequest.onreadystatechange = function() {
    if (this.readyState == 4 && this.status == 200) {
      document.getElementById("myDiv").innerHTML =
                                        this.responseText;
    }
  };
  document.getElementById("myButton").onclick = function() {
    myRequest.open("GET", "greetings.txt", true);
    myRequest.send();
  };
};

  </script>
</head>
<body>
  <form>
    <div id="myDiv">This will change forever.</div>
    <input type="button" value="HTTP GET" id="myButton">
  </form>
</body>
</html>
```

Note that this example will only work if you upload it to a web server because XML-HttpRequest is designed to work with an HTTP protocol. You can conveniently use the development server that we set up on page 37. If you test examples in this book using a local file system, then all the URLs are relative to *file://* instead of *http://*, consequently using a file instead of the required HTTP protocol.

Further Reading

A very basic Ajax tutorial with ready-to-run examples is available at *www.w3schools. com/ajax/default.asp*. In Robin Nixon's book *Learning PHP, MySQL, JavaScript, and CSS* you'll find more explanations about how Ajax works together with tips on using XML instead of plain text for conveniently structuring data that you pass back to your Ajax applications. If you're interested in using the jQuery library (briefly introduced in the next section) for programming your Ajax applications, you will probably want to get David Sawyer McFarland's book *JavaScript and jQuery*. If you are up for more technically profound explanation of the topic, you might want to look into *JavaScript: The Definitive Guide*, a book by David Flanagan.

B.4   jQuery

Currently by far the most popular and widely used JavaScript library is jQuery. There are many reasons why you would decide to use jQuery in your own projects, and they can all be brought under the same umbrella with the motto "write less, do more." If nothing else, programmers like to use the jQuery library because it simplifies mundane everyday tasks and hides the differences between browsers, which have made programmers reach for bottles of aspirin from time immemorial. Also important is the fact that jQuery is stable and well documented.

Getting jQuery

Technically, jQuery is simply a file containing an assemblage of JavaScript code, which you include as you would any other external JavaScript file. That is, you use the src attribute of a <script> HTML element. However, you still have two possibilities regarding the library location: either you can download the jQuery file to your own computer or you can simply link to one of the versions hosted on the web. Needless to say, each of these two approaches has its advantages and downsides but I leave them for you to discover on your own. For the sake of this section, we're going to use the jQuery library version hosted on the *jquery.com* CDN (Content Distribution Network) server:

```
<script src = "http://code.jquery.com/jquery-1.11.2.min.js">
</script>
```

The jQuery() Function

There's one single global function named jQuery() defined in the jQuery library. Because this is the central query function for jQuery, it is intensely used and the library also defines the global symbol $ as a shortcut alias for it. The jQuery() function returns a jQuery object, which contains zero or more DOM elements and a bunch of convenience methods for working with those DOM elements. Note that jQuery() is not a constructor but a factory function and does therefore not need the new operator to create a jQuery object. Most jQuery methods return the same jQuery object that they operate on. This allows for so-called *method chaining*, which is quite frequent in jQuery programming. Method chaining is a compact form of calling methods one after the other in a single "chain." For example, if you want to add a class of `hidden` to all `<p>` elements that have a class of `details`, hiding them at the same time by sliding them up, you can use the following chain of methods:

```
$("p.details").addClass("hidden").slideUp();
```

In this example you see how the `$()` function finds and selects elements simply by using CSS selector syntax. Passing a CSS selector as an argument is the most common way of invoking the `jQuery()` function, but it's not the only one. You can also pass an Element, Document, or Window object, and the `$()` function simply wraps a jQuery object around whatever object it is passed, allowing you to use jQuery methods to manipulate the element. For example, you can change the background color of an element inside an event handler attached to it simply by passing `this` to the `$()` function and use the jQuery `css()` method:

```
$(this).css("background-color", "blue");
```

The third way to invoke `$()` is to call it with a string of plain HTML code as an argument. Note that doing this does not select elements, but instead creates new elements. For example, you can create an `<h3>` heading like this:

```
var h3 = $("<h3>Latest Additions</h3>");
```

The created element is not automatically added to the document, of course. You can, for example, use the `appendTo()` method to append the newly created element as the last child of an existing element:

```
h3.appendTo("#about");
```

The above created `<h3>` heading thus becomes the last child of the element with the id of `about`.

The last way to invoke `$()` is with a function as an argument. When you do that, the passed function will be called after the document has completed loading. This is in fact the same thing as using the `onload` property of the Window object to make sure that JavaScript starts executing only after the document is fully loaded and the DOM is ready to be manipulated. Knowing that, it comes as no surprise that most jQuery programs are written as an anonymous function passed as an argument to `$()`:

```
$(function() {
  //JavaScript and jQuery code
});
```

Instead of the above form, you'll sometimes come across the older and more involved, but equivalent code:

```
$(document).ready(function() {
  //JavaScript and jQuery code
});
```

Iterating Selected Elements

You have just learned about different ways to select HTML elements using the global
jQuery() (a.k.a. $()) function. It is important to understand that $() always returns
a jQuery object containing zero or more elements. Because jQuery objects are array-
like (they have the length property as well as numeric properties from zero to length
- 1) it is possible to iterate selected elements using the for loop, but in practice you'll
seldom need to do this. For one thing, you can use the each() method to loop over
the elements in a jQuery object, but even this is not very commonly used. Most of the
time you'll use jQuery methods, which automatically iterate all the elements within
the jQuery object on which they are invoked. For example, you can change font color
of *all* the <h1> elements on the page like this:

```
$("h1").css("color", "darkblue");
```

Or, you can underline text within any element having a class of important:

```
$(".important").css("text-decoration", "underline");
```

Getting and Setting Values

The most frequent operations on jQuery objects are those that manipulate different
values connected with elements stored in jQuery objects, like getting or setting values
of HTML attributes or CSS styles. There are some general truths about jQuery getter
and setter methods worth noting:

- jQuery uses a single method for both getting and setting a value. If you invoke
  the method with a new value as an argument, then that value is set. If, however,
  you do not provide a value, the method returns the current value, if any.

- If a jQuery object contains more than a single element, then a value is set on all
  the elements in the jQuery object.

- If, however, you want to get a value, then only the first element in the jQuery
  object will be queried for the value.

- You can set more values with a single call to a method if you pass as an argument
  an object whose properties specify names and values to be set.

The examples that follow are intended to give you the basic feel of how the jQuery
getters and setters work.

The next lines of code show how to get and set HTML attributes using the attr()
method. You can also completely remove an attribute from the selected elements using
removeAttr():

```
var src = $("img").attr("src");   //Gets the value of the src
                                   //attribute of the first image on
                                   //the page.
$("a").attr("target", "_blank");  //Sets the target attribute of all
                                   //the <a> elements to _blank. All
                                   //the hyperlinks on the page will
                                   //now load in a new window.
$("a").removeAttr("target");       //Removes the target attribute from
                                   //all the <a> elements on the page.
                                   //All the hyperlinks will load in
                                   //the current window again.
```

The following are examples of getting and setting CSS properties and the class HTML attribute. The css() method is used to get and set individual CSS properties, while addClass(), removeClass(), and hasClass() are meant for manipulating the class HTML attribute:

```
//Queries the background color of the first <div>:
var color = $("div").css("background-color");

//You can use camelCase:
var color = $("div").css("backgroundColor");

//Sets the font color of all the <div> elements.
$("div").css("color", "yellow");

$("div").css({         //Sets multiple properties at once. The
  width: 50,            //argument of css() is a JavaScript object
  height: 50,           //literal.
  borderRadius: 5
});

//Adds a class to the element with id of notes:
$("#notes").addClass("highlight");

//Removes a class from all the <p> elements.
$("p").removeClass("highlight");

//Does the element with id of notes have a class of empty?
var empty = $("#notes").hasClass("empty");
```

Another set of methods allow you to manipulate HTML form values and elements' content, either in plain text or HTML markup:

```
var addr = $("#address").val();   //Gets a value from the text box
                                   //with id of address.
$("#getAmount").val(usd * rate);   //Sets a value of a text box.

var title = $("title").text();     //Gets the title of the document.
```

```
$("title").text("Working Title"); //Sets the title of the document.

var div = $("div").html();          //Gets the HTML contained in the
                                    //first <div> in the document.
//Sets the HTML content of all of the <div> elements in the
//document:
$("div").html("<h1>Z-Files</h1>");
```

Custom Element Data

jQuery has a powerful capability to associate arbitrary data with document elements or with Document or Window objects. The `data()` method allows you to attach data to the elements contained in a jQuery object. You can either pass a name and value as two arguments of the method or you can pass a single object, whose properties will be associated with the elements of the jQuery object.

You can also use `data()` to get values, of course. If you pass a single argument to it, that argument is interpreted as a name whose value is to be returned. When, however, you call the method without any argument, an object is returned that contains all the name/value pairs attached to the first element inside the jQuery object.

If you want to completely remove all or some of the data associated with an element or elements, then use the `removeData()` method.

These are some examples:

```
$("#sprite").data("counter", 10); //Sets counter on sprite to zero
console.log($("#sprite").data("counter")); //Writes 10

$("#sprite").data({x: 13, y: -8});    //Sets x and y on sprite
console.log($("#sprite").data());     //Writes {counter: 10, x: 13,
                                      // y: -8}

$("#sprite").removeData("counter");   //Completely removes counter
console.log($("#sprite").data());     //Writes {x: 13, y: -8}

$("#sprite").removeData();            //Removes all previously set data
```

Note that you can also use the `data()` method to return values set by an HTML5 `data-*` custom data attribute, where the same name transformations apply as with the JavaScript `dataset` property described on page 398. It is, however, not possible to add new custom data attributes to HTML elements using this method.

Manipulating Document Structure

You can use `html()` and `text()` to alter the document structure. However, an HTML document is not represented as a linear sequence of elements, but rather as a tree of nodes. Insertions, deletions, and replacements of elements are therefore not trivial operations, and the jQuery library defines quite a number of handy methods to carry them out. The most essential are the `append()` and `remove()` methods, which are described in this section.

As its name suggests, the `append()` method appends an element at the end of an existing one, so that the appended element becomes the last of its children. The method accepts as an argument the content that is to be appended. You can specify a string of plain text or HTML markup, but you can also pass a jQuery or Element object. Note that, if you attempt to append an element that already exists in the document, that element is removed from its current position to be inserted to a new one. If `append()` is invoked on a jQuery object that contains more than one element, then the element to be appended is cloned as necessary and appended to all the elements within that jQuery object.

For instance, to append a `<p>` element to all the `<div>` elements in the document, you invoke `append()` in the following manner:

```
$("div").append("<p class='answer'>42</p>");
```

Note that `append()` returns the jQuery object on which it is invoked. Sometimes, however, it is more convenient to get the jQuery object that contains the newly added content instead. In this case you can use the `appendTo()` method like this:

```
$("<p class='answer'>42</p>").appendTo("div");
```

Although both of the above lines have exactly the same effect on the document, there are two important differences between them. We have already mentioned that they return quite different jQuery objects. The second difference is connected with the behavior of the `$()` function. If you tried to append plain text using `appendTo()`, then that text would be interpreted as a selector that identifies the target element and not as a string of text to be appended. It is therefore not possible to append plain text using the `appendTo()` method.

When you want to remove one or more elements from a document, you can use the `remove()` method. This method simply removes all the elements contained in the jQuery object on which it is invoked. For example, to remove all the `<h3>` elements from the document, you write:

```
$("h3").remove();
```

jQuery Events

To register an event handler in jQuery, you simply call one of its event-handler registration methods, passing to it a reference to the event-handler function definition. For example, you can register an event handler for click events like this:

```
$("img").click(myClickHandler);
```

Note that the above call to `click()` registers the `myClickHandler()` function on all of the `<img>` elements found in the document.

Here are some of the event handler registration methods defined by jQuery:

```
click()      dblclick()   keydown()
keypress()   keyup()      load()
mouseover()
```

These are common event types and you can find their descriptions in the JavaScript reference on page 402.

jQuery also defines its own Event object, which is always passed as an argument to a jQuery event handler. You can find additional information about the event that you are handling by querying the Event object's properties. These are some of the properties of the jQuery Event object:

```
altKey    button    clientX
clientY   ctrlKey   shiftKey
which
```

You can find descriptions of these properties in the JavaScript reference on page 402. Note that `which` is no longer supported by Web standards but it is emulated by the jQuery library to carry additional information about keyboard and mouse events. It seems that in JavaScript, at the time of writing this book, there is no standardized property to work with keyboard events. The properties `charCode` and `keyCode` are deprecated, while the `key` property, which replaces them, is not yet adequately supported by browsers.

Let's put it all together in a simple example that waits for mouseover and keydown events, and writes the value of the `which` property to the JavaScript Console window each time an event occurs.

```
var writeWhich = function(event) {
  console.log(event.which);
};

$(function () {
  $("p").mouseover(writeWhich);   //Listens to mouseover events over
                                  //each <p> element
  $(document).keydown(writeWhich); //Listens to keydown events
});
```

There are some remarks we should make about the above example. First of all, in order for the mouseover event to be triggered on a particular paragraph, that paragraph should have a nonzero size. This means you have to put at least one printable character inside your sample paragraph if you want to test the example. Next, keyboard events can only be triggered on elements that have focus. Few elements other than form elements can receive focus, so keyboard events are often attached to the document if they are not to be handled by forms. In the above example, the keydown event is used, because this event is also triggered by pressing non-character keys such as arrow or function keys. On the other hand, if you use the keypress event and want to detect whether any of the modifier keys (Shift, Ctrl, or Alt) is also pressed, then you must also query the corresponding shiftKey, ctrlKey, or altKey properties, which return a Boolean value. Note that most of the key combinations using the modifier keys Ctrl and Alt represent shortcuts for browser commands and will therefore be intercepted by the browser. For example, the key combination Ctrl+T will open a new browser tab and will not fire a keypress event on the document.

One more useful feature of the jQuery event API is that it allows you to trigger events manually. You already know that event handlers are normally invoked automatically when certain events occur, but sometimes it is useful to be able to trigger an event from within your code. This is made possible with the trigger() method, which accepts as the first argument a string specifying the type of event. For example, you can trigger a click event on all the elements with a class of special like this:

```
$(".special").trigger("click");
```

You can also pass arguments to the event-handler function when you manually trigger an event. One way of doing this is to pass an object as the first argument to the trigger() method, instead of just passing a string specifying the type of event. If you do this, then a new jQuery Event object will be created and the properties of the passed object will be added to it. You must still specify the event type, of course, which you do by setting the type property of the passed object. For example, you can trigger the keydown event like this:

```
$("p").trigger({type: "keydown", which: 42});
```

Animation

When you add or remove an element from a page, or otherwise change the page appearance, you can do that either instantaneously or you can animate the change by gradually modifying certain properties such as size or opacity. It is often not a very pleasing experience for a visitor if an element just disappears into thin air. It is much nicer if it slowly slides up or fades out, for example. jQuery makes such animations easy by defining some convenient animation methods.

There are some general facts about jQuery animations that you should know:

- Every animation has a duration, which a programmer can change. You can specify a duration either as a number of milliseconds or by providing a string of predefined text. You can use the strings "fast" and "slow", which stand for 200 ms and 600 ms, respectively. If no duration is specified, then the animation lasts for 400 ms, which is default.

- All the jQuery animated effects are asynchronous. When you invoke an animation method, it returns immediately and the animation is performed in the background. This means that the code which follows will be executed while the animation has not yet necessarily completed.

- If there's some code that you don't want to execute while the animation still lasts, you can register it as a callback function. The callback function will be invoked automatically when the effect completes. You cannot pass any arguments to the callback function, but the `this` keyword will be accessible inside the function referring to the element that was animated. So if you need to pass additional information to the callback function, you can do that using the `data()` method, which allows you to associate arbitrary data with an element.

- If you invoke an animation method on an already animated element, then the new animation effect is put into an animation queue and is not animated until all the effects before it are completed. Note, however, that each element has its own animation queue.

There exist nine basic effects that jQuery defines for hiding and showing elements, and one method intended for custom animations. The simplest are effects that only animate the CSS `opacity` property: `fadeIn()`, `fadeOut()`, and `fadeTo()`. These three effects change the element's visibility but don't remove the element from the page even if it becomes completely invisible (when `opacity` becomes zero). The `fadeIn()` and `fadeOut()` methods animate `opacity` towards one and zero, respectively. However, if you want to animate `opacity` to any other value, use the `fadeTo()` method, which accepts the target `opacity` value as the second argument. The first argument is the animation duration, which you must always specify when calling `fadeTo()`.

If you want to remove the element from the page layout (i.e., set the CSS `display` property to `none`), then call `hide()`. This method gradually shrinks the element's `width` and `height` to zero and reduces its `opacity` to zero at the same time. If you want to show the element again, then call `show()`, which reverses the process. By default, `hide()` and `show()` do not animate the change, but hide and show selected elements instantly. If you want to get an animated effect, then pass an optional duration argument. You can also toggle an element from visible to hidden and vice versa by simply calling `toggle()`. Toggle effectively calls `hide()` if the element is visible, and it calls `show()` if the element is hidden.

The last three of the predefined effects are `slideUp()`, `slideDown()`, and `slide Toggle()`. `slideUp()` is similar to `hide()` but it only hides the element by animating its `height` to zero and then setting its CSS `display` property to `none`. The element's width and opacity don't change. The `slideDown()` property shows the element again by reversing the process. `slideToggle()`, as its name suggests, toggles between `slideUp()` and `slideDown()`.

These are some examples of the basic animation methods in action:

```
$("p").fadeTo(500, 0.5); //Changes opacity to 0.5 in 500 ms
$("p").hide(500);        //Hides all the paragraphs in 500 ms
$("p").show();           //Shows them back immediately
$("p").slideUp();        //Slides them up
$("p").slideToggle(function(){   //Registers a callback function
  $(this).text("That's all folks!"); //and slides the paragraphs
});                                   //back down
```

You can produce more universal animated effects with the `animate()` method. In its simplest form, the method accepts a so-called *animation properties object*, which is no more than a plain JavaScript object whose properties are CSS property names. The specified values of these properties are target values towards which the animation will run. Note, however, that only numeric properties can be animated and you cannot animate things like colors and fonts. If you use a number as a property value, then pixels are assumed. You may also specify different units, which you may do if you use a string as a property value. There's another option you have if you use a string: you can specify a relative value by prefixing a value with either "+=" to increase or "-=" to decrease the value of the specified property. For example, you can animate moving all the selected elements to a single absolute position like this:

```
$("p").animate({left: 100, top: 100});
```

In order for this to work, the CSS `position` property of the selected elements must be set to `absolute`, of course.

The next example inflates the width and height of the selected elements by 10 pixels:

```
$("p").animate({width: "+=10", height: "+=10"});
```

Apart from numeric values, it is also possible to use the values `"hide"`, `"show"`, and `"toggle"`. The value `"hide"` first saves the current numeric state of the property and then animates it towards zero. At the end, it sets the `display` property of the animated element to `none`. To animate a property back to its saved value, you use `"show"`. Not surprisingly, `"toggle"` performs either a show or a hide, depending on the current element state.

As an example, the next code mimics an effect similar to that produced by the `slide Toggle()` method except that elements slide to the left:

```
$("p").animate({
    width: "toggle",
    "padding-left": "toggle",
    "padding-right": "toggle"
});
```

Note the quotes around the hyphenated property names. Because they contain hyphens, they are not legal JavaScript identifiers and have to be quoted. You can, of course, also use the camelCased alternatives `paddingLeft` and `paddingRight` if you don't like to use quotes.

Sliding Puzzle

At the end of this concise review of the jQuery library, we look at a somewhat more sophisticated example of a 15-puzzle. The 15-puzzle is a sliding puzzle composed of 15 numbered square tiles, arranged inside a square frame in random order as shown in this picture.

The objective of the puzzle is to place the tiles in order by making sliding moves to the empty space.

The example defines a single object named SlidingPuzzle, which contains the following properties:

- tiles: a two-dimensional array of tiles.
- empty: a plain object that holds the position of the empty space.
- tileSize: holds the display size (in pixels) of one tile.
- numbers: a two-dimensional 4×4 array of numbers from one to 16. This array is used to mark the starting positions of tiles. Because not all starting positions are solvable (in fact, half of the starting positions are impossible to solve), the program starts with an ordered array of 16 numbers and shuffles them before initializing the graphical interface of the puzzle. The position of number 16 is initialized as an empty space.

The SlidingPuzzle object also has some methods:

- setTileSize(): sets the tileSize property.
- shuffle(): shuffles the two-dimensional numbers array.
- start(): initializes the graphical interface of the puzzle.
- translateMove(): a static method that translates an integer from zero to three into an actual move towards one of the four directions (left, right, up, or down). If a move is not possible (because it would fall off the frame), the method returns false.

The example allows the player to interact with the puzzle using either a mouse or a keyboard. A mouse click on a tile that can possibly move (i.e., has empty space at one of its sides) moves that tile. Alternatively, pressing one of the arrow keys moves whichever tile is free to move in that direction. The following are the two event-handler functions to handle mouse and keyboard events:

- moveTile(): handles mouse clicks. If the clicked tile can move, then the handler moves it.
- findAndMoveTile(): handles key presses. It searches for the tile that can move and if such a tile exists, it fires a click event on that tile to actually move it.

Here's the complete code for the 15-puzzle. The code is well commented so you should have no difficulties understanding it:

```html
<!DOCTYPE html>
<html lang="en">
<head>
  <meta charset="utf-8">
  <title>15 Sliding Puzzle</title>
  <style>
/*
 * Tiles must be absolutely positioned so you have better control
 * of their position. Note that all size-related properties are
 * computed and set dynamically using the jQuery css() method.
 */
.tile {
  border-style: outset;
  background-color: #bbb;
  color: #777;
  text-align: center;
  font-family: fantasy;
  position: absolute;
}

/*
 * This is the puzzle frame. It should have position other than
 * static if its absolutely positioned children (tiles) are to
 * respect its position.
 */
#puzzle {
  position: relative;
}
  </style>

  <script src = "http://code.jquery.com/jquery-2.1.1.min.js">
  </script>

  <script>
/*
 * The constructor
 */
var SlidingPuzzle = function() {
  this.tiles = [];
  this.empty = {};
  this.numbers = [[1,  2,  3,  4], [ 5,  6,  7,  8],
                  [9, 10, 11, 12], [13, 14, 15, 16]];
  this.tileSize = null;
};

/*
 * Sets the display size of a tile. All size related CSS properties
 *   will be set according to this value.
```

```
 * Parameters:
 *    s: size of a tile in pixels.
 */
SlidingPuzzle.prototype.setTileSize = function(s) {
  this.tileSize = s;
};


/*
 * Shuffles the numbers. Call this method before initializing the
 * graphical interface with start(). The method tries to move the
 * number 16 (which is a placeholder for the empty space) in random
 * directions in the two-dimensional numbers array 100000 times.
 * If a call to translateMove() returns true, then it is possible
 * to move the empty space from its current position (x, y) in the
 * direction stored in the move object.
 */
SlidingPuzzle.prototype.shuffle = function() {
  var n;
  var x = 3, y = 3;
  var move = {x: 0, y: 0};

  for (n = 1; n < 100000; n++) {
    if (SlidingPuzzle.translateMove(Math.floor(Math.random() * 4),
                                                x, y, move)) {
      this.numbers[x][y] = this.numbers[x + move.x][y + move.y];
      x += move.x;
      y += move.y;
      this.numbers[x][y] = 16;
    }
  }
};


/*
 * Initializes the graphical interface of the puzzle.
 * Parameters:
 *    id: the ID of an HTML element to hold the puzzle.
 */
SlidingPuzzle.prototype.start = function(id) {
  var x, y;
  for (x = 0; x < 4; x++) {
    this.tiles[x] = [];
    for (y = 0; y < 4; y++)
    {
      if (this.numbers[x][y] < 16) {
//If the number does not represent the empty space, then create a
//tile and append it to the puzzle frame. Next, attach a click
//event handler to the created tile. Then set the position of the
//tile. Using data(), associate the jQuery object representing
//the tile with the position of the tile and the reference to the
//puzzle frame. All these data will be used by event handlers
//because it is not possible to pass arguments to them. The
//mouseover event handler is defined so that the mouse cursor will
```

```
//not change to the text select form when over text, but rather
//keep its normal arrow appearance. Note that 'this' inside the
//event-handler definition refers to the tile over which the mouse
//cursor is placed and not to the SlidingPuzzle object instance.
        this.tiles[x][y] = $("<div class='tile'>" +
                    this.numbers[x][y] + "</div>").appendTo("#" + id);
        this.tiles[x][y].click(this.moveTile);
        this.tiles[x][y].css("top", x * 1.2 * this.tileSize);
        this.tiles[x][y].css("left", y * 1.2 * this.tileSize);
        this.tiles[x][y].data("x", x);
        this.tiles[x][y].data("y", y);
        this.tiles[x][y].data("puzzle", this);
        this.tiles[x][y].mouseover(function(){
          $(this).css("cursor", "default");
        });
      }
      else {
//If the number represents the empty space, then store its position.
        this.empty.x = x;
        this.empty.y = y;
      }
    }
  }
//Set the dimensions of the puzzle frame and tiles:
  $("#" + id).css("width", 4 * 1.2 * this.tileSize);
  $("#" + id).css("height", 4 * 1.2 * this.tileSize);
  $(".tile").css("width", this.tileSize);
  $(".tile").css("height", this.tileSize);
  $(".tile").css("border-radius", 0.2 * this.tileSize);
  $(".tile").css("border-width", 0.1 * this.tileSize);
  $(".tile").css("fontSize", 0.8 * this.tileSize);
//Attach the keydown event handler to the document. Associate also
//a reference to the puzzle with the document. This is necessary
//because we need to refer to the puzzle from within the
//findAndMoveTile() event-handler function but it is not possible
//to pass arguments to it.
  $(document).keydown(this.findAndMoveTile);
  $(document).data("puzzle", this);
};


/*
 * A keyboard event handler, which is fired on the document. 'this'
 *    therefore refers to the document object. If a move in the
 *    desired direction is possible, then it triggers a click event
 *    on the tile that can move. Since arrow keys have codes from 37
 *    to 40 (stored in the which property of the Event object), we
 *    need to subtract 37 before passing the desired move to
 *    translateMove().
 * Parameters:
 *    event: an Event object holding the additional information
 *           about the actual event that triggered findAndMoveTile().
 */
```

```
SlidingPuzzle.prototype.findAndMoveTile = function(event) {
  var move = {x: 0, y: 0};
//Retrieve a reference to the puzzle stored in the document object:
  var pzzl = $(this).data("puzzle");
  if (SlidingPuzzle.translateMove(event.which - 37, pzzl.empty.x,
                                  pzzl.empty.y, move)) {
    pzzl.tiles[pzzl.empty.x + move.x][pzzl.empty.y +
                                  move.y].trigger("click");
  }
};


/*
 * A mouse click event handler. It is triggered on the tile that
 * should move. 'this' therefore refers to that specific tile.
 */
SlidingPuzzle.prototype.moveTile = function() {
  var i;
//Fetch the tile's current position and a reference to the puzzle:
  var x = $(this).data("x");
  var y = $(this).data("y");
  var pzzl = $(this).data("puzzle");
  if (Math.abs(x - pzzl.empty.x) + Math.abs(y - pzzl.empty.y) == 1){
//Move only if the tile is currently beside the empty position.
//First, insert the tile to the place of the empty position in the
//"tiles" array. Next, use "data()" to update the position
//information of the tile. Then animate the tile to its new
//position. Finally, update the "empty" object to hold the new
//position of the empty space.
    pzzl.tiles[pzzl.empty.x][pzzl.empty.y] = $(this);
    pzzl.tiles[pzzl.empty.x][pzzl.empty.y].data("x", pzzl.empty.x);
    pzzl.tiles[pzzl.empty.x][pzzl.empty.y].data("y", pzzl.empty.y);
    pzzl.tiles[pzzl.empty.x][pzzl.empty.y].animate({
      "top": pzzl.empty.x * 1.2 * pzzl.tileSize,
      "left": pzzl.empty.y * 1.2 * pzzl.tileSize
    }, 200);
    pzzl.empty.x = x;
    pzzl.empty.y = y;
  }
};


/*
 * A static method that translates an integer from zero to three to
 *   an actual move.
 * Parameters:
 *   n [in]: an integer between zero and three specifying the
 *           desired move.
 *   x, y [in]: the coordinates specifying the current position of
 *              the tile to be moved.
 *   mov [out]: a reference to the object that will hold the x and
 *              y directions of the move. The values are only valid
 *              if the method returns true at the same time.
```

```
 *  Returns:
 *    If the move is possible (i.e., it is not off the puzzle frame),
 *    then it returns true. The actual direction of the move is
 *    stored in the mov parameter.
 *
 */
SlidingPuzzle.translateMove = function(n, x, y, mov) {
  mov.x = mov.y = 0;
  switch (n) {
    case 0: mov.y = 1; break;
    case 1: mov.x = 1; break;
    case 2: mov.y = -1; break;
    case 3: mov.x = -1; break;
    default: return false;
  }
  if (x + mov.x >= 0 &&
      x + mov.x <= 3 &&
      y + mov.y >= 0 &&
      y + mov.y <= 3) {
    return true;
  }
  else {
    return false;
  }
};

/*
 * This handles the Window's load event.
 */
$(function() {
  var puzzle = new SlidingPuzzle();
  puzzle.shuffle();
  puzzle.setTileSize(80);
  puzzle.start("puzzle");
});
    </script>
  </head>
  <body>
    <div id="puzzle"></div>
  </body>
</html>
```

Further Reading

If you have become a keen programmer (I sincerely hope you have), you will soon discover that we only touched the surface of jQuery in this appendix, and quite understandably so. There is simply too much going on in the wild to be able to even think of including everything in an appendix of an introductory programming book. I tried my best, though, to introduce you the basic principles that perpetuate this extremely popular library, but you will soon be reaching for other sources of information. Probably one of the best sources for beginners is *Learning jQuery*, a book by Jonathan Chaffer and Karl Swedberg. You may also want to go directly to the official jQuery website *jquery.com*, with downloads, examples, blogs, and more.

Unless you like reinventing the wheel, you won't stop at the basic jQuery library. There are countless jQuery plug-ins for you to use (check *plugins.jquery.com*, for example). Plug-ins are modules that add new functionality, and jQuery plug-ins are just plain files of JavaScript code. There's even a whole library of popular jQuery plug-ins distributed under the name *jQuery User Interface* (jQueryUI, available at *jqueryui.com*). With these UI plugins you can make elements on your page draggable, droppable, resizable, selectable, or sortable. You will also find practical widgets like date picker, with a convenient calendar, or progress bar. The jQuery UI library also comes with bountiful CSS themes, and if you don't find one to your taste, you can compose and download your own with their online ThemeRoller.

B.5   Go Mobile

If you haven't already done so, it is probably time to start thinking about programming for mobile devices. The good news is that HTML5 allows you to take advantage of everything you've learned about HTML, CSS, and JavaScript. By using the newest HTML5, CSS3, and JavaScript features, you can build web applications that look and feel like native applications. With the latest browsers on the market, even speed is no longer a concern—most web applications built upon the newest technologies are as fast as native applications. Apart from that, HTML5 allows you to "install" web applications in your browser's memory and run them offline, so your users need not be connected to the Internet to use them.

One big disadvantage of native applications is that they are programmed for a specific target platform, mainly using very different technologies. You need to learn the Objective C and Xcode development environment if you want to program Apple's native iOS applications, or you will have to learn Java along with the Eclipse development environment if you want to program native Android applications, for example. By creating web applications, you can mostly forget about the differences between different platforms. With minimal modifications, your application will run on most, if not all, devices.

Mobile Interaction

It's true that you can also view most web pages and run most web applications designed for desktop computers on your mobile device. Just because you don't have a mouse connected to your mobile, that doesn't mean you're handicapped when trying to follow a hyperlink, for example. In mobile devices, mouse events are simulated when touch events occur. This is essential for keeping the mobile web alive. That said, when you design a page or application with a mobile device already in mind, you can improve your user's experience hugely.

There are some considerations that you should give your design if you really want to think mobile. First, there's the type of web navigation: instead of the mouse and keyboard used with desktop browsers, mobile users navigate the Web with their fingers, tapping, swiping, and rotating the device. For the programmer, this basically means that he/she needs to handle different events, and handle them differently. You cannot hover with your finger over an element like you can with your mouse pointer, for example. Even the boniest of us still have fat fingers compared to the single-pixel precision of a mouse pointer. To make a user interface comfortable for tapping and swiping, you should deal with this fact as well. Another thing to consider is connected with the size of a screen, which is considerably smaller on a mobile device, and the user attention span, which is also smaller. This importantly influences the type and quantity of information to include and also affects the user interface design. The last thing you should be aware of is that mobile devices often access the Web via inconsistent and limited public services, which limits the amount of data they are able to exchange with servers. So, stay away from gigantic JavaScript libraries and images.

Swiping the 15-Puzzle

In order to demonstrate a mobile-specific approach to programming, let's give our 15-puzzle a more appealing feel. The puzzle as we programmed it in the previous section can already be used on a mobile phone since finger taps are automatically translated to mouse clicks on a mobile device. It is, however, quite unnatural and uncomfortable to tap tiles in order to slide them. Rather, we want our user to be able to swipe with a finger to slide a tile. There's a jQuery Mobile library available, which defines a user interface for touchscreens. Unfortunately, the library cannot handle vertical swipes directly—it is designed only to handle horizontal swipes. An easy way to circumvent this problem is to use an appropriate plugin, and so I picked up the jQuery TouchSwipe plugin by Skinkers Technology Services. Their website doesn't seem to work any more, but you can get the TouchSwipe plugin and all the documentation at *labs.rampinteractive.co.uk/touchSwipe*

To use the plugin in your code, you first include it as you would include any other external JavaScript file:

```
<script src="jquery.touchSwipe.min.js">
/* Important: include after the basic jQuery library */
</script>
```

Then, you insert the following code just at the end of the `start()` method of the `SlidingPuzzle` object:

```
$(document).swipe({
  swipe: function(event, direction) {
    var dir;
    switch (direction) {
      case "left": dir = 37; break;
      case "up": dir = 38; break;
      case "right": dir = 39; break;
      case "down": dir = 40;
    }
    $(this).trigger({type: "keydown", which: dir});
  },
```

```
    threshold: 30
});
```

You instantiate the plugin and apply a TouchSwipe behavior to a jQuery object by calling the `swipe()` method on that object. Notice that I applied the swiping behavior to the document rather than to individual tiles. This way, a user is free to swipe anywhere over the puzzle in order to slide a tile instead of needing to worry about whether he/she will hit the right tile. The `swipe()` method accepts as the first argument an object containing the event-handler definition (`swipe` in our case, but there are also other events defined by the library) and some properties controlling the handler behavior. One of these properties is `threshold`, which specifies a minimum distance in pixels that a user must swipe in any one of the four directions in order for the event to get fired. The `swipe` event handler also accepts, apart from the usual Event object, some other arguments. I used only the second argument, which holds a string naming the direction of the swipe (`"left"`, `"right"`, `"up"`, or `"down"`). Because a swipe over the puzzle gives exactly the same information as pressing one of the arrow keys, all we have to do is translate a swipe into a corresponding key press. If you look inside the event-handler body, you'll see that the four possible swipe directions are first translated into appropriate key codes and then the keydown event is triggered.

Further Reading

If you would like to know more about mobile web design, you may first want to explore the jQuery Mobile website *jquerymobile.com* with demos, downloads, documentation, recommended resources, and more. As mouse and touch events are simulated on mobile and desktop applications, you don't actually need a touch screen but can test some of the touch and swipe events on your desktop using a mouse pointer. If you prefer books, there's quite an enthusiastic piece of writing available in the form of Estelle Weyl's book *Mobile HTML5*. The book is definitely worth reading if you are concerned about principles, which are at least as important as technology itself.

HTML Mini Reference

This appendix is a list of selected HTML elements and attributes meant mainly as a convenient reference to use with this book. Included are also some of the more important HTML5 elements and properties. For a more exhaustive list, see, for example, *developer.mozilla.org/en-US/docs/Web/HTML/Reference*. You may also want to check the official specification by the World Wide Web Consortium at *www.w3.org/TR/html5*.

C.1 Root Element

`<html>`

Description	Represents the root element of an HTML document.
Content	Contains a `<head>` element followed by a `<body>` element.

Example

```
<!DOCTYPE html>
<html lang="en">
  <head><title>...</title></head>
  <body></body>
</html>
```

C.2 Document Metadata

`<base>`

Description	Defines the document base URL used for resolving relative URLs.

Context	May be used in a `<head>` element that contains no other `<base>` elements.
Content	This is a void element.

Attributes

`href`

This attribute defines a base URL used by relative URL addresses throughout the document. If a `<base>` element has an `href` attribute, it must appear before any other elements that have attributes whose values are URLs.

`target`

This attribute specifies the default target for all hyperlinks and forms in the page. If the `_blank` keyword is used as the value of this attribute, the linked content will be displayed in a new tab or window.

Examples

```
<!-- Define the  base URL. -->
<base  href="http://www.theendisnear.org/docs/">

<!-- With documents opening in a new tab or window. -->
<base  href="http://www.theendisnear.org/docs/" target="_blank">

<!-- This link points to
                     www.theendisnear.org/docs/prophecy.html. -->
<a href="prophecy.html">Read more</a>

<!-- This displays an image from www.theendisnear.org/imgs. -->
<img src="../imgs/planets.jpg" alt="The Solar System">
```

`<head>`

Description	Represents a collection of metadata for the document.
Context	May appear as the first child of an `<html>` element.
Content	One or more metadata elements, exactly one of which should be a `<title>` element.

Example

```
<head>
  <title>Yellow Submarine: Was it Really Yellow?</title>
</head>
```

`<link>`

Description	Used to link the document to other resources, e.g., external JavaScript or CSS.
Context	May appear inside any element that accepts metadata elements.
Content	This is a void element.

Attributes

`href`

This attribute specifies the (absolute or relative) URL of the linked resource and must be present.

`rel`

This attribute names the type of the linked resource and must be present. Most commonly, this attribute is used for specifying a link to an external style sheet, in which case its value is set to `stylesheet`.

Example

```
<!-- Include an external style sheet. -->
<link href="superfancy.css" rel="stylesheet">
```

`<meta>`

Description	Represents various kinds of metadata that cannot be defined using other metadata elements.
Context	Usually used within a `<head>` element.
Content	This is a void element.

Attributes

`charset`

Specifies the character encoding that is used in the document. The `<meta>` element with this attribute should be inside the `<head>` element and within the first 512 bytes of the page, since certain browsers only check these first bytes before deciding on a character set to be used. It is advisable that authors use the UTF-8 encoding.

`content`

This attribute holds the value connected with the name specified by the `name` attribute.

`name`

This attribute allows authors to set the document metadata expressed in terms of name/value pairs. The `name` attribute gives the name, while the corresponding value is given by the `content` attribute on the same `<meta>` element. Some of the possible values for the `name` attribute are:

- `author`: Specifies the name of the author.
- `description`: Contains a short summary of the document. This is one of the more important identifiers that helps search engines, as well as visitors/users, fathom what your page is all about. Also, a lot of browsers use this data as the default description of the page when it is bookmarked.
- `keywords`: A comma-separated list of important words describing the content of the document. This used to be a crucial identifier for early search engines but it is no longer so. Today, the best choice for authors is to avoid using it.

Examples

```
<meta charset="utf-8">
<meta name="description" content="A gang of hyper-intelligent
    pan-dimensional beings request to find out the Answer to the
    Ultimate Question of Life, The Universe, and Everything from the
    Deep Thought supercomputer exclusively designed for this
    purpose.">
<meta name="author" content="Douglas Adams">
```

`<style>`

Description	Used for embedding style information in a document.
Context	Normally used within a `<head>` element.
Content	Depends on the value of the `type` attribute.

Attributes

`type`

Specifies the styling language. If this attribute is absent, then the default CSS (i.e., `text/css`) is assumed.

Example

```
<style>
body {
  color:blue;
}
</style>
```

<title>

Description	Represents the title or name of the document, which is shown in the browser's title bar.
Context	May appear in a `<head>` element that contains no other `<title>` elements.
Content	It can only contain text.

C.3 Scripting

<noscript>

Description	Defines a portion of HTML code to be interpreted if scripting is unsupported or is currently turned off in the browser.
Content	Code that is to be interpreted if scripting is unsupported or turned off.
Display	Inline

Example

```
<noscript>
  <p>If you want to see some mind-blowing stuff, please enable
```

```
      JavaScript in your browser.</p>
   </noscript>
```

`<script>`

Description	Allows authors to include dynamic scripts in their documents. The scripts may either be embedded inline or imported from an external file by means of the `src` attribute.
Content	Without the `src` attribute, content depends on the value of the `type` attribute. With the `src` attribute, the element should either have no content or contain script documentation.

Attributes

`src`

If this attribute is specified, then it gives the name of the external script to be used in the form of a valid URL. Only one `src` attribute is permitted per element. If the `src` attribute is present, no script should be embedded inside the `<script>` element.

`type`

This attribute defines the type of scripting language. If this attribute is absent, then the default JavaScript (`text/javascript`) is used.

Examples

```
<!-- Include code from an external file. -->
<script src="/libs/slideshow.js"></script>


<!-- Include an external file with comments on its usage. -->
<script src="/libs/slideshow.js">
/*
   Set the class attribute of <img> elements that you
   want to include into a slide show to "slide".
*/
</script>


<!-- Inline code -->
<script>
   document.write("Whatever happened to heroes?");
</script>
```

C.4 Sections and Structure

`<article>`

Description	Represents a section of a page that forms a separate part of a document, page, or site, which could also be used independently. This could be a magazine or newspaper article, as well as a forum post, user-submitted comment, or any other independent content.
Category	Sectioning element
Display	Block

Example

```
<article>
  <header>
    <h1>About Kangaroos</h1>
    <h2>Australia's most iconic animals</h2>
  </header>
  <section>
    <h1>The Red Kangaroo</h1>
    <p>A large male can weigh up to 90 kilos and be 2 meters
    high.</p>
  </section>
  <section>
    <h1>The Western Grey Kangaroo</h1>
    <p>Inhabits the southern part of Western Australia, as well
    as coastal South Australia and the Darling River basin.</p>
  </section>
  <footer>
    <p>This article is a part of the series Indigenous Australia</p>
    <p>Copyright (C) 2015</p>
  </footer>
</article>
```

`<aside>`

Description	Represents a part of a page containing information that is tangentially related to the content surrounding the `<aside>` element and could be viewed separately from that content. Such information is often displayed in the form of sidebars or inserts, and includes more loosely related material, like background material, a glossary definition, or advertise-

ments. If an `<aside>` element is removed, the remaining content should still make sense.

Category	Sectioning element
Display	Block

Example

```
<!-- Mark up background material on koalas in a much longer story
on Australia. -->
<aside>
  <h1>Koalas</h1>
  <p>Koalas get very little energy from their diet so they have to
  limit their energy use and sleep 20 hours a day. They spend only
  4 minutes a day in active movement.</p>
</aside>
```

`<body>`

Description	Holds the main content of the document. There may only be one `<body>` element in a document.
Context	As the second child of an `<html>` element.
Category	Sectioning root

`<footer>`

Description	Represents a footer for a section. The section could be implicitly derived from the page structure or explicitly specified by one of the elements, `<article>`, `<aside>`, `<nav>`, or `<section>`. The `<footer>` element typically holds information about the author, copyright data, links to related material, and the like.
Content	Cannot contain any of the following elements: `<h1>` to `<h6>`, `<hgroup>`, `<article>`, `<aside>`, `<nav>`, `<section>`, `<header>`, and `<footer>`.
Display	Block

`<h1>`, `<h2>`, `<h3>`, `<h4>`, `<h5>`, `<h6>`

Full name	Heading
Description	These elements represent six levels (ranks) of headings for their sections. The `<h1>` element has the highest rank, while the `<h6>` element has the lowest rank.
Content	Can only contain text and inline elements.
Display	Block

Example

```
<body>
  <h1>Life, the Universe and Everything</h1>
    <h2>Lord's Cricket Ground</h2>
      <p>...</p>
    <h2>The Starship Bistromath</h2>
      <p>...</p>
    <h2>The Infinite Improbability Drive</2>
      <p>...</p>
      <h3>Theory</h3>
        <p>...</p>
      <h3>Application</h3>
        <p>...</p>
    <h2>The planet Krikkit</h2>
      <p>...</p>
</body>
```

`<header>`

Description	Specifies a header of a page or section, which contains items like introductory information or aids for site navigation. The `<header>` element is meant primarily to include headings (`<h1>` to `<h6>`) or heading groups (`<hgroup>`) but that is not required. It can also contain things like a table of contents, a search form, or different logos.
Content	Cannot contain any `<header>` or `<footer>` elements.
Display	Block

`<hgroup>`

Full name	Headings Group
Description	It represents a heading of a section. The `<hgroup>` element's main purpose is to associate lower rank headings with the one with the highest rank, where the lower rank headings act as secondary titles, like, for example, subheadings, alternative titles, or even bylines or taglines, which shouldn't appear in the outline of the document.
Content	One or more heading elements (`<h1>` to `<h6>`).
Display	Block

Examples

```
<!-- Add a secondary title to the main heading. -->
<hgroup>
  <h1>The Lord of the Rings</h1>
  <h2>The Two Towers</h2>
</hgroup>

<!-- Add a byline to the main heading. -->
<hgroup>
  <h1>The Meaning of Liff</h1>
  <h2>By Douglas Adams and John Lloyd</h2>
</hgroup>
```

`<nav>`

Full name	Navigation
Description	This element represents a section containing links to other documents or to parts within the page. Not all links need to be in the `<nav>` element but only sections of major navigation blocks are suitable for it. Footers may as well contain a list of links, in which case the `<footer>` element is a better choice. Note that no `<nav>` element is allowed inside the `<footer>` element, though.
Category	Sectioning element

Example

```
<nav>
  <ul>
    <li><a href="#">Somewhere</a></li>
    <li><a href="#">Anywhere</a></li>
    <li><a href="#">Nowhere</a></li>
  </ul>
</nav>
```

`<section>`

Description	Defines a document's section representing a thematic grouping of content, typically with a heading. An example of a section would be a chapter.
Category	Sectioning element
Display	Block

Example

```
<section>
  <h1>General Description</h1>
  <p>Dinosaurs have always been extremely fascinating animals.
  Especially ...</p>
</section>
```

C.5   Grouping

`<br>`

Full name	Break
Description	Introduces a line break. This element should only be used for line breaks that are really part of the content, such as line breaks in poems or addresses.
Content	This is a void element.

Example

```
<p>
    In Xanadu did Kubla Khan <br>
    A stately pleasure dome decree: <br>
    Where Alph, the sacred river, ran <br>
    Through caverns measureless to man <br>
    Down to a sunless sea. <br>
    ...
</p>
```

`<div>`

Full name	Division
Description	Defines a generic container, which does not inherently represent anything. It is normally used to group elements for styling purposes. Do not use this element if other semantic elements (such as `<section>` or `<article>`) are available.
Display	Block

Example

```
<div class="FortuneCookie">
  <h1>Your fortune cookie for today</h1>
  <p>You're just another brick in the wall.</p>
</div>
```

`<figcaption>`

Full name	Figure Caption
Description	Represents a caption or legend for the content of its parent `<figure>` element.
Context	The element can only be used as a first or last child of a `<figure>` element.
Display	Block

Example

```
<figure>
  <img src="before.jpg" alt="Old-Looking">
  <img src="after.jpg" alt="Young-Looking">
  <figcaption>
    Before and after our amazing Drop-Pounds-and-Lose-Inches Diet.
  </figcaption>
</figure>
```

`<figure>`

Description	Represents some self-contained content, usually with a caption specified by the first `<figcaption>` element inside the `<figure>` element. The `<figure>` element can be used to add explanations to illustrations, charts, photos, computer code listings, or the like, which are referenced from the text of the main document. They could be moved away from that primary text without affecting the flow of the document. The outline of the `<figure>` element is not a part of the main document outline.
Category	Sectioning root
Display	Block

`<hr>`

Full name	Horizontal Rule (historic name)
Description	Represents a thematic break between paragraphs, such as a transition to another topic within a section or a change of scene in a story.
Content	This is a void element.
Display	Block

Example

```
<p>...until the last one of them fell asleep.</p>
<hr>
<p>Early next morning, a magnificent nuclear explosion
on a nearby meadow woke them up for a fleeting instant...</p>
```

<li>

Full name	List Item
Description	Represents a list or menu item.
Context	May be used inside <ol> or <ul> elements.
Display	Block

Example

```
<!-- NOTE: since <ol> and <ul> elements can only contain <li>
     elements, you can make a sublist by nesting another list
     inside an <li> element. -->

<ul>
  <li>Immortals
    <ul>
      <li>Aphrodite</li>
      <li>Ares</li>
      <li>Hermes</li>
      <li>Zeus</li>
    </ul>
  </li>

  <li>Mortals
    <ul>
      <li>Achilles</li>
      <li>Electra</li>
    </ul>
  </li>
</ul>
```

<ol>

Full name	Ordered List

Description	Represents a list of items where the order of items is important to the meaning of the document.
Content	Zero or more `<li>` elements
Display	Block

Attributes

`start`

This attribute must be a valid integer. It gives the ordinal number of the first item.

Example

```
The Ten Commandments (Exodus 20:1-17 NKJV continued):
<ol start="6">
  <li>You shall not murder.</li>
  <li>You shall not commit adultery.</li>
  <li>You shall not steal.</li>
  <li>You shall not bear false witness against your neighbor.</li>
  <li>You shall not covet your neighbor's house; you shall not
      covet your neighbor's wife, or his male servant, or his
      female servant, or his ox, or his donkey, or anything that
      is your neighbor's.</li>
</ol>
```

`<p>`

Full name	Paragraph
Description	Represents a paragraph.
Content	Can only contain text and inline elements.
Display	Block

`<pre>`

Full name	Preformatted

Description	Introduces a block of preformatted text, which is displayed exactly as it is typed, typically in a monospaced font and including all whitespaces. However, HTHL elements within this element are still interpreted. As the structure within this element is represented by typographic conventions, it can be used for displaying things like, for example, fragments of computer code, or ASCII art.
Display	Block

Examples

```
<!-- Computer code -->
<pre>
  .panic {
    color: red;
  }
</pre>


<!-- ASCII art -->
<pre>
        _
      (. .)
      ( V )
   ------m m--------

</pre>
```

`<ul>`

Full name	Unordered List
Description	Represents a list of items whose order is not important.
Content	Zero or more `<li>` elements.
Display	Block

Example

See `<li>` example on page 318.

C.6   Text-Level Semantics

<a>

Full name	Anchor
Description	Represents a hyperlink or the named target destination for a hyperlink.
Display	Inline

Attributes

`href`

If the <a> element is used as a hyperlink, this is a required attribute. Its value specifies the link target in the form of a URL. Apart from the usual HTTP protocol for transferring hypertext documents, other protocols can be used as well (e.g., HTTPS, FILE, or FTP). One can even establish a link to an internal target location within a document (marked by another <a> element with an `id` attribute). This can be done by means of a URL fragment, which is a name preceded by a hash mark (#).

`target`

This attribute specifies where to display the linked content. If you use the `_blank` keyword as a value of this attribute, the content will be displayed in a new tab or window.

Examples

```
<!-- A link to a hypertext document. -->
<a href="http://www.someweirdstuff.org/memberlogin.html">Let's go
    somewhere</a>


<!-- Open the document in a new tab/window. -->
<a href="http://www.someweirdstuff.org/memberlogin.html"
    target="_blank">Let's go somewhere</a>


<!-- A link to a PDF document. -->
<a href="http://www.someweirdstuff.org/appform.pdf">Application
    form</a>


<!-- Mark a specific location within a page
                    www.someweirdstuff.org/benefits.html. -->
<a id="somewhere">This is totally amazing</a>

<!-- Link to that specific location from within the same page. -->
```

```
<a href="#somewhere">Go there</a>

<!-- Link to that specific location from some other page. -->
<a href="http://www.someweirdstuff.org/benefits.html#somewhere">
   Go there</a>


<!-- Link that opens an email client. -->
<a href="mailto:ceo@nobigdeal.com">ceo@nobigdeal.com</a>

<!-- With the subject line already filled in. -->
<a href="mailto:ceo@nobigdeal.com?subject=News">
   ceo@nobigdeal.com</a>
```

\<b\>

Full name	Bold (historic name)
Description	Represents a part of text of a different meaning than the rest of the content without conveying any additional importance. The examples are: keywords in an abstract of a document, product names, the first sentence or paragraph of an article, or some other content that is typically represented in bold. It is advisable to use the class attribute on this element to identify why the element is being used.
Display	Inline

Example

```
<!-- From "Master of All Masters", an English Fairy Tale
     by J. Jacobs. -->
<p><b class="firstsentence">A girl once went to the fair to
hire herself out for a servant.</b> At last a funny-looking old
gentleman engaged her, and took her home to his house. When
she got there, he told her that he had something to teach her,
for that in his house he had his own names for things.</p>
```

\<em\>

Full name	Emphasized

Description	Represents stress emphasis of its contents. Although many browsers display the contents of an <em> element in italics, the element is not intended to represent a generic italics element. The <i> element is more appropriate for that. Also, the <em> element should not be used to express importance. The <strong> element is more appropriate for that purpose.
Display	Inline

Examples

```
<p><em>Pizza</em> is a healthy food.</p>
<p>Pizza <em>is</em> a healthy food.</p>
<p>Pizza is a <em>healthy</em> food.</p>
<p>Pizza is a healthy <em>food</em>.</p>
<p><em>Pizza is a healthy food!</em></p>
```

<i>

Full name	Italics (historic name)
Description	Defines a fraction of text in an alternate mood or otherwise deviated from normal prose, such as a technical term, an idiomatic phrase from another language, a thought, or some other content that is typically represented in italics. It is advisable to use the class attribute on this element to identify why the element is being used.
Display	Inline

Example

```
<p>The domestic cat (<i class="taxonomy">Felis catus</i>) is a
small, usually furry and cute, domesticated carnivorous mammal.
</p>
```

<q>

Full name	Quote

Description	Represents some content quoted inline from another source.
Display	Inline

Attributes

`cite`

Names the source URL of the quoted information.

Examples

```
<p>Tom said, <q>That is really theoretically impossible but turned
out to be possible in practice</q>.</p>
```

```
<p>In his speech at The University of California DNA stated that
<q cite="http://www.youtube.com/watch?v=_ZG8HBuDjgc">we don't
have to save the world. The world is big enough to look after
itself. What we have to be concerned about is whether or not
the world we live in will be capable of sustaining us in it.</q></p>
```

`<small>`

Description	Defines small print or other side notes. Typically, it is used for disclaimers, legal restrictions, or copyrights, as well as for attribution or meeting licensing requirements.
Display	Inline

Examples

```
<small>(C) Copyright 1454 Gutenberg Ltd.</small>
<small>All terms subject to change without notice.</small>
```

`<span>`

Description	Does not convey any meaning but can be useful in combination with certain attributes. It is normally used in the same way as the `<div>` element, only `<div>` is a block-level element, while `<span>` is an inline element.

Display	Inline

`<strong>`

Description	Gives its contents strong importance. Although many browsers display the contents of the `<strong>` element in bold, it is not intended to represent a generic bold element. The `<b>` element is more appropriate for that.
Display	Inline

Example

```
<p><strong>Warning:</strong>Hot beverages are hot!</p>
```

C.7   Embedded Content

`<canvas>`

Description	Defines a bitmap area that scripts can use for rendering graphics on the fly. To draw on the canvas, you must first obtain a reference to a context using the `getContext()` method of the Canvas object. You can read more about the canvas in section B.1 on page 271.
Content	Alternative content to be used if the `<canvas>` element is not supported or if scripting is disabled in a browser.
Display	Inline

Attributes

`height`

This is an integer attribute that specifies the height of the canvas in CSS pixels. If this attribute is absent, then a default value of 150 is used.

```
width
```

This is an integer attribute that specifies the width of the canvas in CSS pixels. If this attribute is absent, then a default value of 300 is used.

Example

```
<canvas id="clock" width="250" height="250">
  Your browser has disabled scripting or it doesn't support
  &lt;canvas&gt;.
</canvas>
```

`<img>`

Full name	Image
Description	Represents an image. The HTML standard doesn't prescribe the formats that should be supported, so different user agents support different sets of formats. Among the safest to use are JPEG and GIF (including animated GIF).
Content	This is a void element.
Display	Inline

Attributes

```
alt
```

Specifies the alternative text to be displayed if the external image file cannot be used. For example, if the file format is unsupported or the image URL is wrong. Users can also see this text until the image is downloaded.

Omitting this attribute suggests that the image is an essential part of the content.

```
height
```

This is an integer attribute that specifies the height of the image in CSS pixels.

```
src
```

This is an obligatory attribute and gives the URL of the image file.

```
width
```

This is an integer attribute that specifies the width of the image in CSS pixels.

Example

```
<img src="groundhog.jpg" width="200" alt="A Male Groundhog Feeding">
```

C.8 Tabular Data

`<caption>`

Description	Defines the title of the table.
Context	May appear as the first child of a `<table>` element. A `<table>` element is permitted to contain only one `<caption>` element.

Example

See `<table>` example on page 328.

`<table>`

Description	Organizes data systemized in more than one dimension. Data are represented in the form of a table. Historically, tables were sometimes misused in HTML as a way to control a page layout, but tables should not be used as layout tools.
Content	A `<caption>` element (optional), followed by one or more `<tr>` elements.
Display	Block

Attributes

`border` **(Obsolete)**

This attribute specifies whether a border is to be displayed around the table and between its cells. Historically, this attribute was used for specifying the border width, but has now become obsolete for that purpose, so don't use it for border styling. Instead, use CSS properties like `border-width` or `border-style`. You can, however, use this attribute to display table and cell borders to make a table legible in environments when CSS support is absent, such as text-based browsers, and in situations where CSS is turned off or the style sheet cannot be found. In HTML5, the only allowed values for the `border` attribute are 1 and an empty string (`""`).

Example

```
<table border="1">
  <caption>
    <p>Approximate color temperatures of the light emitted from
    different light sources.</p>
  </caption>
  <tr>
    <th>Temperature (K)</th><th>Source</th>
  </tr>
  <tr>
    <td>1,700</td><td>Match flame</td>
  </tr>
  <tr>
    <td>1,850</td><td>Sunset/sunrise</td>
  </tr>
  <tr>
    <td>4,100</td><td>Moonlight</td>
  </tr>
  <tr>
    <td>6,500</td><td>Overcast daylight</td>
  </tr>
</table>
```

<td>

Full name	Table Data
Description	Represents a single data cell in a table.
Context	May be used inside a `<tr>` element.
Category	Sectioning root

Attributes

`colspan`

Determines over how many columns the cell extends.

`rowspan`

Determines over how many rows the cell extends.

Example

See `<table>` example on page 328.

`<th>`

Full name	Table Header
Description	Represents a single header cell in a table.
Context	May be used inside a `<tr>` element.

Attributes

`colspan`

Determines over how many columns the header extends.

`rowspan`

Determines over how many rows the header extends.

Example

See `<table>` example on page 328.

`<tr>`

Full name	Table Row
Description	Holds a row of cells in a table.
Context	May be used as a child of a `<table>` element and it must appear after the `<caption>` element, if the table has one.

Example

See `<table>` example on page 328.

C.9 Forms

`<form>`

Description	Represents an assemblage of interactive form-control elements that collect data from a user. The collected data can be sent to a server for further processing.
Content	Cannot contain any other `<form>` elements.

Display	Block

Attributes

`action`

This attribute is used at form submission. It specifies a server-side program that should process submitted data. In HTML5, the `action` attribute is no longer required as it was in HTML4.

`autocomplete`

Determines whether a browser can automatically complete the values of the controls in this form. Using the `autocomplete` attribute on a specific element can override this value. This attribute can have one of two possible values:

- on: The browser is allowed to automatically complete values according to the values that have been entered during previous uses of the form.
- off: The user must explicitly input every piece of information each time, or the document will provide its own method of autocompletion. The browser will not automatically complete entries.

If this attribute is not set, then the default value on is assumed.

`method`

Specifies the HTTP method the browser will use to submit the form. The two most commonly used methods are:

- get: By this method, the form data is URL encoded with a question mark (?) as a separator.
- post: If the form uses this method, then the entered data will be sent in the message body.

Example

```
<form action="eat_this.asp" method="get">
  <!-- Here come different input controls such as buttons, text
      boxes, radio buttons, etc. -->
</form>
```

`<input>`

Description	Represents a typed data field, most often with a form control for the user to edit the data.
Context	Usually, it is used as a descendant of a `<form>` element. If not, then it should have specified the `form` attribute.
Content	This is a void element.
Display	Inline

Attributes

autocomplete

This attribute determines whether the browser can automatically complete the value of the control. It has one of two possible values:

- on: The browser is allowed to automatically complete values according to the values that have been entered during the previous uses of the form.
- off: The user must enter a complete value into this field each time, or the document will provide its own method of autocompletion. The browser will not automatically complete entries.

If this attribute is not set, then the browser uses the `autocomplete` attribute value of the `<form>` element to which this `<input>` element belongs. The `<input>` element belongs to a form that is either its parent, or the `<input>` element's `form` attribute equals the form's `id` attribute value.

checked

This Boolean attribute is used only when the `type` attribute is set to `checkbox` or `radio`, and it is specified without a value. If the `checked` attribute is present, then the control is selected by default.

form

This attribute connects an `<input>` element with a specific form, which thus becomes the `<input>` element's owner. The value of the `form` attribute must be equal to the value of the `id` attribute of the form. If this attribute is omitted, then the `<input>` element must be a descendant of a `<form>` element.

max

Defines the maximum possible value for this control, which must not be smaller than the value of the `min` attribute.

min

Defines the minimum possible value for this control, which must not be greater than the value of the `max` attribute.

name

Specifies the name of the control, which is used in form submission.

readonly

This is a Boolean attribute and is specified without a value. If it is present, then the visitor cannot modify the value of the control.

step

This attribute is used together with the `min` and `max` attributes, and specifies the increment at which one can set a numeric value.

type

Defines the type of the form control element. If this attribute is not given, then the default value `text` is assumed. Some of the possible values are:

- `button`: A push button.
- `checkbox`: A check box. If the `type` attribute is set to `checkbox`, then the `value` attribute should also be used in order to specify the value submitted with this item. The `checked` attribute can be used to indicate that the item is initially selected.
- `number`: A field for entering a floating point number. With the `type` attribute set to `number`, the attributes `value`, `min`, `max`, and `step` are often used. If they are not explicitly set, the default values `min = 0`, `max = 100`, and `step = 1` are used.
- `password`: A text field with no line breaks that obscures data entry.
- `radio`: A radio button. The value that will be submitted by this element is specified by the `value` attribute. Authors can make the item selected by default by including the `checked` attribute. The radio buttons with the same value of the `name` attribute form a group, and only one button in a group can be selected at a time.
- `reset`: A push button that resets the values of all the control elements of the owner form to default values.
- `submit`: A push button that submits the form.
- `text`: A field for entering text with no line breaks.

value

Specifies an initial value of a control. It is optional except when the `type` attribute is set to `radio` or `checkbox`.

Examples

```
<!-- A push button triggering a JavaScript alert box. -->
<input type="button" value="Click me" onclick="alert('Hi there!')">


<!-- Select zero or more from the list. -->
<label><input type="checkbox" name="goodies" value="lamp" checked>
  Fruits Table Lamp
</label>
<label><input type="checkbox" name="goodies" value="clock">
  Kit-Cat Clock
</label>
<label><input type="checkbox" name="goodies" value="tea_set">
  Ceramic Tea Set
</label>


<!-- Decide for exactly one of the options. -->
<label><input type="radio" name="offer" value="yes" checked>
  Yes, please.
</label>
<label><input type="radio" name="offer" value="no">
  No, thanks.
</label>


<!-- Fetch some generic text input. -->
<input type="text" name="opinion" value="Enter your opinion...">


<!-- A button to set all the form controls to default values. -->
<input type="reset">


<!-- Some numeric input -->
<label>Rate this seller
  <input type="number" name="rate" min="1" max="10" value="5">
</label>
```

`<label>`

Description	Represents a caption within a form, which can be associated with a specific form-control element. The association can be established either by using the `<label>` element's `for` attribute, or by putting the form-control element inside the `<label>` element itself.

Context	Usually, it is used as a child of a `<form>` element. If not, then it should have specified the `form` attribute.
Content	Can contain no `<label>` elements and no labelable form-associated elements other than the one that it labels.
Display	Inline

Attributes

`for`

With this attribute, authors indicate a form control element with which they want to associate the caption. The value of this attribute should be equal to the value of the `id` attribute of the associated element. Alternatively, if there is no `for` attribute, the association can be made by simply putting the form control element inside the `<label>` element.

`form`

This attribute should be specified if a `<label>` element is located outside of any `<form>` element. It should be given the value of the `id` attribute of the `<form>` element to which it belongs. This attribute enables authors to put `<label>` elements anywhere in the document.

Examples

```
<label for="first_name">First Name:</label>
<input type="text" id="first_name" name="first_name">
```

```
<label>Middle Name:
  <input type="text" id="middle_name" name="middle_name">
</label>
```

`<option>`

Description	Represents an item within a `<select>` element.
Context	It can be used as a child of a `<select>` element.
Content	Can only contain text.

Attributes

`label`

Provides a caption for an element. If this attribute is not used, then it assumes the value of the content of the `<option>` element.

`selected`

This is a Boolean attribute and is specified without a value. If it is present, then the option is initially put into the selected state.

`value`

Provides a value for an element. If this attribute is not used, then it assumes the value of the content of the `<option>` element.

Examples

```
<!-- Offer to select one of the options with the Super King Size
     Bed preselected. -->
<select name="select_a_bed">
  <option value="compact">Compact Single Bed</option>
  <option value="single">Single Bed</option>
  <option value="compact_double">Compact Double Bed</option>
  <option value="double">Double Bed</option>
  <option value="king">King Size Bed</option>
  <option value="super_king" selected>Super King Size Bed</option>
</select>


<!-- Offer to select any number of options. -->
<select size="3" name="order" multiple>
  <option>Double Down</option>
  <option>Pumpkin Spice Latte</option>
  <option>Cherry Limeade</option>
  <option>Beefy Crunch Burrito</option>
  <option>Sweet Potato Fries</option>
</select>
```

`<select>`

Description	Represents a control for selecting between a set of possible choices.
Content	Zero or more `<option>` elements.
Display	Inline

Attributes

`form`

This attribute connects a `<select>` element with a specific form, which thus becomes the `<select>` element's owner. The value of the `form` attribute must be equal to the value of the `id` attribute of the form. If this attribute is omitted, then the `<select>` element must be a descendant of a `<form>` element.

`multiple`

This is a Boolean attribute and does not require a value. If it is present, then multiple options can be selected at the same time. Multiple options can be selected by holding the `Ctrl` key and simultaneously clicking on desired options.

`name`

Specifies a name for a control, which is used at form submission.

`size`

This attribute gives the number of options that are visible at the same time. If the `multiple` attribute is present, then the default value of the `size` attribute is 4, otherwise it is 1.

Examples

See the `<option>` examples on page 335.

`<textarea>`

Description	Represents a multi-line control for entering plain text.
Context	Usually, it is used as a descendant of a `<form>` element. If not, then it should have specified the `form` attribute.
Content	It can contain character data.
Display	Inline

Attributes

`cols`

The `cols` attribute is a non-negative integer specifying the expected maximum number of characters in a line. If this attribute is not specified, it defaults to 20.

form

This attribute connects a `<textarea>` element with a specific form, which thus becomes the `<textarea>` element's owner. The value of the `form` attribute must be equal to the value of the `id` attribute of the form. If this attribute is omitted, then the `<textarea>` element must be a descendant of a `<form>` element.

rows

The `rows` attribute is a non-negative integer specifying the number of text lines for the control. If this attribute is not specified, it defaults to 2.

Example

See the `localStorage` example on page 281.

C.10   Global Attributes

The attributes in this section may be specified on any HTML element.

class

This attribute represents a list of one or more class names separated by spaces. Classes offer CSS and JavaScript access to specific HTML elements by means of *class selectors* or special functions like the `document.getElementsByClassName()` method.

id

This attribute defines an identifier, which must be *unique* in the whole document, so it exclusively identifies an element on which it is used. IDs offer CSS and JavaScript access to specific HTML elements by means of *id selectors* or special functions like the `document.getElementById()` method. The `id` attribute on an `<a>` element can also be used to name a URL fragment marking an internal hyperlink target location within a document.

In HTML5, there are no limits on what form a value of an ID may assume other than that it must contain at least one character and must not contain any space characters.

lang

With this attribute, authors specify the language for the content of the element. Its value must be a valid ISO 639-x/IETF BCP 47 language code, which is en for English.

style

This attribute allows authors to write inline CSS code.

```
title
```

This attribute represents textual information describing the element it belongs to, typically in the form of a tooltip.

C.11 Event-Handler Attributes

One way that HTML elements can get notified of events (e.g., mouse clicks or key presses) is to specify an event handler using the corresponding HTML attribute on an element that needs to respond to the event. Event-handler attribute names begin with on and are the same as the names of the event-handler properties defined by the JavaScript Element object. You can find some of them in the table on page 402.

Using HTML event-handler attributes to specify event handlers is not in line with good practice. Namely, it requires mixing structure (HTML) and behavior (JavaScript). A much better approach is using the analogous properties of the Element JavaScript object, or even the addEventListener() method of the EventTarget object. The latter is not covered in this book, though.

Example

```
<!-- A click on the following <p> element will trigger the
     JavaScript alert() method. -->
<p onclick="alert('Don\'t you ever listen?')">Don't click me!</p>
```

CSS Mini Reference

This appendix is a handy list of some of the most useful CSS properties, pseudo-elements, and pseudo-classes. Mainly they are older and well-established properties, but included are also some of the CSS3 properties that are well supported by modern browsers. If you want to plunge deeper into CSS, there exist many exhaustive sources that you might want to look at. For example, *CSS3* by D. S. McFarland, or *developer.mozilla.org/en-US/docs/CSS/CSS_Reference*. For full details about the latest CSS specification right from the horse's mouth, visit the World Wide Web Consortium at *www.w3.org/Style/CSS/current-work*.

D.1 CSS Data Types

Each CSS declaration consists of a property name and a corresponding value. Values are categorized according to CSS data types, some of which are described in this section.

\<color\> Data Type

The \<color\> CSS data type is used to specify a color. Most usually, a color is assigned to fonts, borders (including shadows), and backgrounds. A color can be specified either by using one of the available keywords, or by using the RGB model, which allows you to mix red, green, and blue components to get a color of your liking.

Color Keywords

You can specify color directly by its name using a color keyword. Color keywords are case-insensitive. Although there are more of them, only 17 color names are currently understood by all browsers: `aqua`, `black`, `blue`, `fuchsia`, `gray`, `green`, `lime`, `maroon`, `navy`, `olive`, `orange`, `purple`, `red`, `silver`, `teal`, `white`, and `yellow`. Note that none of the colors defined by keywords can have any transparency. If you want to use transparent colors, then you must use the `rgba()` function.

A special color keyword is the `transparent` keyword, which represents a fully transparent color. Technically, this is a black color having the alpha channel set to its lowest value and it is equivalent to `rgba(0, 0, 0, 0)`.

RGB Values

Colors can be specified using the RGB model in either a hexadecimal or functional notation.

The hexadecimal notation is composed of a hash sign (#) followed by either six or three hexadecimal numbers (i.e., decimal digits 0–9 and letters A–F), specifying the amount of red (R), green (G), and blue (B) light to use:

```
#RRGGBB
#RGB
```

Both notations are equivalent and you can get the longer one simply by duplicating each of the three hexadecimal digits of the shorter one. The next two lines both represent exactly the same screaming green color:

```
#66ff11
#6f1
```

Lower- as well as upper-case letters can be used as hexadecimal digits A–F.

The other method of specifying RGB colors is a functional notation. This notation uses the rgb() function with three <integer> values between 0 and 255, or three <percentage> values between 0 and 100% as arguments:

```
rgb(<integer>, <integer>, <integer>)
rgb(<percentage>, <percentage>, <percentage>)
```

Using the functional notation, the same screaming green color can be expressed like this:

```
rgb(102, 255, 17)
rgb(40%, 100%, 6.7%)
```

The integer number 255 is identical to FF in hexadecimal notation, or to 100%. Note that it is not possible to use hexadecimal numbers with the rgb() function.

RGBA Values

The RGBA color model extends the RGB model with the alpha channel, which specifies the color opacity. The opacity is specified by a real number between 0 (meaning completely invisible) and 1 (totally opaque) as a fourth argument of the rgba() function. You can use either three <integer> or three <percentage> values for a color, and a <number> value for opacity:

```
rgba(<integer>, <integer>, <integer>, <number>)
rgba(<percentage>, <percentage>, <percentage>, <number>)
```

For example, the following two lines both create a semi-transparent green:

```
rgba(0, 255, 0, 0.5)
rgba(0%, 100%, 0%, 0.5)
```

<integer> Data Type

The <integer> CSS data type is used to specify an integer. An integer is composed of any number of decimal digits 0–9, preceded by an optional + or – sign. Any <integer> value is also a <number> value, while the opposite is not true.

<length> Data Type

The <length> CSS data type is used for specifying distance measures. It is composed of a <number> directly followed by a length unit. No space is allowed between the number and the unit literal. Note that for certain CSS properties, negative lengths are also allowed. For example, by specifying either a negative or positive <length> for the text-indent property, you specify whether text is indented to the left or right.

Some properties that accept <length> values also accept <percentage> values. Note, however, that <percentage> values are not <length> values.

Units

CSS provides a number of different units to measure lengths. Although there is no limit on which units can be used with which property, you will generally not use units like centimeter (cm) or inch (in) for an on-screen display. Those units are more appropriate for printing on paper. For display on screen you will almost exclusively use ems (em) if you want relative units or pixels (px) if you want absolute units. You can read more about ems on page 67, and about CSS pixels on page 47.

For measurements of zero length it is not necessary to use units. For example, instead of 0px you can simply write 0.

<number> Data Type

The <number> CSS data type is used to specify a real number. A <number> is composed of an <integer> part, possibly extended with a fractional part—a decimal point (.) followed by any number of decimal digits 0–9. The integer part can also be omitted. Note that any <integer> value is also a <number> value, while the opposite is not true.

These are examples of valid <number> values:

```
42
3.14159
-273.15
.99
```

<percentage> Data Type

The <percentage> CSS data type is used to specify percentages. Percentages are composed of a <number> directly followed by a percentage sign (%). There is no space between the number and the percentage sign.

Even if many length properties accept percentages as relative units, the <percentage> data type is not the <length> data type. You can read more about how to use percentages as length measurements on page 68.

<url> Data Type

The <url> CSS data type represents a reference to an external resource. It is expressed exclusively by means of the url() function, which accepts as an argument the URL of the resource. The URL does not need to be quoted unless it contains spaces. Allowed are single as well as double quotes. Specified URLs may be absolute or relative and when they are relative, they are relative to the URL of the style sheet and not the URL of the web page.

Here are some examples:

```
url(/bullets/funnybullet.gif)
url('/bullets/another funny bullet.gif')
url("/bullets/another funny bullet.gif")
url(http://fancybullets.com/animals/mosquito.gif)
```

D.2 inherit keyword

The inherit keyword is common to all CSS properties. With this keyword you can force a property that isn't normally inherited to inherit a value from the parent. For example, if you want elements of the important class to inherit a border from their parents, then you declare:

```
.important {
  border: inherit;
}
```

Note that you cannot use the inherit keyword to inherit individual property values through shorthand properties. You can apply the inherit keyword to a shorthand property only as a whole.

D.3 Text Properties

Using properties in this section you can decide how text should be formatted on a web page. Most of the text properties are inherited, so it is not necessary to use them directly with text-level elements, or elements like <h1> to <h6> or <p>. It is often better to define an overall font, color, alignment, or similar for a page or section. For example, you can apply font properties to a <body> or <article> element to create a uniform overall appearance.

color (Inherited)

The color CSS property is used to set the foreground color of the element's text together with its decorations (e.g., underlining). This property is inherited, so the color property applied to the document body, for example, will also be applied to text inside all descendants of the <body> element.

From CSS3 onwards it is possible to use transparency with text colors.

Values

Any valid <color> value.

Default Value

Depends on the browser.

Examples

```
color: blue;
color: #0000FF;
color: rgba(0, 0, 255, 0.5);
```

font (Inherited)

This is a shorthand property that sets several font properties in one declaration. The following properties can be set: font-style, font-variant, font-weight, font-size, line-height, and font-family. The values must be separated by a space, except font-size and line-height, which must be separated by a slash (/). The font-size and font-family values are required and should be the last two items in a declaration.

Like with any shorthand property, the values that are not explicitly set revert to their individual default values, even if they were previously set to different values using other properties. You can find out what the default values are from descriptions of respective properties.

Examples

```
font: 1.25em Arial, Helvetica, sans-serif;
font: italic small-caps bold 2em/150% Georgia, serif;
font: italic; /* Error: font-size and font-family
                  are missing. */
font: italic serif 18px; /* Error: font-size and font-family
                             are specified in wrong order. */


font-style: italic;
/* The following declaration overrides the above specified
   italic style with the default one (normal): */
```

```
font: 20px Helvetica, sans-serif;
```

font-family (Inherited)

The font-family CSS property allows you to specify the font in which the browser will display text. It can happen that a precise font is not installed on a viewer's computer—that's why designers usually specify a list of two or more options. At the end of the list there should always be one generic font family in case more specific fonts are not available. In contrast to most other CSS properties, values of the font-family property are delimited by commas to suggest that they are alternatives. The browser will use the first of the fonts in the list that it finds on the computer.

Values

Any font family name can be used, such as, for example, "Helvetica" or "Courier New". Quotes are not necessary unless names include spaces.

The generic font families serif, sans-serif, monospace, fantasy, or cursive can be used. Generic names act as a fallback system: if the browser does not find any font on the list, then it will use a generic font, which should always be specified last. Generic font family names should not be quoted because they are keywords.

Examples

```
font-family: "Times New Roman", "Georgia", serif;
font-family: "Comic Sans MS", fantasy;
```

font-size (Inherited)

The font-size CSS property sets the size of the font—precisely the height of the glyphs of the font. Because the actual font size is used for computing the value of the em length unit, setting it may in turn affect the sizes of other objects.

While keywords such as xx-small or large can also be used with the font-size property, many designers use only pixels, ems, or percentages because the interpretation of these keywords can vary between browsers.

Values

A positive <length> or <percentage> value. If a relative unit or <percentage> is used, then the font size is computed relative to the size of the font of the parent element.

xx-small, x-small, small, medium, large, x-large, xx-large

Default Value

medium

Example

```
font-size: 2em;
```

`font-style` (Inherited)

The `font-style` CSS property permits you to select an italic or oblique font. There is a slight difference between italic and oblique fonts in that that an italic font is normally cursive while an oblique font is merely a slanted version of the normal font. However, in most browsers there is no difference between the values `italic` and `oblique`. The `normal` value returns the font to its normal (roman) state.

Values

`italic`, `oblique`, `normal`

Default Value

`normal`

Example

```
font-style: italic;
```

`font-variant` (Inherited)

The `font-variant` CSS property allows a small-caps font to be selected, which is all upper-case letters except that the letters that were originally lower case are smaller. The `normal` value returns the font to its normal state.

Values

`small-caps`, `normal`

Default Value

`normal`

Example

```
font-variant: small-caps;
```

`font-weight` (Inherited)

The `font-weight` CSS property allows you to make text bold. The `normal` value returns the font to its normal state.

Values

`bold`, `normal`

Default Value

`normal`

Example

```
font-weight: bold;
```

letter-spacing (Inherited)

The `letter-spacing` CSS property defines the space between text characters.

Values

Positive or negative <length> values are allowed that specify the amount of space between characters *in addition to* the initial character spacing. Therefore, specifying a <length> value of 0 is equivalent to the `normal` value.

`normal`

Default Value

`normal`

Examples

```
letter-spacing: normal;
letter-spacing: -2px;
```

line-height (Inherited)

The `line-height` CSS property adjusts line spacing, i.e., space between lines of text. The default value for this property is `normal`. Although the amount of normal line spacing depends on the user agent, you can expect it to be about 120 percent of the font size for desktop browsers.

Values

A positive <number> value, which is multiplied by the element's font size to get the height of the line. This is normally the preferred way of setting the `line-height` property as it has no unexpected results.

A positive <length> or <percentage> value. If a relative unit or <percentage> is used, then `line-height` is computed relative to the size of the element's font. Note that using relative units like ems or percentages may yield unexpected results because of their poor inheritance behavior.

`normal`

Default Value

`normal`

Examples

```
line-height: normal;
line-height: 1.4;
line-height: 25px;
```

text-align (Inherited)

The `text-align` CSS property specifies how text or other inline content is aligned inside its parent block element. Note that it is not possible to control the alignment of a block element with the `text-align` property.

Because browsers do not hyphenate text, the alignment to both the left and right ends of an element (obtained by applying the `justify` value) may cause an uneven word spacing if lines of text are not long enough, which makes text difficult to read.

Values

`left`, `right`, `center`, `justify`

Default Value

`left`

Example

```
text-align: right;
```

text-decoration (Inherited)

The `text-decoration` CSS property permits authors to draw lines under, above, or through text. The color of the line is the same as the font color of the element being styled. The `none` value can be used to remove the underline that usually appears under hyperlinks.

Values

`none`, `underline`, `overline`, `line-through`

Default Value

Examples

```
text-decoration: none;
text-decoration: line-through;
text-decoration: underline overline;
```

text-indent (Inherited)

The text-indent CSS property specifies how much the first line in a block of text should be indented.

Values

A positive or negative absolute <length> value can be used. Negative values indent to the left so that the first line hangs over the left edge of the block of text.

A <percentage> value can also be used, which specifies the amount of indent relative to the containing block width.

Default Value

0

Examples

```
text-indent: 20px;
text-indent: 10%;
text-indent: 0;
```

text-transform (Inherited)

The text-transform CSS property controls the capitalization of text, which can be all upper case, all lower case, or have each word capitalized. When the capitalize value is used, each word is capitalized regardless of accepted title casing conventions, like the English convention where secondary words are not capitalized.

Values

uppercase, lowercase, capitalize, none

Default Value

Example

```
text-transform: uppercase;
```


word-spacing (Inherited)

The word-spacing CSS property works the same as the letter-spacing property, only that it regulates space between tags and words instead of individual characters.

Values

Positive or negative <length> values are allowed that specify the amount of space between words *in addition to* the initial word spacing. Therefore, specifying a <length> value of 0 is equivalent to the normal value.

normal

Default Value

normal

Examples

```
word-spacing: 1em;
word-spacing: -2px;
```


D.4   List Properties

The properties in this section influence the formatting of lists with bullets (<ul>) and numbered lists (<ol>).

list-style (Inherited)

The list-style CSS property is a shorthand property for specifying three list properties on a single line: list-style-type, list-style-image, and list-style-position. Values are separated by a space and can be provided in any order. If you specify list-style-type and list-style-image at the same time, then the bullet type will be used only if the image is not found. That way you are out of danger of ending up with a list without bullets, if the path to your bullet image doesn't work.

Like with any shorthand property, the values that are not explicitly set revert to their individual default values, even if they were previously set to different values using other properties. You can find out what the default values are from descriptions of respective properties.

Examples

```
list-style: square url(/images/mysquarebullet.gif) inside;
list-style: inside;


list-style-position: inside;
/* The following declaration overrides the above specified
   position with the default one (outside):              */
list-style: lower-roman;
```

list-style-image (Inherited)

The list-style-image CSS property defines an image that will be used as a bullet in a bulleted list.

Values

<url> of the image to be used as a bullet or the none keyword.

Default Value

Example

```
list-style-image: url(/images/mysquarebullet.gif);
```

list-style-position (Inherited)

The list-style-position CSS property decides whether the bullets or numbers should appear inside or outside of the content flow. If they are positioned outside of the content flow, then they hang off to the left, and if they are positioned inside the content flow, then they are put at the position where the first letter of the first line normally sits.

Values

inside, outside

Default Value

outside

Example

```
list-style-position: inside;
```

list-style-type (Inherited)

The `list-style-type` CSS property specifies the type of bullet for an unordered list (`<ul>`) or type of numbering for an ordered list (`<ol>`). Use the `none` value to completely remove the bullets.

Values

`disc`, `circle`, `square` (used with unordered lists)

`decimal`, `decimal-leading-zero`, `lower-alpha`, `upper-alpha`, `lower-roman`, `upper-roman`, `lower-greek` (used with ordered lists)

`none`

Default Value

`disc` (for unordered lists), `decimal` (for ordered lists)

Example

```
list-style-type: lower-roman;
```

D.5   Borders

These properties allow you to add and design borders of your elements.

border

The `border` CSS property is a shorthand method of defining the `border-width`, `border-style`, and `border-color` properties using a single declaration. The values of the `border` property are delimited with spaces and can be listed in any order. Like with any shorthand property, the values of properties that are omitted are set to their default values.

Example

```
border: 1px solid brown;
```

border-color

The `border-color` CSS property gives a color to all four edges of the border. You can also assign a different color to each of the four edges by specifying a space-delimited list of values in the following order: top, right, bottom, and left. If you list two colors, then the first will be used for the top and bottom edges, and the second for the left and right edges.

Note that the border line will not show as long as you don't declare the `border-style` property as well.

Values

Any valid <color> value can be used.

Default Value

The default value is equal to the current foreground color of the element.

Examples

```
/*Must specify border-style, or the border will not show:     */
border-style: solid;
border-color: #FFA000;
/* Black vertical and red horizontal borders:                */
border-color: black red black red;
border-color: black red;    /* The same as the previous line. */
```

border-radius

The border-radius CSS property rounds the corners of an element. The property is only effective with an element that has a border, or background color or image. It accepts one, two, or four space-separated values that define the radii of the circles used to round the corners. If only one value is given, it is used for all four corners. If two values are given, the first is used for top left and bottom right corners, and the second is used for top right and bottom left corners. If four values are given, then they are interpreted in this order: top left, top right, bottom right, and bottom left.

Elliptical corners can be specified by adding a slash sign (/) and a second set of values defining the vertical radius of corresponding ellipses.

Values

Nonnegative <length> or <percentage> values can be used. When percentages are used, the horizontal radius is relative to the width of the element, while the vertical radius is relative to the height of the element. That means that using percentages will generally produce elliptical corners unless the element has a shape of a square.

Default Value

0

Examples

```
border-radius: 20px;
border-radius: 50%; /* Full ellipse or full circle is produced. */
border-radius: 3em / 1em; /* Elliptical corners.                */
```

border-style

The `border-style` CSS property specifies the line style used to draw the border. You can also define a different style for each of the four edges by specifying a space-delimited list of values in the following order: top, right, bottom, and left. If you list two styles, then the first will be used for the top and bottom edges, and the second for the left and right edges.

Note that the default value of the `border-style` property is `none`. That means that even if you set the `border-width` and `border-color` properties, the border will not show unless you specify a value for this property different from `none` or `hidden`.

Values

`solid`, `dotted`, `dashed`, `double`, `groove`, `ridge`, `inset`, `outset`, `none`, `hidden`

Default Value

`none`

```
border-style: solid;
```

border-top, border-right, border-bottom, border-left

These are shorthand properties that work in the same way as the `border` property except that each sets a specific edge.

Example

```
border-bottom: 2px dashed rgb(250, 120, 0);
```

border-top-color, border-right-color, border-bottom-color, border-left-color

These properties set color for only one border. They are usually used to override the basic color defined by more generic `border` or `border-color` properties, when a specific color is needed for just one border.

Example

```
/* Must specify border-style, or the border will not show: */
border-bottom-style: dashed;
border-bottom-color: green;
```

border-top-style, border-right-style, border-bottom-style, border-left-style

These properties behave the same as the `border-style` property, only that each operates just on one edge.

Example

```
border-top-style: double;
```

border-top-width, border-right-width, border-bottom-width, border-left-width

These properties behave just like the `border-width` property, only that they apply just to one edge.

Example

```
/* Must specify border-style, or the border will not show: */
border-bottom-style: solid;
border-bottom-width: 1px;
```

border-width

The `border-width` CSS property specifies the thickness of the line used to draw the border. You can also assign a different thickness to each of the four edges by specifying a space-delimited list of values in the following order: top, right, bottom, and left. If you list two values, then the first will be used for the top and bottom edges, and the second for the left and right edges.

Note that the border line will not show as long as you don't declare the `border-style` property as well.

Values

A nonnegative <length> value or one of the keywords `thin`, `medium`, or `thick`.

Default Value

`medium`

Examples

```
/* Must specify border-style, or the border will not show:   */
border-style: solid;
border-width: thin;
```

```
/* Different vertical and horizontal borders:          */
border-width: .2em .1em .2em .1em;
border-width: .2em .1em;   /* The same as the previous line. */
```

D.6   Spacing

These properties control the space around and inside an element.

`margin`

The `margin` CSS property defines the amount of empty space between the element's border and the margin of a neighboring element. You can also assign a different amount to each of the four margins by specifying a space-delimited list of values in the following order: top, right, bottom, and left. If you list two values, then the first will be used for the top and bottom margins, and the second for the left and right margins.

Note that in some situations vertical margins can collapse, leaving a smaller gap between elements than anticipated. You can read about collapsing margins on page 79.

Values

Positive or negative <length> or <percentage> values can be used. If percentages are used, they are relative to the width of the containing block.

A special keyword `auto` can be used, which is replaced by the browser with some convenient value. It can be used, for example, for centering block elements. The next CSS rule will center `<div>` elements horizontally:

```
div {
  width: 50%;
  margin: 0 auto;
}
```

Default Value

0

Examples

```
/* Set all the margins to 10 px:                      */
margin: 10px;
/* Set top, right, bottom, and left margins individually: */
margin: 1em 0.5em 2em 0.5em;
```

margin-top, margin-right, margin-bottom, margin-left

These properties act just like the margin property, only that each sets the margin for its own edge.

Example

```
margin-bottom: 20px;
```

padding

The padding CSS property specifies the amount of empty space between the element's content (e.g., an image or block of text) and its border. You can also assign a different amount to each of the four paddings by specifying a space-delimited list of values in the following order: top, right, bottom, and left. If you list two values, then the first will be used for the top and bottom paddings, and the second for the left and right paddings.

Values

Nonnegative <length> or <percentage> values can be used. If percentages are used, they are relative to the width of the containing block.

Default Value

0

Examples

```
/* Set all the paddings to 10 px:                    */
padding: 10px;
/* Set top, right, bottom, and left paddings individually: */
padding: 1em 0.5em 2em 0.5em;
```

padding-top, padding-right, padding-bottom, padding-left

These properties act just like the padding property, only that each sets the padding for its own edge.

Example

```
padding-top: 5px;
```

D.7   Background Properties

With these properties, you can control the appearance of the background of an element by specifying a color or including an image.

background-color

The `background-color` CSS property gives a color to the background of an element. Note that background lies underneath the border, so if you use a non-solid border such as dotted or dashed, the background color will show through the spaces between the dots or dashes. Background also lies underneath a background image and will not be seen if a background image is specified unless the image is not fully opaque.

Values

Any valid <color> value can be used.

Default Value

`transparent`

Examples

```
background-color: magenta;
background-color: transparent;
```

background-image

The `background-image` CSS property enables authors to put an image into the background of an element. All the other parts of the element (including the border) sit on top of the background image except for the background color, which hides underneath the image. If either the width or height (or both) of the image are smaller than the width and height of the element, then the image will tile. If the image doesn't fit, it will be clipped.

Values

A <url> of the image, or the `none` keyword.

Default Value

`none`

Example

```
background-image: url(/backgrounds/sky.jpg);
```

D.8 Table Properties

The properties in this section are used exclusively for work with HTML tables, the look of which can be significantly improved by using CSS. Note that, by default, HTML tables and table cells do not have borders. Use the `border-style` CSS property found on page 353 of this appendix to make borders visible.

border-collapse

The `border-collapse` CSS property specifies whether space should be left between cell and table borders, or if borders should collapse. Namely, even if you set the `border-spacing` CSS property to zero, there will still remain some space between borders of neighboring cells or between cells and table borders, which produces double border lines. They will turn into single lines if you set `border-collapse` to `collapse`.

Values

`collapse`, `separate`

Default Value

`separate`

Example

```
border-collapse: collapse;
```

border-spacing

The `border-spacing` CSS property specifies how much space should be left between cells of a table. When two values are provided, they are used as two separate values for horizontal and vertical spacing. This property is also used in calculating space between table cells and the outside edge of a table.

Note that the `border-spacing` property does not apply when the `border-collapse` property is set to `collapse`.

Values

Nonnegative <length> values can be used.

Default Value

0

Examples

```
border-spacing: 0.1em;
border-spacing: 5px 0; /* Different horizontal and vertical
                          spacing. */
```

caption-side

The `caption-side` CSS property decides whether a table caption should appear above or under the table. This attribute can only be applied to the `<caption>` HTML element.

Values

`top`, `bottom`

Default Value

`top`

Example

```
caption-side: bottom;
```

empty-cells

The `empty-cells` CSS property specifies the look of empty table cells, i.e., cells that have no visible content. When the `hide` value is used, then the place for an empty cell will still be reserved but nothing will be displayed there, neither border nor background. Note, however, that placing a nonbreakable space character (i.e., ` `) inside a cell makes that cell nonempty. In that case, even if a cell looks empty, the `hide` value will not hide the border and background. The `empty-cell` property can only be applied to the `<table>` HTML element.

Values

`show`, `hide`

Default Value

`show`

Example

```
empty-cells: hide;
```

D.9 Size Properties

The two properties in this section allow you to set specific dimensions of your elements.

height

The `height` CSS property defines the height of the element's content area. The actual height of the element can be taller than that specified by the `height` property since the actual height also includes both (top and bottom) vertical paddings, both vertical border edges, and both vertical margins. If you are not familiar with these terms, you can read about the CSS box model on page 74.

The `auto` keyword lets the browser compute and select the most appropriate height for the element.

Note that the actual height of the content area on the screen can show taller than the set height. That can happen, for example, if there's too much text inside the content box or if the user increases the font size. Browsers don't all handle this situation in the same way. Some will increase the height of the content box while others will simply overflow text outside of the box.

Values

Nonnegative <length> and <percentage> values can be used. When percentages are used, they are relative to the height of the containing block. Note that percentages may not have an effect if the containing element's `height` property is set to `auto`.

auto

Default Value

auto

Example

```
height: 200px;
```

width

The `width` CSS property defines the width of the element's content area. Note that the actual width of the element can be wider since it also includes the widths of both (left and right) horizontal paddings, both horizontal border edges, and both horizontal margins. If you are not familiar with these terms, you can read about the CSS box model on page 74.

The `auto` keyword lets the browser compute and select the most appropriate width for the element.

Values

Nonnegative <length> and <percentage> values can be used. When percentages are used, they are relative to the width of the containing block.

```
auto
```

Default Value

```
auto
```

Example

```
width: 80%;
```

D.10   Positioning

The properties in this section control the layout of elements on a web page. They are among the most difficult to understand and use correctly.

bottom

The `bottom` CSS property is used with positioned elements to specify part of their position. The property has no effect on elements that are not positioned (i.e., have their `position` property set to `static`).

If this property is used on an element whose `position` property is set to `absolute` or `fixed`, then it sets the position of the bottom margin edge of the element relative to the bottom edge of its containing block.

If the `position` property is set to `relative`, then the element is simply moved above its normal position.

Values

Negative or positive <length> or <percentage> values can be used. Positive values move the element up and negative down. If percentages are specified, they are relative to the height of the containing block.

```
auto
```

Default Value

```
auto
```

Example

```
bottom: 5%;
```

clear

The `clear` CSS property defines whether an element can be placed next to floated elements (i.e., elements that have their `float` property different from `none`) that are placed *before* it in the original HTML code or an element has to move down (clear) below the floated elements that are placed before it.

The `clear` property works with floated as well as non-floated elements. When a *non-floated* block is cleared, its top border edge is moved below the bottom margin edge of any floated elements before it. Note that this move collapses margins, although the resulting margin is not the largest of both margins. Rather, it is defined by the position of the lowest bottom margin edge of the relevant floated elements.

When, however, a *floated* block is cleared, its top *margin* (not border) edge is moved below the bottom margin edge of any floated elements before it. This move does not collapse margins.

It is possible to clear an element in three different ways:

- If the `left` value is used, then the element will move down past all the floated elements that precede it in the original HTML document and are floated to the left.

- If the `right` value is used, then the element will move down past all the floated elements that precede it in the original HTML document and are floated to the right.

- If the `both` value is used, then the element will move down past all the floated elements that precede it in the original HTML document.

Values

`left, right, both, none`

Default Value

`none`

Example

```
clear: left;
```

display

The `display` CSS property defines the type of box used for rendering the element. CSS defines many different keyword values to be used with this property, but for the purposes of this book we mention only three of them: `block`, `inline`, and `none`. Basically, they work as follows: The `block` value adds a line break before and after an element, making the element occupy the whole available width of the document. The `inline` value places an element on the same line with the neighboring elements if there is enough space for them. With the `none` value you completely remove the element from the page. It is possible to make the element appear later with the `:hover` pseudo-class or some JavaScript code. That is useful, for example, if you want to animate the opening of sub-menus in a menu system as you can see in the example on page 103. You can find another example of using this property on page 238, where it is used for expanding and collapsing long text passages.

Values

`block, inline, none`

Default Value

Depends on the type of an HTML element.

Example

```
display: inline;
```

float

The `float` CSS property allows authors to take an element from its normal flow and place it by the left or right edge of its container. Text and inline elements that appear in the original HTML code after the floated element move up to fill the empty space to the right or left (depending on the specified direction of the float) and wrap below the floated element. Setting `float` to either `left` or `right` implies the block layout and hence changes the computed value of the element's `display` property to `block`.

Note that the `float` property does not work on absolutely positioned elements, i.e., elements that have their `position` property set to either `absolute` or `fixed`.

Values

`left, right, none`

Default Value

`none`

Example

```
float: left;
```

left

The `left` CSS property is used with positioned elements to specify part of their position. The property has no effect on elements that are not positioned (i.e., have their `position` property set to `static`).

If this property is used on an element whose `position` property is set to `absolute` or `fixed`, then it sets the position of the left margin edge of the element relative to the left edge of its containing block.

If the `position` property is set to `relative`, then the element is simply moved to the right from its normal position. This may seem odd, but the `left` property actually specifies how far an element is positioned from the left edge towards the *inside* of the containing element, which means it is positioned to the right.

Values

Negative or positive <length> or <percentage> values can be used. Positive values offset the element to the right and negative to the left. If percentages are specified, they are relative to the width of the containing block.

`auto`

Default Value

`auto`

Example

```
left: 50px;
```

position

The `position` CSS property specifies the type of positioning used for placing an element on the web page. Any element that has its `position` property set to either `relative` or `absolute` or `fixed` is said to be *positioned*. Conversely, any element that has its `position` property set to `static` is *not positioned*.

If `position` is `static`, then the element is placed to its normal position, which corresponds to its current position in the document flow. Any of the `top`, `right`, `bottom`, `left`, and `z-index` CSS properties set on that element will be ignored.

The `relative` value offsets an element from its normal position, i.e., relative to where it would have been placed if its `position` property was set to `static`. Note, however, that a relatively positioned element is not taken out of the document flow, i.e., it leaves behind a gap as if it was still there, which lets all the other elements stay in their positions.

The `absolute` value takes an element totally out of the document flow. Other elements disregard this element when placing themselves and may appear underneath it. With the `absolute` value you place an element in an exact position relative to its closest ancestor that is positioned or to its containing block. Note that margins of absolutely positioned elements do not collapse with any other margins.

The `fixed` value also takes an element out of the document flow, putting it in an exact position in the browser's document window (i.e., viewport). Other elements place themselves as if the fixed element was not there. If the page is scrolled, the element stays fixed on the screen.

Note that the `position` property itself does not offer you full control over the element's position. To fine-tune the position, you use the `top`, `right`, `bottom`, and `left` CSS properties.

Values

`static`, `relative`, `absolute`, `fixed`

Default Value

`static`

Example

```
position: relative;
```

right

The `right` CSS property is used with positioned elements to specify part of their position. The property has no effect on elements that are not positioned (i.e., have their `position` property set to `static`).

If this property is used on an element whose `position` property is set to `absolute` or `fixed`, then it sets the position of the right margin edge of the element relative to the right edge of its containing block.

If the `position` property is set to `relative`, then the element is simply moved to the left from its normal position. This may seem odd, but the `right` property actually specifies how far an element is positioned from the right edge towards the *inside* of the containing element, which means it is positioned to the left.

Values

Negative or positive <length> or <percentage> values can be used. Positive values offset the element to the left and negative to the right. If percentages are specified, they are relative to the width of the containing block.

`auto`

Default Value

`auto`

Example

```
right: 50px;
```

top

The `top` CSS property is used with positioned elements to specify part of their position. The property has no effect on elements that are not positioned (i.e., have their `position` property set to `static`).

If this property is used on an element whose `position` property is set to `absolute` or `fixed`, then it sets the position of the top margin edge of the element relative to the top edge of its containing block.

If the `position` property is set to `relative`, then the element is simply moved below its normal position.

Values

Negative or positive <length> or <percentage> values can be used. Positive values move the element down and negative up. If percentages are specified, they are relative to the height of the containing block.

`auto`

Default Value

`auto`

Example

```
top: 10%;
```

visibility

The `visibility` CSS property specifies whether or not an element is visible. Unlike the `none` value of the `display` CSS property, this property does not remove the element from the page, i.e., the element still takes up space on the page even if it is made invisible. This property is most often used with the `display` property set to `absolute`. Because just hiding an element isn't very useful, you can make the element visible at any time with some JavaScript programming or the `:hover` pseudo-class.

Values

`visible`, `hidden`

Default Value

`visible`

Example

```
visibility: hidden;
```

z-index

The z-index CSS property controls the position (stacking) of an element on the z axis, which is perpendicular to the screen plane. Normally, if two elements overlap, the one with the higher z-index will appear on top. This may not be true if one of the elements is a descendant of another positioned element. Namely, each positioned element creates its own local stacking context for its descendants, whose z-index values are not compared to the z-index values of elements outside of this context.

Note that this property only works on positioned elements, i.e., elements that have their position property set to either absolute or relative or fixed.

Values

Nonnegative <integer> values can be used.

```
auto
```

Default Value

```
auto
```

Example

```
z-index: 42;
```

D.11   Pseudo-Classes and Pseudo-Elements

In this section you will find some predefined selectors (called pseudo-classes and pseudo-elements) that let you select parts of a web page that are not equipped with tags but are still easy to identify. Names of pseudo-classes begin with a colon (:) while names of pseudo-elements begin with two colons (::).

:active

The :active CSS pseudo-class matches an element that is being activated by the user: for example, when the user is pressing the mouse button while pointing to the element.

When using the :active pseudo-class selector with other link-related pseudo-classes on the same element, :active must be put after any other link-related rules respecting the so-called *LVHA order* (i.e., :link, :visited, :hover, and :active).

Example

```
a:link { color: blue; }      /* Unvisited links */
a:visited { color: black; } /* Visited links   */
a:hover { color: orange; }  /* Hovered links    */
a:active { color: red; }     /* Active links     */
```

:first-child

The :first-child CSS pseudo-class matches the first of the group of siblings (children of the same parent).

Examples

```
/* Select any paragraph that is the first child of its parent: */
p:first-child
/* Select the first child of any paragraph:                     */
p :first-child
```

::first-letter

The ::first-letter CSS pseudo-element matches the first letter in the first line of text in a block, if there is no other content before it. This pseudo-element can be used, for example, to create a decorative initial capital letter, also called a *drop cap* or *initial*—the first letter of a paragraph that is enlarged to drop down a few lines. Only a subset of all CSS properties works with this pseudo-element.

Note that it is not always trivial to identify the first letter. For example, any punctuation before or immediately after the first letter is also selected by this pseudo-element.

Example

```
/* Create a drop cap for all paragraphs. */
p::first-letter {
  font-size: 3em;
  float: left;
}
```

::first-line

The ::first-line CSS pseudo-element matches the first line of text of an element. Some designers like to style the first line of a paragraph or section in a different font or color. That way, text can appear more lively and appealing. Only a subset of all CSS properties works with this pseudo-element.

Example

```
p::first-line {
  font-weight: bold;
}
```

:hover

The :hover CSS pseudo-class matches an element when the mouse cursor passes over it. The resulting effect can provide useful visual feedback about where the mouse cursor is. Note, however, that on touch screens this selector can be problematic. It is therefore important that you do not make any content accessible exclusively through the use of this selector, or touch-screen users won't be able to see it.

When using the :hover pseudo-class selector with other link-related pseudo-classes on the same element, :hover must be put after :link and :visited and before :active, respecting the so-called *LVHA order* (i.e., :link, :visited, :hover, and :active).

Examples

```
/* Change color of elements of class clickable
   when hovered over. */
.clickable:hover {
  color: white;
  background-color: black;
}
```

:last-child

The :last-child CSS pseudo-class matches the last of the group of siblings (children of the same parent).

Examples

```
/* Select any paragraph that is the last child of its parent: */
p:last-child
/* Select the last child of any paragraph: */
p :last-child
```

:link

The :link CSS pseudo-class matches any hyperlink that the user hasn't visited yet, and is not being activated or hovered over. This pseudo-class selector only works with hyperlinks.

When using the :link pseudo-class selector with other link-related pseudo-classes on the same element, :link must be put before any other link-related rules respecting the so-called *LVHA order* (i.e., :link, :visited, :hover, and :active).

:nth-child()

The :nth-child() CSS pseudo-class allows authors to select any combination of alternating child elements. An extra piece of information in parentheses is needed in order for this pseudo-class selector to work. The simplest option is to use one of the keywords odd or even, which select alternating odd or even elements. Every ath child can also be selected, optionally starting with an arbitrary chosen child (bth child), using this syntax:

```
:nth-child(an + b)
```

Constants a and b must be integers. If b is omitted, it is assumed to be 1. Note that child counting begins with one so that the element's first child has an index value of 1.

If n and b are omitted, then only a single child will be selected, which is the ath child.

Examples

```
/* Select every second child, starting with the first one: */
:nth-child(2n+1)
:nth-child(odd) /* The same as above.                      */

/* Select every third child, starting with the first one:  */
:nth-child(3n)

/* Select every third child, starting with the second one: */
:nth-child(3n + 2)

/* Select the first child of the body:                      */
body :nth-child(1)

/* Color every second row of a table:                       */
tr:nth-child(even) {
  background-color: lightgray;
}
```

:visited

The :visited CSS pseudo-class matches a hyperlink that a visitor has already visited and is saved in the web browser's history. Using this pseudo-class selector, you can show your visitor the links that he/she has already followed in a different color. The :visited pseudo-class only works with hyperlinks.

When using the `:visited` pseudo-class selector with other link-related pseudo-classes on the same element, `:visited` must be put after `:link` and before `:visited` and `:active`, respecting the so-called *LVHA order* (i.e., `:link`, `:visited`, `:hover`, and `:active`).

Owing to privacy concerns, serious limits have been posed as to which CSS properties are allowed with the `:visited` pseudo-class. Namely, some JavaScript functions exist that can easily figure out what sites the user has visited, which allows some uninvited iniquitous person to learn a lot about a user's identity. Hence, the limitations have been made to the amount of styling that can be applied to visited links. Basically, you are allowed to style `color`, `background-color`, and `border-color` properties with this selector.

Example

```
/* Style all visited hyperlinks inside any <nav> element: */
nav :visited {
  color: black;
}
```

JavaScript Mini Reference

In this appendix, you will find a convenient list and descriptions of the core and client-side JavaScript classes, methods, and properties that are used in this book, and also some others that are not but may come in handy while taking the first steps through the language. However, if you're looking for a more complete and elaborate reference, you can find one in David Flanagan's must-have book *JavaScript: The Definitive Guide*. Another quite detailed and comprehensive language reference is available on-line at *developer.mozilla.org/en-US/docs/Web/JavaScript/Reference*. If you have the nerve, you may even want to look at the latest (at the time of writing this book) published language specification, ECMAScript 5.1 at *www.ecma-international.org/ecma-262/5.1*.

E.1 Operator Precedence and Associativity

The following table summarizes JavaScript operators by their precedence and associativity. The operators with higher precedence are listed before the operators with lower precedence. Operators between two horizontal lines have equal precedence order.

Operators	Associativity
++, -, - (sign) , + (sign), !, typeof	right-to-left
*, /, %	left-to-right
+, -	left-to-right
<, <=, >, >=	left-to-right
==, !=, ===, !==	left-to-right
&&	left-to-right
||	left-to-right
? :	right-to-left
=, *=, /=, %=, +=, -=	right-to-left
,	left-to-right

E.2 arguments[] (Core JavaScript)

Syntax:

```
arguments
arguments[n]
```

The arguments object is an array-like object of the Arguments type. It is defined only within a body of a function and holds all the values that were passed as arguments to the function.

Object Instance Creation

Each time a function is called, an Arguments object is created, which holds all the arguments passed to the function. Technically, arguments is a local variable of the function, which is automatically declared and initialized to refer to the created Arguments object.

Properties

length

Syntax:

```
arguments.length
```

This property defines the number of arguments passed to the function. It is important to note that this is the number of arguments actually passed and not the number expected. The property enables a programmer to check whether all the parameters have been passed or to fetch additional, unnamed arguments of the function.

There's an example of using arguments on page 194.

E.3 Array (Core JavaScript)

The Array object acts as a constructor for arrays and therefore defines an object class. An array is a list-like object with prototyped methods suited for list manipulations. Basically, you can think of an array as a collection of different objects—called elements—which you can traverse, add, delete, search, sort, and so on.

Object Instance Creation

Array() //Constructor

Syntax:

```
new Array(element1, element2, ..., elementN)
```

```
new Array(arrayLength)
new Array()
```

Creates, initializes and returns a new array. The created array is initialized with the values passed as arguments to a constructor (*element1*, *element2*, ..., *elementN*). If only a single numeric argument is passed to the Array() constructor (*arrayLength*), which is an integer between 0 and 2^{32-1} inclusive, then the constructor creates an array with the given number of undefined elements. When the Array() constructor is invoked without any arguments, it returns an empty array, whose length property has a value of 0.

Examples:

```
var a = new Array(42, 4); //An array with two elements: 42 and 4
var a = new Array();      //An empty array
var a = new Array(42);    //An array with 42 undefined elements
var a = new Array("q");   //An array with a single element "q"
```

```
[] //Literal syntax
```

Syntax:

```
[element1, element2, ..., elementN]
```

An alternative way of creating an array is to place a comma-separated list of expressions inside square brackets. The values of the supplied expressions become the array elements. Note that a single numeric value is treated as a single element with which to initialize the array and not as the number of undefined elements like it is the case with the Array() constructor.

Examples:

```
var a = [42, 4];          //An array with two elements: 42 and 4
var b = [];               //An empty array
var c = [42];             //An array with a single element 42
var d = [a[1], b.length]; //An array with two elements: 4 and 0
```

Properties

```
length
```

Syntax:

```
array.length
```

The `length` property is an integer whose value is one bigger than the index of the last element of *array*. If *array* has contiguous elements, then `length` tells the number of elements in *array*. You can also directly change the value of the `length` property, which truncates or lengthens *array*.

Examples:

```
var a = [5, 2, 9];
console.log(a.length);   //The length of a is 3
a.length = 2;            //Truncates a to two elements
console.log(a[2]);       //The third element of a is now undefined
```

Methods

`indexOf()`

Syntax:

```
array.indexOf(searchElement)
array.indexOf(searchElement, searchFrom)
```

The `indexOf()` method searches *array* for the first element that is equal to *search Element* and returns the index of the found element. By default, the search begins at the first element (with the index of 0), unless the optional *searchFrom* argument specifies a different start index. If no matching element is found, then `indexOf()` returns -1. Note that the method uses the `===` operator for testing equality.

Examples:

```
var arr = [5, 40, "9", 40];
arr.indexOf(40)      //Returns 1
arr.indexOf(40, 2)   //Returns 3
arr.indexOf(9)       //Returns -1
arr.indexOf("9")     //Returns 2
arr.indexOf(5, 1)    //Returns -1
```

`lastIndexOf()`

Syntax:

```
array.lastIndexOf(searchElement)
array.lastIndexOf(searchElement, searchFrom)
```

The `lastIndexOf()` method searches *array* backwards for the first element that equals *searchElement* and returns the index of the found element. By default, the search begins at the last element (with the index of *array*.length − 1), unless the optional *searchFrom* argument specifies a different start index. If no matching element is found, then indexOf() returns -1. Note that the method uses the === operator for testing equality.

Examples:

```
var arr = [5, 40, "9", 40];
arr.lastIndexOf(40)       //Returns 3
arr.lastIndexOf(40, 2)    //Returns 1
arr.lastIndexOf(5, 1)     //Returns 0
arr.lastIndexOf("9", 1)   //Returns -1
```

pop()

Syntax:

```
array.pop()
```

The pop() method returns the value of the last element in *array* and removes that element from the array. In order to reflect the change in the size of the array, the length property is decremented accordingly. If pop() is invoked on an empty array, then it returns the undefined value and the array is not changed.

Examples:

```
var arr = [2, 4, 11];
arr.pop();    //Returns 11, arr is now [2, 4]
arr.pop();    //Returns 4, arr is now [2]
arr.pop();    //Returns 2, arr is now []
arr.pop();    //Returns undefined, arr is still []
```

push()

Syntax:

```
array.push(element1, element2, ..., elementN)
```

This method appends one or more elements (*element1*, *element2*, ..., *elementN*) to the end of *array*. The method also increments the length property to reflect the new size of the array and returns it.

Examples:

```
var arr = [];
arr.push(2, 3, 5);  //Returns 3, arr is now [2, 3, 5]
arr.push(7);        //Returns 4, arr is now [2, 3, 5, 7]
arr.push(11);       //Returns 5, arr is now [2, 3, 5, 7, 11]
```

```
reverse()
```

Syntax:

```
array.reverse()
```

The `reverse()` method reverses the order of the array elements, so that the first element of the array becomes the last and vice versa. The rearrangement of the elements is done *in place*, i.e., the elements of the array on which the method is invoked are reordered without creating a new array. The method returns the reference to the reversed array, so that the method chaining is possible.

Example:

```
var arr = [1, 1, 2, 3, 5, 8];
arr.reverse(); //arr is now [8, 5, 3, 2, 1, 1]
```

```
sort()
```

Syntax:

```
array.sort()
array.sort(compareFunction)
```

The `sort()` method sorts the elements of an array, which is done *in place* (i.e., no copy of the original array is made), and returns the reference to the sorted array so that the method chaining is possible. If no additional argument is supplied, then the array elements are sorted alphabetically (more specifically, in an ascending order according to the character codes determined by the character encoding).

Because the array elements are sorted alphabetically, numbers are not arranged as you might expect. For example, 2 comes before 42 in a numeric sort, but since numbers are converted to strings before sort is applied, `"42"` comes before `"2"`.

If you want to use some other criteria to sort elements, then you can supply a custom comparison function (*compareFunction*). This function takes two arguments and must return one of the following values:

- A value less than zero if the first argument should appear before the second one in the sorted array.
- A value greater than zero if the second argument should appear before the first one in the sorted array.
- Zero if both arguments are considered equal in the scope of this sort and their order doesn't matter.

If there are any undefined elements, they always come at the end of the sorted array. This is not possible to change even if you provide your own comparison function because undefined values are never passed to the *compareFunction* function.

Example:

```
//Sort numbers alphabetically and then numerically.
var arr = [700, 900, 6000, 80];
var numeric = function(x, y) {return x - y;};
arr.sort();       //arr is now [6000, 700, 80, 900]
arr.sort(numeric); //arr is now [80, 700, 900, 6000]
```

toString()

Syntax:

```
array.toString()
```

The method converts *array* to a string, which it returns. The original array is not changed.

The conversion is performed in two steps: First, every array element is converted to a string. Second, all thus obtained strings are concatenated into a single string with additional commas placed between them.

Example:

```
var arr = [1, 2, 3, 4];
var str = arr.toString(); //str is now "1,2,3,4".
str[3]                    //Returns ","
str[4]                    //Returns "3"
```

E.4   Boolean (Core JavaScript)

Boolean is one of the JavaScript fundamental data types. The Boolean object is a wrapper around the primitive boolean value. The essential role of this object is to provide a `toString()` method, which converts boolean values `true` and `false` to strings.

Object Instance Creation

```
Boolean() //Constructor
```

Syntax:

```
new Boolean(value) //Constructor
Boolean(value)     //Type conversion function
```

When a boolean to string conversion is necessary, JavaScript often implicitly creates a temporary Boolean object in order to be able to call its `toString()` method. Although you will rarely need to create a Boolean object for yourself, you can do that by using the `Boolean()` function. The `Boolean()` function can either be invoked as a constructor (with the `new` operator) or as a conversion function (without the `new` operator). In both cases its first step is to convert *value* to a boolean value, where the values 0, `NaN`, `null`, `undefined`, `false`, and the empty string `""` are all converted to `false`. All other values (primitive and object)—including the string `"false"`—are converted to `true`. After the conversion, the `Boolean()` constructor returns a Boolean object containing the converted value. The `Boolean()` conversion function, however, simply returns a primitive boolean value.

Methods

`toString()`

Syntax:

```
boolean.toString()
```

Returns a string representation of the value stored in *boolean*, which can be either `"true"` or `"false"`. The method can also be invoked on a primitive boolean value, in which case JavaScript creates a temporary wrapper Boolean object in order to be able to call the method. However, this is rarely needed in practice because JavaScript calls the `toString()` method automatically when the automatic conversion rules demand that a boolean value be converted to a string (e.g., in a string concatenation with the concatenation operator (+)).

Examples:

```
var b = true;
var B = new Boolean(true);
B.toString()        //Returns "true"

//In the following two examples JavaScript automatically creates
//temporary wrapper objects to be able to call toString():
b.toString()        //Returns "true"
false.toString()    //Returns "false"
```

`valueOf()`

Syntax:

```
boolean.valueOf()
```

Returns the primitive boolean value stored in the object.

E.5 `console` (Client-Side JavaScript)

Modern browsers have integrated a debugging console (e.g., the JavaScript Console in Google Chrome, or the Browser Console in Firefox, both accessible through the

---

`Ctrl+Shift+J` shortcut) and implement a Console object that is accessible through the `console` global property. The Console object defines an API for basic debugging tasks such as displaying different messages to a console window. Note that there's no formal standard defining the Console API, but vendors follow a de facto standard set up by the Firebug debugging extension of Firefox. While implementations may differ from vendor to vendor, support for the essential `log()` method is almost universal.

Methods

`dir()`

Syntax:

```
console.dir(object)
```

Displays *object* in the Console window. The programmer can examine the properties, methods, or elements of *object*, and inspect any nested objects interactively by clicking disclosure triangles.

`log()`

Syntax:

```
console.log(obj1, obj2, ...,objN)
```

Outputs the string representations of its arguments to the console.

Consider, for example, the next code:

```
var t = "Hello there!";
var o = {a: 10, b: 20};
console.log(t, o);
```

Inside the Console window, you can see the following information produced by the above call to the `log()` method.

Observe how the string is displayed as normal text while the o object is displayed with the names and values of both its properties.

You can also examine programmatically generated DOM elements in the Console window. Take, for example, this code:

```
var table = document.createElement("table");
var row = document.createElement("tr");
var cell = document.createElement("td");
cell.innerHTML = "Howdy!"
table.appendChild(row);
row.appendChild(cell);
console.log(table);
```

Feeding the code to the browser produces the following situation.

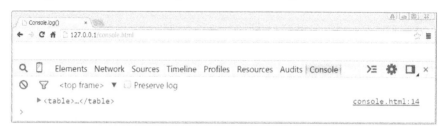

Inspecting the Console window only shows you the `<table>` element but you will notice an additional triangle in front of it, which indicates that this item is expandable. If you click it, the element expands and reveals the following structure.

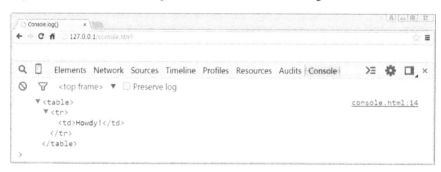

time()

Syntax:

```
console.time(name)
```

The `time()` method starts a timer and assigns *name* to it. The timer's name is later used by the `timeEnd()` method to stop the timer.

timeEnd()

Syntax:

```
console.timeEnd(name)
```

The `timeEnd()` method stops the timer with the specified *name*. It also displays in the Console the name and the time passed since the last call to the `time()` method with the same name. The `time()` and `timeEnd()` pair can be used to establish how much time a certain code fragment in your program takes to execute.

E.6   Date (Core JavaScript)

The Date object is a JavaScript built-in data type allowing you to work with date and time. Internally, the object stores time value as a number of milliseconds since midnight on January 1, 1970 in Universal Coordinated Time (which is the same as Greenwich Mean Time). This time stamp is also known as the *UNIX epoch*. According to the ECMAScript standard requirement, the Date object should be able to represent any moment, to milliseconds precision, in a period of 100 million days before or after the UNIX epoch. That's more than a quarter million years to the past and future.

Object Instance Creation

`Date() //Constructor`

Syntax:

```
new Date()
new Date(milliseconds )
new Date(datestring )
new Date(year , month [, day [, hrs [, mins [, secs [, millis ]]]]])
//Called as a function:
Date() //Returns a string representation of current date and time
```

Constructs a Date object pointing to a certain moment in time. If you supply no arguments to the constructor, then the created Date object is set to the current date and time (more precisely, the local date and time as set by the operating system of the computer running the script). When you pass a single numeric argument (*milliseconds*), this argument is understood as the number of milliseconds since midnight on January 1, 1970 (UTC), which is the object's internal representation of date and time.

The third of the above constructors allows you to use a string representation (an argument *datestring*) of date and time, which is specified in local time, not UTC.

When two or more (up to seven) numeric arguments are passed to the constructor, they specify respective fields of the date and time, again in local time. Note that all but the first two fields are optional as indicated by the square brackets in the above syntax. Square brackets should not be used in a constructor, they just signal which arguments are optional. For example, if you specify *mins*, then you must specify *day* and *hrs* as well.

The next table summarizes ranges of possible values for each of the seven date and time fields.

Argument	Values
year	The year should be specified in full format. For backwards compatibility, if this argument is between 0 and 99 inclusive, 1900 is automatically added to it.
month	An integer from 0 (January) to 11 (December).
day	An integer from 1 to 31.
hrs	Hours, an integer from 0 (midnight) to 23 (11 p.m.).
mins	Minutes, an integer from 0 to 59.
secs	Seconds, an integer from 0 to 59.
millis	Milliseconds, an integer from 0 to 999.

Finally, you can use Date() as an ordinary global function (without the new operator) to get a string representation of the present time and date.

Examples:

```
new Date(3600000)       //1 a.m. on January 1, 1970 (UTC)
new Date(2016, 0, 1, 12)   //Noon on January 1, 2014 (local time)
new Date("1/1/2016 12:00") //Same as above
Date()                  //Returns human readable local date and time
```

Out-of-Range Values of Time and Date Fields

When a value provided to the Date() constructor is out of its valid range, it rolls over and the adjacent value is adjusted accordingly. For example, the expression Date(2016, 0, 1, 26) is equivalent to calling Date(2016, 0, 2, 2), and Date (2016, 0, 2, -4) is the same as Date(2016, 0, 1, 20). The same happens when you use out-of-range values with methods that set individual time and date fields of the Date object like, for example, setHours(), or setDate().

For example, the next code fragment checks whether a year is a leap year by attempting to set date to February 29:

```
dt = new Date(year, 1, 29);
if (dt.getMonth() == 1) {
  //year is a leap year
}
```

Methods

Since the internal date and time representation of the Date object is the number of milliseconds since the UNIX epoch, no properties exist that could be read and written directly. Instead, there are methods for setting and getting individual date and time fields. They take care of conversion to milliseconds, taking into account a time difference between UTC and local time where needed. Most methods are designed in pairs to work with both, local time and UTC, which is evident from their names—if there is "UTC" in the method's name, then that method operates using universal time. In method descriptions below, notation [UTC] means that a method exists with and without "UTC" in its name.

get[UTC]Date()

Syntax:

```
date.getDate()
date.getUTCDate()
```

Returns the day of the month specified by *date*, using local (getDate()) or universal (getUTCDate()) time. The method returns values between 1 and 31.

get[UTC]Day()

Syntax:

```
date.getDay()
date.getUTCDay()
```

Returns the day of the week specified by *date*, using local (getDay()) or universal (getUTCDay()) time. The method returns values from 0 (Sunday) and 6 (Saturday).

get[UTC]FullYear()

Syntax:

```
date.getFullYear()
date.getUTCFullYear()
```

Returns the year as specified by *date*, using local (getFullYear()) or universal (getUTCFullYear()) time. The method always returns a full year value, without any abbreviations.

Example:

```
var dt = new Date(-1000); //1 second before midnight 1/1/70
dt.getFullYear()     //Returns 1970 in Paris or Sydney (local time)
dt.getUTCFullYear()  //Returns 1969 (UTC)
```

`get[UTC]Hours()`

Syntax:

```
date.getHours()
date.getUTCHours()
```

Returns the hour of the day as defined by *date*, using local (getHours()) or universal (getUTCHours()) time. The return value follows a 24-hour system, where 0 is midnight and 23 is 11 p.m.

Example:

```
var dt = new Date(0); //Midnight 1/1/70
dt.getHours()         //Returns 19 in New York (local time)
dt.getUTCHours()      //Returns 0 (UTC)
```

`get[UTC]Milliseconds()`

Syntax:

```
date.getMilliseconds()
date.getUTCMilliseconds()
```

This method returns the milliseconds in the second as specified by *date*, using local (getMilliseconds()) or universal (getUTCMilliseconds()) time. The return value is between 0 and 999.

`get[UTC]Minutes()`

Syntax:

```
date.getMinutes()
date.getUTCMinutes()
```

Returns the minutes in the hour as specified by *date*, using local (getMinutes()) or universal (getUTCMinutes()) time. The return value is between 0 and 59.

`get[UTC]Month()`

Syntax:

```
date.getMonth()
date.getUTCMonth()
```

This method returns the month as defined by *date*, using local (getMonth()) or universal (getUTCMonth()) time. The return value is from 0 (January) to 11 (December).

get[UTC]Seconds()

Syntax:

```
date.getSeconds()
date.getUTCSeconds()
```

The method returns the seconds in the minute as specified by *date*, using local
(getSeconds()) or universal (getUTCSeconds()) time. The return value is between
0 and 59.

getTime()

Syntax:

```
date.getTime()
```

Returns the number of milliseconds between midnight on January 1, 1970 (UTC) and
the date and time stored in *date*. This method is useful, for example, if you want
to establish how much time has elapsed between two moments in date and time, or
simply to compare two Date objects.

Note that the number of milliseconds stored inside a Date object is always measured
from the UNIX epoch in universal time and is independent of the location, so there is
no getUTCTime() method.

Examples:

```
var today = new Date();                    //Local time
var theMoment = new Date(2017, 2, 13); //Local time
if (today.getTime() < theMoment.getTime()) {
  //theMoment is yet to come.
}


var utc = new Date(3600000);           //1 am on 1/1/70 (UTC)
var local = new Date(1970, 0, 1, 1); //1 am on 1/1/70 (local time)
//The next expression is true or false, depending on a location.
//It is true in London and false in Tokyo, for example.
utc.getTime() == local.getTime()
```

getTimezoneOffset()

Syntax:

```
date.getTimezoneOffset()
```

Returns the difference between UTC and local time. Since some countries have time zones that are not an integer number of hours offset from UTC, the method returns minutes rather than hours. Essentially, this method gives you information about current offset from UTC, which depends on the time zone that JavaScript is running and whether or not daylight savings time is (or would be) effective at the moment specified by *date*.

Example:

```
//Set the date to July (when daylight savings time is in effect):
var summer = new Date(2014, 6);
//Set the date to January (when standard time is in effect):
var winter = new Date(2014, 0);
summer.getTimezoneOffset() //Returns -120 in Vienna, -60 in London
winter.getTimezoneOffset() //Returns -60 in Vienna, 0 in London
```

now() //Static method

Syntax:

```
Date.now()
```

This is a static method that returns the current time in milliseconds since the UNIX epoch. Because the method is static, it is invoked directly through the Date object itself, rather than through particular Date object instances.

Example:

```
//The following two expressions return equivalent values:
Date.now()
(new Date()).getTime()
```

set[UTC]Date()

Syntax:

```
date.setDate(dayOfMonth)
date.setUTCDate(dayOfMonth)
```

Sets the day of the month of the *date* object, using local (setDate()) or universal (setUTCDate()) time. The *dayOfMonth* argument should be an integer between 1 and 31.

set[UTC]FullYear()

Syntax:

```
date.setFullYear(year[, month[, day]])
date.setUTCFullYear(year[, month[, day]])
```

This method sets the year field of *date*, using local (setFullYear()) or universal (setUTCFulYear()) time. The *year* argument must be an integer specifying a full year, and not an abbreviation such as 84 instead of the full 1984. Optionally, you can specify *month*, which is an integer between 0 (January) and 11 (December), and *day*, which is an integer between 1 and 31. Note that square brackets should not be used in a method. They just indicate which arguments are optional.

set[UTC]Hours()

Syntax:

```
date.setHours(hours[, minutes[, seconds[, milliseconds]]])
date.setUTCHours(hours[, minutes[, seconds[, milliseconds]]])
```

This method sets the hours (and, optionally, minutes, seconds, and milliseconds) field of *date*, using local (setHours()) or universal (setUTCHours()) time. The *hours* argument should be an integer between 0 and 23, the *minutes* and *seconds* arguments should be integers between 0 and 59, and the *milliseconds* argument should be an integer between 0 and 999. Note that square brackets should not be used in a method. They just indicate which arguments are optional.

set[UTC]Milliseconds()

Syntax:

```
date.setMilliseconds(milliseconds)
date.setUTCMilliseconds(milliseconds)
```

Allows you to set the milliseconds field of *date*, using local (setMilliseconds()) or universal (setUTCMilliseconds()) time. The *milliseconds* argument should be an integer between 0 and 999.

set[UTC]Minutes()

Syntax:

```
date.setMinutes(minutes)
date.setUTCMinutes(minutes)
```

This method sets the minutes field of *date*, using local (setMinutes()) or universal (setUTCMinutes()) time. The *minutes* argument should be an integer between 0 and 59.

set[UTC]Month()

Syntax:

```
date.setMonth(month)
date.setUTCMonth(month)
```

This method sets the month field of *date*, using local (setMonth()) or universal (setUTCMonth()) time. The *month* argument should be an integer between 0 (January) and 11 (December).

set[UTC]Seconds()

Syntax:

```
date.setSeconds(seconds)
date.setUTCSeconds(seconds)
```

This method sets the seconds field of *date*, using local (setSeconds()) or universal (setUTCSeconds()) time. The *seconds* argument should be an integer between 0 and 59.

setTime()

Syntax:

```
date.setTime(milliseconds)
```

Sets the number of millisecond between the wanted date and midnight on January 1, 1970 (UTC). Because this number is independent of the time zone, no "UTC" counterpart of this method exists.

to[UTC]String()

Syntax:

```
date.toString()
date.toUTCString()
```

Returns a human-readable string representing the date and time stored in *date*. The string is expressed in either local (toString()) or universal (toUTCString()) time.

Example:

```
var d = new Date();
d.toString()    //Returns "Fri Jan 09 2015 16:12:22 GMT+0100
                //          (Central European Standard Time)"
```

```
d.toUTCString() //Returns "Fri, 09 Jan 2015 15:12:22 GMT"
```

UTC() //Static method

Syntax:

```
Date.UTC(year, month[, day[, hrs[, mins[, secs[, millis]]]]])
```

This is a static method that converts the specified universal time to the milliseconds since the UNIX epoch. Because the method is static, it is invoked directly through the Date object itself, rather than through individual Date object instances. For the description of arguments, see the table summarizing the Date() constructor arguments on page 384. Note that square brackets should not be used in a method. They just indicate which arguments are optional.

Although the Date.UTC() static method and the Date() constructor can accept the same parameters and transform them to a single number of milliseconds since the UNIX epoch, there are two important differences between them: First, Date() creates a Date object, while Date.UTC() only returns an integer. Second, the arguments passed to Date() are understood to specify local time, and arguments passed to Date.UTC() are assumed to represent universal time. By combining Date() and Date.UTC(), you can easily create a Date object using universal time:

```
//Creates a Date object set to 2 p.m. on August 1, 2016 (UTC):
var dt = new Date(Date.UTC(2016, 7, 1, 14));
```

valueOf()

Syntax:

```
date.valueOf()
```

Returns the same millisecond value as the getTime() method.

E.7   document (Client-Side JavaScript)

Every web page that is loaded in a browser owns one Document object accessible through the global document property. Document implements an interface that acts as an entry point into the content of the web page. Many of the Document's properties and methods offer access to elements and other important objects contained in the document. Apart from that, the Document object defines a number of so-called "factory methods" used for creating elements and other relevant objects.

Technically, a Document object is a subclass of Node. That means that it inherits all the properties and methods of Node, which you can use in addition to the Document's own properties and methods described in this section. Note that a Document itself is not an Element. Rather, it contains a single element (i.e., an <html> element) accessible through its documentElement property.

Properties

`body`

Syntax:

```
document.body
```

This property refers to the Element object representing the <body> element.

`charset`

Syntax:

```
document.charset
```

Defines the character encoding for the document.

`documentElement`

Syntax:

```
document.documentElement
```

This read-only property refers to the Element object representing the <html> element, which is the root element of the document.

`head`

Syntax:

```
document.head
```

This read-only property refers to the Element object representing the <head> element.

`images`

Syntax:

```
document.images
```

This read-only property is a reference to an array-like object storing all images inside the document.

The next example searches for the image element with id of logo and replaces the image that the element refers to with the one stored in the *newlogo.gif* file.

```
  var i, imgList = document.images;

  for(i = 0; i < imgList.length; i++) {
    if(imgList[i].id == "logo") {
      imgList[i].src = "newlogo.gif";
    }
  }
```

lastModified

Syntax:

```
  document.lastModified
```

The `lastModified` read-only property holds the date and time of the latest change made to the document in the visitor's local time zone.

You can use `lastModified` directly as in the next example:

```
  console.log(document.lastModified); //Writes "08/19/2014 10:25:59"
```

If you want to have some control over how the date and time are output, then you can construct a Date object by passing the `document.lastModified` to the `Date()` constructor:

```
  var lastChange = new Date(document.lastModified);
```

links

Syntax:

```
  document.links
```

This read-only property is a reference to an array-like object storing all hyperlinks in the document.

title

Syntax:

```
  document.title
```

This property is a string containing the text found inside the `<title>` element of the document.

Example:

```
//Changes the document title if a video, or a game on the page is
//paused:
document.title += " (paused)";
```

Methods

createElement()

Syntax:

```
document.createElement(elementName)
```

Creates and returns an empty Element object with *elementName*, which must be an HTML element name. You can then fill the created element's content using the Element's innerHTML property or the Node's appendChild() method, for example. You can use the latter because Element is a subclass of Node and thus inherits all the Node's properties and methods.

The next example creates an empty <h1> element, fills its content with "Howdy," and appends it to the document body:

```
var mainHeading = document.createElement("h1");
mainHeading.innerHTML = "Howdy";
document.body.appendChild(mainHeading);
```

You can find more examples of using createElement() on pages 252 and 258.

getElementById()

Syntax:

```
document.getElementById(elementId)
```

The getElementById() method searches the document for an element that has an id attribute with the value *elementId* and returns it in the form of an Element object. If no such element exists, then it returns null.

Elements that are not part of the document are not searched by getElementById(). After you have created a new element, you should insert it into the document tree (e.g., using the Element's appendChild() method) or getElementById() won't be able to find it.

Note that the *elementId* argument is case-sensitive. For example, if there is an element with id of "first", then document.getElementById("First") will return null because "f" and "F" are different as far as the getElementById() method is concerned.

Note also the capitalization of "Id" in the name of the method. New programmers sometimes mistakenly write getElementByID() (capital "D") instead of the correct form getElementById().

You can find examples of using getElementById() on pages 237 and 241.

getElementsByClassName()

Syntax:

```
document.getElementsByClassName(classNames )
```

The getElementsByClassName() method returns a reference to a read-only array-like object of elements with a class attribute whose value contains all of the specified classNames. classNames can either be a single class name or a space-separated list of class names. When you supply a list of class names, they should still be specified as a single string argument like, for example, "my-border blue fancy". The returned list is live, which means that it is automatically updated whenever the document changes. The order in which elements appear in the returned list is the same as the order in which they appear in the document. Note that you can also invoke this method on any Element object, in which case only the descendants of that element are included in the search.

getElementsByTagName()

Syntax:

```
document.getElementsByTagName(elementName )
```

The getElementsByTagName() method returns a reference to a read-only array-like object of elements with the specified elementName. The method compares names in a case-insensitive manner. You can also use an asterisk (*) as an element name, in which case all elements will be returned. The returned list is live, which means that it is automatically updated whenever the document changes. The order in which elements appear in the returned list is the same as the order in which they appear in the document. Note that you can also invoke this method on any Element object, in which case only the descendants of that element are included in the search.

There are examples of using this method on pages 248 and 254.

write()

Syntax:

```
document.write(markup )
```

This method writes text to the document while the document is loading, rendering any HTML tags that may appear in the markup argument. If the document has already finished loading, a call to `write()` will reopen the document, erasing all its contents. In the old days of JavaScript, the `write()` method used to be the only way of generating the document's content dynamically, but in modern web programming this method is seldom used.

Events

There are few events that browsers fire directly at Document objects. Most events that a document will receive are events that have been fired at the elements it contains and have bubbled up to the document. Hence, the Document object supports all of the Element's event-handler properties, some of which are listed on page 402.

Event bubbling, or *event propagation*, is a mechanism that fires an event not only at the element on which it occurs but also on all its ancestors up the document tree. Event bubbling allows you to register a single event handler on a document or some other container element, which is sometimes more convenient than to do that on each element separately.

E.8   Element (Client-Side JavaScript)

Each element in an HTML document can be represented by an Element object in JavaScript code. The properties and methods of Element allow you to query and manipulate elements positioned on a web page. Since Element is a subclass of Node, you can use the properties and methods of Node on Element objects as well. A very important subject connected with Element are events. Web browsers fire numerous kinds of events on HTML elements, and Element has defined different event-handler properties to handle those events.

Object Instance Creation

The Element object defines no constructor of its own, but you can create elements using the `createElement()` method of Document. After creating a new element, you must insert it into the document using methods like `appendChild()` of the Node object.

You can also obtain the elements that already exist on a web page. Listed under the Document object, you will find properties like `head` and `body`, which return specific elements, and also many methods that allow you to obtain elements according to different criteria like `getElementById()`, `getElementsByTagName()`, or `getElementsByClassName()`, for example. Note that some of those methods are defined not only on Document but also on Element.

Properties

In this section, you will find listed some of the properties that Element defines. Apart from these, you can also use the properties listed under Node (as Element is a subclass of Node) and properties which, for the most part, have the same names as attributes of corresponding HTML elements.

HTML Attributes

HTML attributes of an HTML element are accessible as additional properties of the Element object representing that HTML element.

Imagine, for example, that myPicture refers to an Element object representing an <img> HTML element on your web page,[1] and you want to change the image displayed by that <img> element. You can do that by setting its src HTML attribute, which is accessible as the src property of myPicture from JavaScript:

```
myPicture.src = "sunset.jpg";
```

Some HTML attributes are reserved words in JavaScript and are therefore accessible through different names in JavaScript. For example, the class HTML attribute is named className in JavaScript, and the for attribute of the <label> element is named htmlFor in JavaScript.

children

Syntax:

```
element.children
```

This read-only property holds a reference to an array-like object of the Element children of *element*. Other children, such as text or comment nodes, are excluded from the array.

clientHeight, clientWidth

Syntax:

```
element.clientHeight
element.clientWidth
```

These read-only properties specify the dimensions of *element* including the content and padding, and excluding the border and margin. When *element* is the root element (like the <html> element returned by document.documentElement), these properties specify the size of the browser window without scrollbars and other browser user interface elements.

---

[1]To be exact, this is an Image object. While the Image object offers some additional information about an image through its properties, for the most part, you need not treat it any differently from any other Element object.

dataset

Syntax:

```
element.dataset.camelCasedName = string
string = element.dataset.camelCasedName
```

HTML5 allows you to associate your own data with a specific HTML element even though the data have no language-defined meaning. You can do that using so-called *custom data attributes* whose names begin with `data-`. Custom attribute names are mapped to properties of the `dataset` property of the Element object using the following transformation rules:

- The prefix `data-` is removed.
- Any combination of a dash (-) followed by a lower-case letter is replaced by a corresponding upper-case letter.

Consider, for example, the following HTML:

```
<div id="item1" data-max-allowed-price="unlimited">Oh, boy!</div>
```

You can access the above `data-max-allowed-price` attribute from JavaScript like this:

```
//Writes unlimited to the JavaScript Console:
console.log(
  document.getElementById("item1").dataset.maxAllowedPrice
);
```

You can, of course, change the value of a custom data attribute:

```
document.getElementById("item1").dataset.maxAllowedPrice =
                                              "0.99 USD";
```

firstElementChild

Syntax:

```
element.firstElementChild
```

This read-only property returns the first Element child of *element*, or `null` if there is none. Any non-Element child, such as a text or comment node, is ignored.

As an example, consider the next HTML code fragment representing an unordered list:

```
<ul id="some-list">
  <li>First</li>
  <li>Second</li>
  <li>Third</li>
  <li>Fourth</li>
</ul>
```

Suppose you want to traverse all the list items within the above list and get the content of each individual item. This is how you can do it:

```
var list = document.getElementById("some-list"); //Get the list
var item = list.firstElementChild;        //Get the first item
while (item) {                            //While item is not null
  console.log(item.innerHTML);            //Write the item's content
  item = item.nextElementSibling;         //Get the next item
}
```

Note that `firstElementChild` is used on the list while `nextElementSibling` is used on the list items.

When running the above code, you will get the words "First," "Second," "Third," and "Fourth" written in the debugging console.

innerHTML

Syntax:

element .innerHTML

This property is a string specifying the HTML markup contained between the start and end tags of *element*. Note that the tags themselves are not included in the string. When you set this property to a string of HTML, the specified text will replace the content of *element* with the parsed form of the text. When you query this property, you get the content of *element* in the form of a string of HTML.

lastElementChild

Syntax:

element .lastElementChild

This read-only property returns the last Element child of *element*, or `null` if there is none. Any non-Element child, such as a text or comment node, is ignored.

nextElementSibling

Syntax:

```
element.nextElementSibling
```

This read-only property returns the first Element sibling that follows `element` in the `childNodes` array of its parent node. If there aren't any more elements following this one, then the `nextElementSibling` property returns `null`. Any non-Element sibling, such as a text or comment node, is ignored.

For an example of using `nextElementSibling`, refer to the `firstElementChild` property on page 398.

outerHTML

Syntax:

```
element.outerHTML
```

This property is like `innerHTML`, only the start and end tags of `element` are included in the string specified by `outerHTML`. Therefore, using `outerHTML` you do not change only the content of `element` but the whole element, including its name and attributes.

previousElementSibling

Syntax:

```
element.previousElementSibling
```

This read-only property returns the Element sibling that directly precedes `element` in the `childNodes` array of its parent node. If there aren't any elements preceding this one, then the `previousElementSibling` property returns `null`. Any non-Element sibling, such as a text or comment node, is ignored.

tagName

Syntax:

```
element.tagName
```

This read-only property returns the name of `element`. Note that the name is always returned in upper case, regardless of how it appears in the document source. For example, the value of the `tagName` property of an Element object representing a `<table>` HTML element will be `"TABLE"`.

Methods

Apart from the methods listed in this section, you can also use the methods defined by Node on any Element object because Element is a subclass of Node.

getElementsByClassName()

Syntax:

```
element.getElementsByClassName(classNames)
```

See the description of the getElementsByClassName() method of the Document object on page 395.

getElementsByTagName()

Syntax:

```
element.getElementsByTagName(elementName)
```

See the description of the getElementsByTagName() method of the Document object on page 395.

removeAttribute()

Syntax:

```
element.removeAttribute(attrName)
```

This method removes the attribute with the name *attrName* from *element*. Note that attribute names are case insensitive.

Events

The Element object defines a number of different event-handler properties representing various events that the browser fires on HTML elements. If you want to handle a particular event on a particular element, you simply set the corresponding event-handler property to the function you want to execute as a response to the event. Not all events listed here are triggered on all elements, but because events bubble up the document tree, the event-handler properties are universally defined for all elements and even for Document and Window objects referred to by document and window.

The next table sums up some of the event-handler properties defined by the Element object.

Event-Handler Property	Invoked When...
onchange	The content or state of a form control changes. This event is not fired for individual keystrokes but for complete edits.
onclick	A mouse click occurs.
ondblclick	Two consecutive mouse clicks (i.e., double click) occur.
oninput	A form control receives some input. This event is similar to onchange but is fired more frequently.
onkeydown	A key is pressed.
onkeypress	A key is pressed that generates a printable character.
onkeyup	A key is released.
onload	Resource loading finishes (e.g., loading an image referred to by the src attribute of <img>). If this event is triggered on window, it is triggered after the document and all its external resources have been fully loaded.
onmousedown	A mouse button is pressed.
onmousemove	The mouse moves.
onmouseout	The mouse leaves the element.
onmouseover	The mouse enters the element.
onmouseup	A mouse button is released.
onmousewheel	The mouse wheel is rotated.

E.9   Event (Client-Side JavaScript)

Every time an event handler is invoked, an Event object is passed to it that holds detailed information about the event. For example, an Event object may tell you which key has been pressed or the exact location of a mouse click. Although the standard event model associates different types of event objects with different types of events, a simplified model used by the jQuery library is described here. This model uses a single Event object, which gathers properties for all possible events under the same umbrella, even when they do not apply. Event properties that hold information about mouse events, for example, won't have a meaningful value when a keyboard event occurs, but those properties will nevertheless be defined. Note that all properties of the Event object are read-only.

Properties

altKey

Syntax:

> *event*.altKey

This is a boolean property, which is set to `true` if the Alt key was held pressed when *event* occurred. Otherwise, it is set to `false`. This property is set for mouse and keyboard events.

button

Syntax:

> *event*.button

This property has a meaningful value for a mousedown, mouseup, or click event. It tells you which of the mouse buttons caused the event. If `button` has the value of 0, the left button was pressed, and if `button` has the value of 2, the right button was pressed. The value of 1 signals the middle button. Note that some browsers only generate events for the left mouse button.

If you are using jQuery, then it is better to use `which` instead of `button`.

charCode

Syntax:

> *event*.charCode

This property is used for keypress events (but not for keydown and keyup events) and specifies the Unicode encoding of the generated printable character. This property has been removed from web standards and should not be used in new programs. The property has been replaced by the `key` property, which is not yet fully supported by browsers at the time of writing this book.

clientX, clientY

Syntax:

> *event*.clientX
> *event*.clientY

These properties specify the *x*- and *y*-coordinates of the mouse pointer relative to the browser window. Note that they do not take into account document scrolling—if the mouse pointer is 10 pixels from the top of the window, for example, `clientY` will be set to 10 regardless of how far the document has been scrolled. The `clientX` and `clientY` properties are specified for all kinds of mouse events.

Example:

```
//Log window coordinates of a mouse click when the user clicks
//the element with id of "klik":
document.getElementById("klik").onclick = function(e) {
  console.log(e.clientX, e.clientY);
};
```

ctrlKey

Syntax:

event.ctrlKey

This is a boolean property, which is set to true if the Ctrl key was held pressed when *event* occurred. Otherwise, it is set to false. This property is set for mouse and keyboard events.

key

Syntax:

event.key

This property is used by all kinds of keyboard events and it specifies the purpose of the pressed key. Its value can either be text or a predefined function key value. The key property is not yet fully supported by browsers at the time of writing this book.

keyCode

Syntax:

event.keyCode

This property is used by all kinds of keyboard events. It specifies the virtual keycode of the pressed key. Keycodes are not standardized and may differ from system to system. However, keyCode for printing characters is most usually set to their Unicode encoding. This property has been removed from web standards and should not be used in new programs. The property has been replaced by the key property, which is not yet fully supported by browsers at the time of writing this book.

pageX, pageY

Syntax:

event.pageX

```
event.pageY
```

These properties are not standardized but are commonly supported by browsers. They are like `clientX` and `clientY`, only that they return a position relative to the document instead of a position relative to the window.

shiftKey

Syntax:

```
event.shiftKey
```

This is a boolean property, which is set to `true` if the Shift key was held pressed when *event* occurred. Otherwise, it is set to `false`. This property is set for mouse and keyboard events.

which

Syntax:

```
event.which
```

This nonstandard property may be supported by some browsers but has been removed from Web standards, so do not use it in your JavaScript programs. However, the property is emulated by the jQuery library. For keyboard events, it is set to the same value as the `keyCode` property. For mouse events, its value is one bigger than that of the `button` property.

E.10 Function (Core JavaScript)

A function is a sequence of JavaScript statements with optional parameters, which you can *call* or *invoke* in order to execute the statements. The statements packed inside the function definition usually perform a specific task. Technically, a JavaScript function is a "callable" object. Like any object, it can have properties and methods, but you can also call a function, something that you cannot do with an ordinary object. A function call returns a value that you specify using a `return` statement. The default return value of JavaScript functions is `undefined`.

You must not confuse function properties with its local variables. Local variables can only be accessed within the function body and are referenced by their names alone. Function properties, on the other hand, are accessed through the function name, and are also accessible from the outside of the function body.

Example:

```
var justDoIt = function() {
    var x = 1;              //A local variable
```

```
  justDoIt.y = 5;        //A property
};

justDoIt();
console.log(x);        //x is not defined
console.log(justDoIt.y); //Writes 5
```

Object Instance Creation

Syntax:

```
//A function expression:
var fun_name = function(par1, par2, ..., parN) {
  statements
};

//A function definition statement:
function fun_name(par1, par2, ..., parN) {
  statements
}
```

Both a function expression and a function definition statement create *fun_name*, which is a Function object. The main difference between both types of creation is that function definition statements are hoisted while function expressions are not. That means that you can invoke a function before you declare it if you use a function definition statement. However, when you define a function through a function expression, you cannot invoke it before it is defined.

Examples:

```
sayHello();      //sayHello is undefined
var sayHello = function() {
  console.log("Hello!");
};
```

```
sayHello();      //Writes "Hello!"
function sayHello() {
  console.log("Hello!");
}
```

Note that it is also possible to create a Function object with the use of the Function() constructor, which is not very efficient and thus not a common way of doing it.

E.11 Global Variables, Functions, and Objects (Core JavaScript)

Technically speaking, all global variables, functions, and objects (including user-defined ones) are part of the global object, which does not have a name in core

JavaScript. The global object is accessible through the `this` keyword, although you will almost never need to refer to the global object explicitly—you simply refer to all global items using their own names.

Variables

Infinity

Syntax:

```
Infinity
```

`Infinity` is a read-only global variable (property of the global object) that holds positive infinity as a special numeric value. Mathematically, this value behaves like infinity, which means that anything (other than `Infinity`) divided by `Infinity` is zero, and anything multiplied by `Infinity` is `Infinity`.

NaN

Syntax:

```
NaN
```

`NaN` is a read-only global variable (property of the global object) that holds not-a-number as a special numeric value—it represents an illegal number, such as the result of multiplication of strings, for example.

`NaN` is very special in that it does not compare equal to any other value including itself. Therefore, if you want to test whether x is set to `NaN`, you use the expression x `!==` x, which evaluates to `true` if, and only if, x is `NaN`.

Example:

```
var myNumber;
console.log(myNumber - 10);   //Writes NaN
```

undefined

Syntax:

```
undefined
```

`undefined` is a read-only global variable (property of the global object) that stores the `undefined` value. This value is returned by a declared but uninitialized variable. You will also get the `undefined` value if you want to inspect the value of a nonexistent property of an object.

For testing whether a value is undefined, you should use the === operator, because the == operator compares undefined as equal also with the null value.

Examples:

```
var y;
var child = {name: "Tom", age: 7};
y              //Returns undefined
child.surname //Returns undefined


null === undefined //Returns false
null == undefined  //Returns true
```

Functions

parseFloat()

Syntax:

```
parseFloat(string)
```

This function parses *string* and returns the first number found in it. The function stops parsing with the first character that is not a part of the floating point number. If *string* begins with a character or set of characters that cannot be parsed, the NaN (not-a-number) value is returned.

Examples:

```
parseFloat("1.3rm4") //Returns 1.3
parseFloat("1.1e2")  //Returns 110 (E notation)
parseFloat("-r2")    //Returns NaN
parseFloat("11-7")   //Returns 11
parseFloat(".22")    //Returns 0.22
```

parseInt()

Syntax:

```
parseInt(string)
```

This function parses *string* and returns the first integer found in it. The function stops parsing with the first character that is not a part of the integer. If *string* begins with a character or set of characters that cannot be parsed, the NaN (not-a-number) value is returned.

Examples:

```
parseInt("*42")      //Returns NaN
parseInt("-3.14")    //Returns -3
parseInt("  8-) ")   //Returns 8
```

Objects

Apart from global variables and functions, core JavaScript also defines a number of global objects, some of which are listed in the table below for convenience. You will find more detailed descriptions of the individual objects in corresponding sections of this appendix.

Object	Description
Array	The `Array()` constructor.
Boolean	The `Boolean()` constructor.
Date	The `Date()` constructor.
Math	A reference to a built-in object with properties and methods for doing math.
Number	The `Number()` constructor.
String	The `String()` constructor.

E.12   Math (Core JavaScript)

Math is a built-in object that does not define a class of objects like Date and String do, so there is no `Math()` constructor. All properties and methods of a Math object are static and you access and invoke them directly through the Math object. You can also think of expressions like `Math.PI` and `Math.pow()` simply as plain global constants and functions.

Constants

E

Syntax:

```
Math.E
```

This constant holds the mathematical constant e. The number e is the base of the natural logarithm, which is approximately equal to 2.71828.

PI

Syntax:

```
Math.PI
```

This constant holds the mathematical constant π. The number π is the ratio of a circle's circumference to its diameter, which is approximately equal to 3.141592.

Static Functions

abs()

Syntax:

```
Math.abs(x)
```

Returns the absolute value of x.

atan2()

Syntax:

```
Math.atan2(y, x)
```

Returns the counterclockwise angle in radians between the positive x-axis and the point specified by x and y. The return value ranges between $-\pi$ and π radians. Note the unusual order of the function arguments, where the y-coordinate is passed before the x-coordinate.

ceil()

Syntax:

```
Math.ceil(x)
```

Returns the smallest integer that is greater than or equal to x.

Examples:

```
Math.ceil(41.0001)  //Returns 42
Math.ceil(9.99)     //Returns 10
Math.ceil(-9.99)    //Returns -9
```

cos()

Syntax:

```
Math.cos(x)
```

Returns the cosine of the angle x. Note that x is assumed to be given in radians so you must convert degrees to radians before passing an angle argument to `Math.cos()`. To do that, simply multiply the degree value by $\pi/180$.

Example:

```
var angle = 45;                   //Degrees
Math.cos(angle * Math.PI / 180) //Returns 0.7071067811865476
```

exp()

Syntax:

```
Math.exp(x)
```

Returns e raised to the power of x. The number e is the base of the natural logarithm, which is approximately equal to 2.71828.

floor()

Syntax:

```
Math.floor(x)
```

Returns the largest integer that is less than or equal to x.

Examples:

```
Math.floor(41.0001) //Returns 41
Math.floor(9.99)    //Returns 9
Math.floor(-9.99)   //Returns -10
Math.floor(13)      //Returns 13
```

log()

Syntax:

```
Math.log(x)
```

Returns the natural logarithm of x. If the value of x is negative, the function returns NaN. If the value of x is zero, then the function returns -Infinity, which is actually the limit as the logarithm of zero is not defined.

`max()`

Syntax:

```
Math.max(val1, val2, ..., valN)
```

Returns the largest of (zero or more) supplied arguments `val1` to `valN`. If no arguments are supplied, the function returns `-Infinity`. If any of the arguments cannot be converted to a valid number or is NaN, then the function returns NaN.

Examples:

```
Math.max("2", 8, 8, -179) //Returns 8
Math.max("4a", 9001)      //Returns NaN
Math.max()                //Returns -Infinity
```

`min()`

Syntax:

```
Math.min(val1, val2, ..., valN)
```

Returns the smallest of (zero or more) supplied arguments `val1` to `valN`. If no arguments are supplied, the function returns `Infinity`. If any of the arguments cannot be converted to a valid number or is NaN, then the function returns NaN.

Examples:

```
Math.min("2", 8, 8, -179) //Returns -179
Math.min("4a", 9001)      //Returns NaN
Math.min()                //Returns Infinity
```

`pow()`

Syntax:

```
Math.pow(x, y)
```

Computes and returns x to the power of y. If the result is a complex number (when raising a negative number to a fractional power), then the function returns NaN.

Examples:

```
Math.pow(3, 4)    //Returns 81
Math.pow(9, 1/2)  //Returns 3 (the square root of 9)
Math.pow(8, 1/3)  //Returns 2 (the cube root of 8)
```

```
Math.pow(-1, 1/2) //Returns NaN
```

random()

Syntax:

```
Math.random()
```

Returns a floating point pseudorandom number greater than or equal to 0 and less than 1. You can scale the obtained number and round it to an integer if you desire a different range and type of pseudorandom numbers.

Example:

```
//Returns an integer pseudorandom number between 0 and 9
//(inclusive):
Math.floor(Math.random() * 10)
```

round()

Syntax:

```
Math.round(x)
```

This function returns the value of x rounded to the nearest integer. If the fractional part of x is 0.5 or higher, x is rounded up. Otherwise, if the fractional part of x is less than 0.5, x is rounded down.

Examples:

```
Math.round(77.5)   //Returns 78
Math.round(77.499) //Returns 77
```

sin()

Syntax:

```
Math.sin(x)
```

Returns the sine of the angle x. Note that x is assumed to be given in radians so you must convert degrees to radians before passing an angle argument to `Math.sin()`. To do that, simply multiply the degree value by $\pi/180$.

Example:

```
var angle = 90;                    //Degrees
Math.sin(angle * Math.PI / 180) //Returns 1
```

sqrt()

Syntax:

```
Math.sqrt(x)
```

Returns the square root of x. If x is negative, then this function returns NaN. If you want to compute an arbitrary root of a number, you can do that using Math.pow() with a fractional power.

tan()

Syntax:

```
Math.tan(x)
```

Returns the tangent of the angle x. Note that x is assumed to be given in radians so you must convert degrees to radians before passing an angle argument to Math.tan(). To do that, simply multiply the degree value by $\pi/180$.

E.13 Node (Client-Side JavaScript)

The Node object is fundamental to a document tree. It provides the basic properties and methods for traversing and manipulating the tree for all objects in a document tree, including the Document object itself. This book covers only Document and Element objects, which are both subclasses of Node. There are other types of nodes in a document tree, such as text and comment nodes, which are represented by classes of their own. However, they are subclasses of Node just like the Document and Element objects are.

Properties

childNodes

Syntax:

```
node.childNodes
```

This property holds a reference to an array-like object of the child nodes of *node*. Note that childNodes is never null. If *node* has no children, then the length property of childNodes is simply set to zero.

---

firstChild

Syntax:

    node.firstChild

This property is a reference to the first child of *node*. If *node* has no children, firstChild is set to null.

lastChild

Syntax:

    node.lastChild

This property is a reference to the last child of *node*. If *node* has no children, lastChild is set to null.

nextSibling

Syntax:

    node.nextSibling

Returns the first sibling node that follows *node* in the childNodes array of its parent node. If there aren't any more nodes following *node*, then the nextSibling property returns null.

parentNode

Syntax:

    node.parentNode

This property returns a reference to the parent node of *node*. If *node* has no parent, then parentNode is null. The Document node and nodes that have been removed from the document never have a parent. Newly created nodes that have not yet been inserted into the document also have parentNode set to null.

previousSibling

Syntax:

    node.previousSibling

This property returns the sibling node that directly precedes *node* in the childNodes array of its parent node. If there aren't any sibling nodes preceding *node*, then the previousSibling property returns null.

Methods

appendChid()

Syntax:

```
node.appendChild(child)
```

This method inserts the node *child* into the document tree, making it the last child of *node*. If, by any chance, *child* already is in the document, then it is first removed from its current position and reinserted into a new one.

insertBefore()

Syntax:

```
node.insertBefore(child, referenceChild)
```

Inserts into the document tree the node *child* before the node *referenceChild* as a child of *node*. If *referenceChild* is null, then insertBefore() behaves just like appendChild(), inserting *child* as the last child of *node*. If, by any chance, *child* already is in the document, then it is first removed from its current position and reinserted to a new one.

removeChild()

Syntax:

```
node.removeChild(child)
```

This method removes *child* from the document tree. Note that the removed child is not deleted but only removed from the tree. The removeChild() method returns a reference to the removed child, which can be later reinserted into the document tree to a new location. If *child* is not a child of *node*, then the method fails and *child* is not removed from the document tree.

E.14   Number (Core JavaScript)

Numbers are JavaScript's primitive data type, but the language also supports the Number object. Essentially, the Number object is a wrapper object around a primitive numeric value, whose role is to provide some convenience methods and constants to work with numbers.

Object Instance Creation

`Number() //Constructor`

Syntax:

```
new Number(value) //Constructor
Number(value)     //Type conversion function
```

Normally, the Number object is automatically created by JavaScript when needed. For the rare occasion when you want to explicitly create a Number object, there is the `Number()` function available. The function can be used either as a constructor (with the `new` operator) or as a conversion function (without the `new` operator). Used either way, `Number()` first tries to convert *value* to a number. If the conversion is not possible, then *value* is converted to NaN. Note that, unlike the global `parseFloat()` and `parseInt()` functions, `Number()` does not perform partial conversions. If there is a character at the end of a string that cannot be evaluated as a part of a number that precedes it, for example, the conversion still results in NaN. When used as a constructor, `Number()` returns a new Number object holding the converted value. When used as a function, `Number()` simply returns the primitive numeric value obtained from the conversion.

Examples:

```
Number("42a")     //Returns NaN
Number("13")      //Returns a primitive value 13
new Number("13") //Creates and returns a Number object
```

Constants

`MAX_VALUE`

Syntax:

```
Number.MAX_VALUE
```

This constant represents the biggest number representable in JavaScript, which is approximately 1.7977×10^{308}. Anything larger than `Number.MAX_VALUE` is represented as `Infinity`.

Example:

```
Number.MAX_VALUE / Infinity  //Returns 0
```

MIN_VALUE

Syntax:

```
Number.MIN_VALUE
```

The `Number.MIN_VALUE` constant represents the smallest number representable in JavaScript that is still greater than zero, which is approximately 5×10^{-324}. Anything smaller than `Number.MIN_VALUE` is converted to 0.

Methods

`toExponential()`

Syntax:

```
number.toExponential()
number.toExponential(fractionDigits)
```

This method formats *number* to exponential notation, placing the first digit before the decimal point and *fractionDigits* digits after the decimal point. An optional *fractionDigits* argument can be an integer between 0 and 20, and when not specified, *number* is represented with as many digits as needed. When necessary, zeros are appended at the end of the fractional part, or the number is rounded. The formatted number is returned as a string.

Examples:

```
//In the following, implicit Number objects are automatically
//created by JavaScript in order to be able to call
//toExponential():
var num = 1123.456;
var min = Number.MIN_VALUE;
num.toExponential()    //Returns "1.123456e+3"
num.toExponential(5)   //Returns "1.12346e+3"    (rounded)
num.toExponential(8)   //Returns "1.12345600e+3" (padded zeros)
min.toExponential()    //Returns "5e-324"
min.toExponential(20)  //Returns "4.94065645841246544177e-324"
```

`toFixed()`

Syntax:

```
number.toFixed()
number.toFixed(fractionDigits)
```

Formats *number* to fixed-point notation and returns a string containing the formatted number. An optional *fractionDigits* argument can be an integer between 0 and

20, and when not specified, 0 is assumed. When necessary, zeros are appended at the end of the fractional part, or the number is rounded. If *number* is greater than or equal to 1e+21, this method simply calls `toString()` on *number*, which then returns an exponential representation of the number.

Examples:

```
var n = 123.789;
n.toFixed(0)   //Returns "124"        (no fractional part, rounded)
n.toFixed()    //Same as n.toFixed(0)
n.toFixed(1)   //Returns "123.8"      (rounded)
n.toFixed(5)   //Returns "123.78900" (padded zeros)
(1e-5).toFixed(3) //Returns "0.000"
(1e+21).toFixed() //Returns "1e+21"
```

toPrecision()

Syntax:

```
number.toPrecision()
number.toPrecision(precision)
```

This method returns a string representing *number* with *precision* significant digits. An optional *precision* argument should be an integer between 1 and 21, and when omitted, the `toString()` method is used instead to convert the number. If *precision* is large enough to include all the digits in the integer part of *number*, then fixed-point notation is used to format the returned string. Otherwise, exponential notation is used. If necessary, the number is rounded or padded with zeros.

Examples:

```
var num = 123.456;
num.toPrecision(2)    //Returns "1.2e+2"
num.toPrecision(3)    //Returns "123"
num.toPrecision(5)    //Returns "123.46"     (rounded)
num.toPrecision(8)    //Returns "123.45600" (padded zeros)
```

toString()

Syntax:

```
number.toString()
```

This method converts *number* to a string.

```
valueOf()
```

Syntax:

```
number.valueOf()
```

Returns the primitive number value stored in *number*. This method is most often invoked internally by JavaScript when needed and you will rarely need to call it explicitly.

E.15   Object (Core JavaScript)

The Object class object provides a superclass for all other objects in JavaScript. All JavaScript objects inherit properties and methods from the `prototype` property of the Object class.

Object Instance Creation

```
Object() //Constructor
```

Syntax:

```
new Object()
new Object(value)
```

The `Object()` constructor creates an appropriate wrapper object for *value*, which can be of any type. If the constructor is invoked without an argument, then an empty Object instance is created. It is possible to call the `Object()` constructor without the new operator, but the effect is just the same.

Example:

```
new Object(true) //Equivalent to new Boolean(true)
new Object(10)   //Equivalent to new Number(10)
```

```
{} //Literal syntax
```

Syntax:

```
{name1: value1, name2: value2, ..., nameN: valueN}
```

Creates an Object instance with zero or more properties *name1*, *name2*, ..., *nameN* having values *value1*, *value2*, ..., *valueN*, respectively. Values can be of any type while names can either be valid JavaScript identifiers or strings.

Examples:

```
var employeeOfTheMonth = {
  name: "Bob",
  title: "Vice-president",
  age: 45,
  "home address": {street: "Abbey Road", country: "UK"}
};
var point = {x: -3.6, y: 2};    //An object with properties x and y
var o = {};                     //An empty object
```

Properties

prototype

Syntax:

object .prototype

The prototype property is used as part of the JavaScript inheritance mechanism. You will use this property to add properties and methods to Function objects that act as constructors for a class of objects.

Say, for example, that you wish to augment the Date JavaScript class with a method that returns true if the specified year is a leap year, and false if it is not a leap year. In the Gregorian calendar, a year is a leap year if it is evenly divisible by four but not with 100. However, if the year is evenly divisible by 400, it is again a leap year. Here is how you can do it:

```
//Add isLeapYear() to the Date's prototype property:
Date.prototype.isLeapYear = function() {
  var y = this.getFullYear();
  return ((y % 4 == 0) && (y % 100 != 0)) || (y % 400 == 0);
};

var d = new Date();
//d inherits the isLeapYear() method:
if (d.isLeapYear()) {
  console.log(d.getFullYear() + " is a leap year.");
}
else {
  console.log(d.getFullYear() + " is not a leap year.");
}
```

E.16 String (Core JavaScript)

String is a JavaScript's primitive data type and the String class type provides methods for working with primitive string values. For example, you can search a string to find a character or substring in it, or you can convert all string characters to upper case. Keep in mind that strings in JavaScript are *immutable*, which means that you cannot modify the contents of a string. As a consequence, String methods such as `toUpperCase()` don't modify the original string, but instead return a completely new one. Because strings act like read-only arrays of single-character strings, you can use the [ ] operator to extract a character from a string.

Object Instance Creation

`String() //Constructor`

Syntax:

```
new String(val) //Constructor
String(val) //Conversion function
```

`String()` can be used either as a constructor (with the `new` operator) or conversion function (without the `new` operator). Either way, `String()` first converts *val* to string. The `String()` constructor then creates and returns a String object holding the string representation of *val*, while the `String()` conversion function simply returns the converted value as a primitive string.

When you use any of the String properties or methods on a primitive string value, JavaScript automatically creates a temporary wrapper String object in order to be able to use the property or invoke the method.

Properties

`length`

Syntax:

```
string.length
```

This read-only property is an integer that specifies the number of characters in a string. The index of the last character is `length - 1`.

Examples:

```
var s = "abc";
s[s.length - 1]              //Returns "c"
"Count my characters".length //Returns 19
```

Methods

`charAt()`

Syntax:

> *string*`.charAt(`*n*`)`

This method returns the *n*th character of *string*. Note that the first character of a string has the index of zero and the last character has the index `length - 1`. If *n* does not specify a valid character index, then `charAt()` returns an empty string. Because JavaScript does not define a special character data type, the returned character is a string of length 1.

Strings act like read-only arrays of single-character strings, which means that instead of `charAt()`, you can also use the [ ] operator to extract a character from a string.

Examples:

```
var s = "Something blue";
s.charAt(2)   //Returns "m"
s[2]          //Same as s.charAt(2)
```

`charCodeAt()`

Syntax:

> *string*`.charCodeAt(`*n*`)`

This method is like `charAt()`, only it returns the Unicode character encoding of the character at a given location, instead of the string containing the character proper. If *n* does not specify a valid character index (i.e., is negative or greater than `length - 1`), then `charCodeAt()` returns NaN.

Examples:

```
var s = "abc";
s.charCodeAt(2)    //Returns 99
s.charCodeAt(3)    //Returns NaN
"W".charCodeAt(0)  //Returns 87
```

`fromCharCode()` `//Static method`

Syntax:

> `String.fromCharCode(`*char1*`, `*char2*`, ..., `*charN*`)`

This method creates a string from one or more individual characters, specified by their numeric Unicode encodings (*char1* to *charN*).

Example:

```
//Creates the string "What's up?":
var greeting = String.fromCharCode(87, 104, 97, 116, 39, 115, 32,
                                   117, 112, 63);
```

indexOf()

Syntax:

```
string.indexOf(str)
string.indexOf(str, begin)
```

This method searches *string* for the first occurrence of *str*. The search starts at the beginning of *string* or, optionally, at position *begin* within *string*. The method returns the index of the first character of the first occurrence of *str* in *string*. If *str* is not found, then -1 is returned. The search is case sensitive, but you can make it case insensitive by using toUpperCase() or toLowerCase(). Because these methods return a modified copy of the string, you can chain them in a single expression with indexOf(). Don't forget that the position of the first character of the string is numbered 0.

Examples:

```
var magic = "Abracadabra";
magic.indexOf("abra")                  //Returns 7
magic.toLowerCase().indexOf("abra")    //Returns 0
magic.indexOf("abra", 8)               //Returns -1
```

lastIndexOf()

Syntax:

```
string.lastIndexOf(str)
string.lastIndexOf(str, begin)
```

This method works like indexOf(), except that it searches *string* in the opposite direction so that the first occurrence of *str* found is in fact the last one within *string*. Note that, in spite of the reversed search direction, character positions are still numbered from the beginning. The first character within a string has the position 0, while the last one has the position length - 1.

Examples:

```
var magic = "Abracadabra";
magic.lastIndexOf("abra", 8)              //Returns 7
magic.lastIndexOf("abra", 6)              //Returns -1
magic.toLowerCase().lastIndexOf("abra", 6) //Returns 0
magic.lastIndexOf("abra")                 //Returns 7
```

slice()

Syntax:

```
string.slice(begin)
string.slice(begin, end)
```

This method returns a portion (substring) of *string* containing all characters from and including the character at the position *begin* and up to but not including the character at the position *end*. The method does not change *string*. The arguments *begin* and *end* can also assume negative values, in which case they specify positions counted from the end of the string. For example, -1 indicates the last character, -2 the second to last character, and so forth. If the *end* argument is omitted, then all characters from *begin* to the end of the string are returned.

It is important to remember that the character at the position *begin* is included in the returned substring while the character at the position *end* is not. This may seem strange, but there are some noteworthy implications of this rule. First, the length of the returned string is always *end* - *begin*. Second, if you want to make several contiguous slices of *string*, which you don't want to overlap, you simply use the *end* argument of the preceding slice as the *begin* argument of the current slice.

Examples:

```
var s = "ABCDE";
s.slice(1)      //Returns "BCDE"
s.slice(-1)     //Returns "E" (the last character)
s.slice(0, 1)   //Returns "A"
s.slice(1, -2)  //Returns "BC"
```

split()

Syntax:

```
string.split()
string.split(delimiter)
```

This method creates and returns an array of strings obtained by splitting *string* into substrings. The split() method first searches for all the occurrences of *delimiter* in *string* and then breaks the string before and after each occurrence of *delimiter*. Thus-obtained substrings (without any delimiting text) are then arranged into an array, which `split()` returns. Note that, if the delimiter occurs at the beginning of *string*, the first element in the returned array will be an empty string. The same goes for the situation when the delimiter occurs at the end of *string*: the last element in the returned array will be an empty string.

The *delimiter* argument can also be an empty string, in which case *string* is broken between each character so that the returned array has the same length as the original string.

If *delimiter* is omitted, the returned array contains a single element that is equal to *string*.

Examples:

```
var commaDelimited = "6, 33, 1, 7,";
commaDelimited.split(",")  //Returns ["6", " 33", " 1", " 7", ""]


var s = "HELP!";
s.split("")  //Returns ["H", "E", "L", "P", "!"]
```

toLowerCase()

Syntax:

```
string.toLowerCase()
```

Returns a copy of *string* with all upper-case letters (if any) replaced by their lower-case counterparts. The method does not modify *string* itself.

toString()

Syntax:

```
string.toString()
```

Returns the primitive string value stored in *string*. This method is usually invoked internally by JavaScript and you will almost never need to call it explicitly.

toUpperase()

Syntax:

```
string.toUpperCase()
```

Returns a copy of *string* with all lower-case letters (if any) replaced by their upper-case counterparts. The method does not modify *string* itself.

trim()

Syntax:

```
string.trim()
```

This method returns a copy of *string* without any leading or trailing whitespace. trim() does not change *string*.

Example:

```
var s = "   Too much whitespace.     ";
s.trim()  //Returns "Too much whitespace."
```

valueOf()

Syntax:

```
string.valueOf()
```

Same as toString().

E.17 window (Client-Side JavaScript)

The window global property is a reference to the Window object, which represents a browser window or tab. It acts as a global object in client-side JavaScript, which means that you do not have to refer to it explicitly. All properties and methods of the Window object can be used as though they were global variables and functions. For example, if you want to invoke the setTimeout() method of Window, you simply use setTimeout() instead of window.setTimeout(). Also, all global variables and functions that you define in your script are in fact properties and methods of Window.

Some of the properties and methods of Window actually operate on the browser window. Others are here merely as generic global variables and functions. Apart from these, Window also implements all the global variables and functions of core JavaScript. Some of them are listed in section E.11 on page 406 of this appendix.

A peculiar thing is that window, which refers to the Window object, is the property of the Window object itself. This makes the expressions window, window.window, window.window.window, and so on all return the same Window object.

Properties

`document`

Syntax:

```
document
window.document
```

A reference to the Document object that holds the content of `window`. For a detailed description of the Document object, refer to section E.7 on page 391.

`innerHeight, innerWidth`

Syntax:

```
innerHeight
innerWidth
window.innerHeight
window.innerWidth
```

These properties specify the pixel dimensions of the document display area of `window`.

`localStorage`

Syntax:

```
localStorage
window.localStorage
```

A reference to a Storage object that implements client-side data storage. The Storage object provides means of storing name/value pairs on a local computer and recalling them any time in the future.

`location`

Syntax:

```
location
window.location
```

The `location` property holds the URL of the document that is currently displayed within this window. If you set `location` to some other URL, the browser will load and display the document from that new URL.

```
window
```

Syntax:

```
window
```

The `window` property holds a reference to the Window object representing the current browser window. It is rarely necessary to use this property explicitly because the Window object is the global object of client-side JavaScript. This makes all the properties and methods of Window accessible without explicitly referring to `window`. However, the `window` property is often used explicitly when specifying the `onload` event handler on Window.

Methods

```
alert()
```

Syntax:

```
window.alert(message)
```

Displays *message* in plain text inside an alert dialog box. The dialog has a single OK button and is modal, which means that a call to `alert()` blocks the code execution until the user closes the dialog.

```
clearInterval()
```

Syntax:

```
window.clearInterval(intervalID)
```

This method cancels the repeating of the code execution that was scheduled by a previous call to `setInterval()`. The *intervalID* argument should have the same value as returned by a corresponding call to `setInterval()`.

Example:

```
h1 = setInterval(f1, 100);
h2 = setInterval(f1, 5000);
//Stops the repeating of the execution of f1:
clearInterval(h1);
```

```
clearTimeout()
```

Syntax:

```
window.clearTimeout(timeoutID)
```

This method cancels the pending code execution that was scheduled by a previous call to `setTimeout()`. The *timeoutID* argument should have the same value as returned by a corresponding call to `setTimeout()`.

```
confirm()
```

Syntax:

```
window.confirm(question)
```

Displays *question* in plain text inside a confirm dialog box. The dialog has OK and Cancel buttons for the user to answer the question. If the user selects OK, then `confirm()` returns `true`, otherwise it returns `false`. The confirm dialog is modal, which means that the call to `confirm()` blocks the code execution until the dialog closes after clicking OK or Cancel.

```
prompt()
```

Syntax:

```
window.prompt(message)
window.prompt(message, defaultText)
```

This method opens a prompt dialog with the indicated plain text *message*, a text input field, and OK and Cancel buttons. Dialog waits until the user types some text into the text input field and clicks either OK or Cancel. If the user clicks the OK button, then `prompt()` returns the string of text that is currently shown in the input field. Otherwise, if the user clicks Cancel, `prompt()` returns `null`.

The optional *defaultText* argument gives the initial value of the text input field.

In modern web programming, this method is seldom used and many consider it bad design. That said, it can be quite convenient for a beginner to capture the user input.

```
setInterval()
```

Syntax:

```
window.setInterval(function, interval, arg1, arg2, ..., argN)
```

This method schedules *function* to be invoked at every *interval* milliseconds. Zero or more optional arguments (*arg1*, *arg2*, ..., *argN*) will be passed to *function*.

Note that *function* is invoked as a method of the Window object, which means that the `this` keyword refers to the Window object when used inside the body of *function*.

The `setInterval()` method returns a value that can later be used as an argument of `clearInterval()` to cancel the repetition of the invocation of *function*.

setTimeout()

Syntax:

```
window.setTimeout(function, timeout, arg1, arg2, ..., argN)
```

This method is like `setInterval()`, except that it schedules a *function* invocation only once: *function* will be invoked after *timeout* milliseconds.

The `setTimeout()` method returns a value that can later be used as an argument of `clearTimeout()` to cancel the pending invocation of *function*.

Events

You can also use all of the event-handler properties that are defined on Element on the Window object. That is possible because most events bubble up the document tree all the way to the Window object. Window also defines some event-handler properties of its own (e.g., the `onstorage` event-handler property) but they are out of the scope of this book.

onload

Syntax:

```
window.onload = handlerFunction;
```

This is an event handler for the load event on the Window object, which is one of the most important handlers. The load event fires after the document, and all the external resources referred to from within the document have been fully loaded and are ready to be manipulated. The specified *handlerFunction* function is where JavaScript code normally begins its execution.

Take, for example, the following code:

```
window.onload = function() {
  document.getElementById("status").innerHTML =
                                    "Boarding completed.";
};
```

Suppose there is an HTML element with ID of `status` in the corresponding document. After the document has loaded, you see the text "Boarding completed." written in it.

Index

bottom CSS property, 361

 element, 11, 315
 as semantic tool in poems, 240
break statement
 terminating loop, 177
 terminating switch, 151
bullets
 image, 350
 numbered, 351
 position, 350
button control, 243, 332
 and value property, 243
byline, 314

C

callback function, 236
camelCase, *see* variable
<canvas> element, 271, 325
caption
 of figure, 316
 of table, 327
  positioning of, 359
<caption> element, 20, 327
caption-side CSS property, 359
cascade
 determining style specificity, 64
 last style wins, 58
 most specific style wins, 64
 specificity of pseudo-classes and
  pseudo-elements, 95
Cascading Style Sheets (CSS)
 browser default, 64
  inspecting, 93
 declaration, 39
 external, 41
 including multiple, 41
 inline, 40
 internal, 40
 overriding properties, 58
 purpose of, 38
 rule, 40
 scripting styles using JavaScript,
  238
 selector, 40
 style attribute, 39
 syntax, 38
case keyword, 150
case sensitivity
 CSS, 42

CSS class names, 55
 entity names, 32
 HTML, 12
 ID selectors, 57
centering
 block elements, 355
 text, 347
chaining, *see* method chaining
character codes in Unicode, 120
character encoding, 6, 308
character entities, 32
 entity numbers, 32
charset attribute, <meta> element, 6,
 308
check box control, 332
checked attribute, <input> element,
 331
child, 5, 62
cite attribute, <q> element, 324
class, *see* class of objects
class global attribute, 16, 21, 337
 className, renamed in
  JavaScript, 238
class of objects, 158
 adding methods to, 212
 constructor
  defining, 210
  overloading, 211
  this, 210
 factory method, 212
 inheritance, 214
  prototype object, 214
 invocation context, this, 215
class selector, *see* CSS selectors
className, *see* class global attribute
clean code, *see also* good style, 42
clean design, 58
clear CSS property, 362
 example of, 87
client, 36
client side JavaScript, *see* JavaScript
closure, 201
 definition of, 204
code indentation, *see* indentation
code validation, *see* validation
collapsing margins, *see* margin
<color> CSS data type, 44, 339
color CSS property, 46, 343

fragment, 30
relative, 28, 29
root-relative, 28, 37
universal time, 159
<url> CSS data type, 43, 342
UTC() method, Date object, 163
UTC, Coordinated Universal Time,
159

V
validation
CSS, 42
HTML, 16
value attribute, <input> element, 332
value attribute, <option> element,
335
value property, retrieving value from
text box, 243
values
number range in JavaScript, 111
out-of-range, Date object, 384
primitive, 110
variable, *see also* scope
absence of value, 113
declaring, 107
global, 196
hoisting, 198
local, 197
difference with object
property, 215
naming of, 108
camelCase, 108
uninitialized, 113
var keyword, 107
vertical rhythm of the page, 70
example of, 89
viewport, 83
visibility CSS property, 366
difference between
visibility:hidden and
display:none, 366
:visited CSS pseudo-class, 93, 370
limitations on, 371
LVHA order, 94

W
WAMP package, 37
EasyPHP, 37
web application, 233

web design, 26
web server, *see* server
web syndication, 26
website root
index.html, 37
location, 37
while loop, 135
whitespace, *see* space
width attribute, <canvas> element,
326
width attribute, <img> element, 326
width CSS property, 360
content box, 76
Window global object, 229, 427
alert() method, 429
clearInterval() method, 429
clearTimeout() method, 430
confirm() method, 430
document property, 428
events, 431
innerHeight property, 428
innerWidth property, 428
localStorage property, 428
location property, 428
onload event handler, 231, 431
prompt() method, 430
bad design, 430
setInterval() method, 236, 430
setTimeout() method, 236, 431
window property, 429
referring to itself, 427
window global property, 427
word-spacing CSS property, 349
World Wide Web Consortium (W3C),
17, 21
wrapper object, temporary, 183

X
x line, 89
XMLHttpRequest object, Ajax, 283

Z
z-index CSS property, 367